Data Reconciliation & Gross Error Detection

To Dave Earp,

With best wishes for success,

Cornel Iordache

1/18/2000

HOUSTON, TX

Data Reconciliation & Gross Error Detection

An Intelligent Use of Process Data

Shankar Narasimhan
and Cornelius Jordache

Gulf Publishing Company
Houston, Texas

Data Reconciliation & Gross Error Detection

Gulf Publishing Company
Book Division
P.O. Box 2608 □ Houston, Texas 77252-2608

10 9 8 7 6 5 4 3 2 1

Library of Congress Cataloging-in-Publication Data

Narasimhan, Shankar.
Data reconciliation and gross error detection : an intelligent use of process data / Shankar Narasimhan and Cornelius Jordache.
p. cm.
ISBN 0-88415-255-3 (alk. paper)
1. Chemical process control—Automation. 2. Automatic data collection systems. 3. Error analysis (Mathematics) I. Jordache, Cornelius. II. Title.
TP155.75.N367 2000
660′.2815—dc21 99-44868
CIP

Printed in the United States of America.
Printed on acid-free paper (∞).

To our guru Professor Richard S. H. Mah, who played the roles of an initiator and a catalyst.

"Since all measurements and observations are nothing more than approximations to the truth, the same must be true of all calculations resting upon them, and the highest aim of all computations made concerning concrete phenomena must be to approximate, as nearly as practicable, to the truth. But this can be accomplished in no other way than by a suitable combination of more observations than the number absolutely requisite for the determination of the unknown quantities."

Gauss, K. G. *Theory of Motion of Heavenly Bodies,* New York, Dover, 1963, p. 249.

Contents

Chapter 3: Linear Steady-State Data Reconciliation, 59

Chapter 4: Steady-State Data Reconciliation for Bilinear Systems, 85

Chapter 5: Nonlinear Steady-State Data Reconciliation, 119

Chapter 6: Data Reconciliation in Dynamic Systems, 142

Chapter 7: Introduction to Gross Error Detection, 174

Chapter 8: Multiple Gross Error Identification Strategies for Steady-State Processes, 226

Chapter 9: Gross Error Detection in Linear Dynamic Systems, 281

Chapter 10: Design of Sensor Networks, 300

Chapter 11: Industrial Applications of Data Reconciliation and Gross Error Detection Technologies, 327

Appendix A: Basic Concepts of Linear Algebra, 373

Appendix B: Graph Theory Fundamentals, 378

Appendix C: Fundamentals of Probability and Statistics, 384

Acknowledgments

The authors are indebted to several people who have contributed to the preparation of this book. The main contributions came from Prof. Narasimhan's students at the Indian Institute of Technology (IIT)—Madras. T. Renganathan and J. Prakash, currently doing their doctoral programs, prepared the solutions for all examples with assistance from Sreeram Maguluri. Murukutla Rajamouli spent hours in the night preparing the tables and figures in different chapters. The successful completion of this book was due to their efforts. Drs. Sachin Patwardhan and S. Pushpavanam, faculty at IIT Madras, provided critical inputs to improve the focus and clarity of the text.

Thanks are also due to the RAGE software development team at Engineers India Limited, R&D Center, consisting of Dr. Madhukar Garg, Dr. V. Ravikumar, and Mr. Sheoraj Singh from whom Prof. Narasimhan gained valuable practical experience in implementing data reconciliation to industrial processes. Prof. Narasimhan also thanks Prof. Jose Ragot and Dr. Didier Maquin at CRAN-INPL, in Nancy, France, for arranging a summer visit for him, during which a significant part of the text was completed.

Dr. Jordache thanks all colleagues and managers from Chemshare Corp., Raytheon Process Automation, and especially Simulation Sciences Inc., for challenging him with practical issues that helped him clarify many implementation details for data reconciliation technology. Dr. Tom Clinkscales provided valuable input on polynomial filters. Drs. Wen-Jing Lin and Ricardo Dunia helped with clarifying some theoretical derivations mentioned in the text.

Special thanks to the R&D management of Simulation Sciences Inc. (SIMSCI®) for the encouragement and support given to Dr. Jordache during the writing of this manuscript and for allowing him to use Datacon™ software and the training book for preparing a detailed industrial example included in this book.

Very special thanks are also due to our respective wives, Jaishree Narasimhan, and Doina Jordache, who shielded us from all the problems on the home front during the past two years.

Moreover, the authors express a heart-felt gratitude to Debbie Markley of Gulf Publishing Company, who patiently reviewed every detail of the manuscript and worked long hours to help with finishing the index and to make sure this book is as accurate as possible.

Finally, the authors want to acknowledge and thank Prof. Miguel Bagajewicz for his excellent review and very useful suggestions on additional material to be included.

Preface

The quality of process data in chemical/petrochemical industries significantly affects the performance and the profit gained from using various software for process monitoring, online optimization, and control. Unfortunately plant measurements often contain errors that invalidate the process model used for optimization and control. Data reconciliation and gross error detection are techniques developed in the past 30 years for improving the accuracy of data so that they satisfy the plant model. During the last decade, they have been widely applied in refineries, petrochemical plants, mineral processing industries, and so forth, in order to achieve more accurate plant-wide accounting and superior profitability of plant operations.

Although commercial software for data reconciliation and gross error detection are available, the accompanying manuals usually give little theoretical background. In order to be able to select the best methods and gain the most benefits from their implementation, one needs a good understanding of the fundamental concepts. This book explains the basic concepts in data reconciliation and gross error detection using many illustrative examples. It also contains descriptions of different techniques that have been developed for these purposes and presents a unified perspective of these diverse methods. Certain criteria for selecting various techniques and guidelines for their practical implementation are also indicated.

The focus in this entire text is on classical methods that make use of process model equations, such as conservation laws, equilibrium relations, and equipment performance equations, to reconcile data and to detect and identify gross errors. In recent years, the use of artificial neural networks for data reconciliation and gross error detection has been proposed. These methods have not been included because they have not attained maturity, although they may become important in the future.

This book is organized in such a way that it is useful for both industrial personnel and academia. The book will be a valuable tool to engineers and managers dealing with the selection and implementation of data reconciliation software, or those involved in the development of such software. The book will also be useful as a supplementary reference for an undergraduate/graduate level course in chemical process instrumentation and control in which basic concepts can be taught or as a text for a full graduate level course in these topics. Unlike this book, the other books that are currently available on these topics do not present an in-depth analysis of the different techniques available, their limitations or their interrelationships.

The book is organized as follows: The first chapter motivates the need for data reconciliation and gross error detection and introduces the major concepts involved. Chapter 2 introduces the statistical characterization of measurement errors and various filtering techniques used for error reduction that can form part of an overall data processing strategy. The next four chapters deal with the subject of data reconciliation in increasing level of complexity. Steady-state linear data reconciliation is described in Chapter 3. Decomposition techniques for linear models with both measured and unmeasured variables are described here. The techniques related to the classification of variables as observable and redundant are also presented. Methods for estimating the measurement error variances from measured data are also described in this chapter.

Chapter 4 deals with steady-state data reconciliation for bilinear systems consisting of component balances and, in certain cases, energy balances. The motivation for considering such processes is illustrated using examples from mineral processing industries as well as utility distribution networks in chemical industries. Chapter 5 treats nonlinear data reconciliation. Nonlinear models are often used to accurately describe chemical processes. The most efficient and widely used solution procedures for the nonlinear reconciliation are presented in this chapter. Handling inequality constraints, such as bounds on variables, is also analyzed in this chapter. Data reconciliation techniques for dynamic systems are discussed in Chapter 6. Both Kalman filtering methods and general optimization techniques designed for dynamic nonlinear problems are presented.

Chapters 7 through 9 deal with the problem of gross error detection. Chapter 7 introduces the issues involved in gross error detection and describes the basic statistical tests that can be used to detect gross errors.

The underlying assumptions, characteristics, and relative advantages and disadvantages of various statistical tests are also discussed. For identifying multiple gross errors, complex strategies are required. A plethora of strategies have been proposed and evaluated in the research literature. A special effort has been made in Chapter 8 to give a unified perspective by classifying the different strategies on the basis of their core principles. We also describe in detail a typical strategy from each of these classes. Chapter 9 treats the problem of gross error identification in dynamic systems.

The efficacy of data reconciliation and gross error detection depends significantly upon the location of measured variables. Recent attempts to optimally design the sensor network for maximizing accuracy of data reconciliation solution are described in Chapter 10.

Several industrial applications and existent software systems for data reconciliation and gross error detection are also discussed in Chapter 11. Various aspects related to the benefits of offline and on-line data reconciliation, the methods mostly used and their performances are analyzed here.

In order to make this book self-sufficient with respect to the mathematical background required for a good understanding, appendices are included that describe the necessary basic concepts from linear algebra, graph theory, and probability and statistical hypothesis testing.

1

The Importance of Data Reconciliation and Gross Error Detection

PROCESS DATA CONDITIONING METHODS

In any modern chemical plant, petrochemical process or refinery, hundreds or even thousands of variables—such as flow rates, temperatures, pressures, levels, and compositions—are routinely measured and automatically recorded for the purpose of process control, online optimization, or process economic evaluation. Modern computers and data acquisition systems facilitate the collection and processing of a great volume of data, often sampled with a frequency of the order of minutes or even seconds.

The use of computers not only allows data to be obtained at a greater frequency, but has also resulted in the elimination of errors present in manual recording. This in itself has greatly improved the accuracy and validity of process data. The increased amount of information, however, can be exploited for further improving the accuracy and consistency of process data through a systematic data checking and treatment.

Process measurements are inevitably corrupted by errors during the measurement, processing and transmission of the measured signal. The total error in a measurement, which is the difference between the measured value and the true value of a variable, can be conveniently represented as the sum of the contributions from two types of errors—*random errors* and *gross errors.*

The term *random error* implies that neither the magnitude nor the sign of the error can be predicted with certainty. In other words, if the measurement is repeated with the same instrument under identical process conditions, a different value may be obtained depending on the outcome of the random error. The only possible way these errors can be characterized is by the use of probability distributions.

These errors can be caused by a number of different sources such as power supply fluctuations, network transmission and signal conversion noise, analog input filtering, changes in ambient conditions, and so on. Since these errors can arise from different sources (some of which may be beyond the control of the design engineer), they cannot be completely eliminated and are always present in any measurement. They usually correspond to the high frequency components of a measured signal, and are usually small in magnitude except for some occasional spikes.

On the other hand, *gross errors* are caused by nonrandom events such as instrument malfunctioning (due to improper installation of measuring devices), miscalibration, wear or corrosion of sensors, and solid deposits. The nonrandom nature of these errors implies that at any given time they have a certain magnitude and sign which may be unknown. Thus, if the measurement is repeated with the same instrument under identical conditions, the contribution of a systematic gross error to the measured value will be the same.

By following good installation and maintenance procedures, it is possible to ensure that gross errors are not present in the measurements at least for some time. Gross errors caused by sensor miscalibration may occur suddenly at a particular time and thereafter remain at a constant level or magnitude. Other gross error causes such as the wear or fouling of sensors can occur gradually over a period of time and so the magnitude of the gross error increases slowly over a relatively long time period. Thus, gross errors occur less frequently but their magnitudes are typically larger than those of random errors.

Errors in measured data can lead to significant deterioration in plant performance. Small random and gross errors can lead to deterioration in the performance of control systems, whereas larger gross errors can nullify gains achievable through process optimization. In some cases, erroneous data can also drive the process into an uneconomic or—even worse—an unsafe operating regime. It is therefore important to reduce, if not completely eliminate, the effect of both random and gross errors. Several data processing techniques can be used together to achieve this objective. In this text, we describe methods which can play an important role as part of an integrated data processing strategy to reduce errors in measurements made in continuous process industries.

Research and development in the area of signal conditioning have led to the design of analog and digital filters which can be used to attenuate the effect of high frequency noise in the measurements. Large gross errors can be initially detected by using various data validation checks. These include checking whether the measured data and the rate at which it is changing is within predefined operational limits. Today, smart sensors are available which can perform diagnostic checks to determine whether there is any hardware problem in measurement and whether the measured data is acceptable.

More sophisticated techniques include *statistical quality control tests* (SQC) which can be used to detect significant errors (outliers) in process data. These techniques are usually applied to each measured variable separately. Thus, although these methods improve the accuracy of the measurements, they do not make use of a process model and hence do not ensure consistency of the data with respect to the interrelationships between different process variables. Nevertheless, these techniques must be used as a first step to reduce the effect of random errors in the data and to eliminate obvious gross errors.

It is possible to further reduce the effect of random error and also eliminate systematic gross errors in the data by exploiting the relationships that are known to exist between different variables of a process. The techniques of *data reconciliation* and *gross error detection* that have been developed in the field of chemical engineering during the past 35 years for this purpose are the principal focus of this book.

Data reconciliation (DR) is a technique that has been developed to improve the accuracy of measurements by reducing the effect of random errors in the data. The principal difference between data reconciliation and other filtering techniques is that data reconciliation explicitly makes use of process model constraints and obtains estimates of process variables by adjusting process measurements so that the estimates satisfy the constraints.

The reconciled estimates are expected to be more accurate than the measurements and, more importantly, are also consistent with the known relationships between process variables as defined by the constraints. In order for data reconciliation to be effective, there should be no gross errors either in the measurements or in the process model constraints. Gross error detection is a companion technique to data reconciliation that has been developed to identify and eliminate gross errors. Thus, data reconciliation and gross error detection are applied together to improve accuracy of measured data.

Data reconciliation and gross error detection both achieve error reduction only by exploiting the redundancy property of measurements. Typically, in any process the variables are related to each other through physical constraints such as material or energy conservation laws. Given a set of such system constraints, a minimum number of error-free measurements is required in order to calculate all of the system parameters and variables. If there are more measurements than this minimum, then redundancy exists in the measurements that can be exploited. This type of redundancy is usually called *spatial redundancy* and the system of equations is said to be *overdetermined.*

Data reconciliation cannot be performed without spatial redundancy. With no extra measured information, the system is *just determined* and no correction to erroneous measurements is possible. Further, if fewer variables than necessary to determine the system are measured, the system is *underdetermined* and the values of some variables can be estimated only through other means or if additional measurements are provided.

A second type of redundancy that exists in measurements is *temporal redundancy.* This arises due to the fact that measurements of process variables are made continually in time at a high sampling rate, producing more data than necessary to determine a steady-state process. If the process is assumed to be in a steady state, then temporal redundancy can be exploited by simply averaging the measurements, and applying steady-state data reconciliation to the averaged values.

If the process state is dynamic, however, the evolution of the process state is described by differential equations corresponding to mass and energy balances, which inherently capture both the temporal and spatial redundancy of measured variables. For such a process, dynamic data reconciliation and gross error detection techniques have been developed to obtain accurate estimates consistent with the differential model equations of the process.

Signal processing and data reconciliation techniques for error reduction can be applied to industrial processes as part of an integrated strategy referred to as *data conditioning* or *data rectification.* Figure 1-1 illustrates the various operations and the position occupied by data reconciliation in data conditioning for on-line industrial applications.

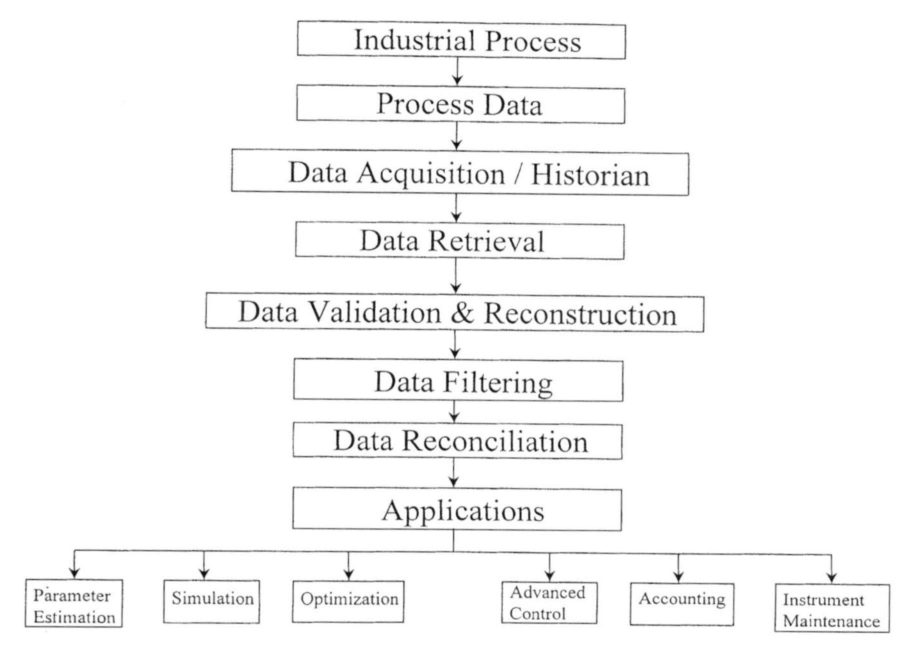

Figure 1-1. Online data collection and conditioning system.

INDUSTRIAL EXAMPLES OF STEADY-STATE DATA RECONCILIATION

Here we will briefly describe two examples of industrial applications of steady-state data reconciliation drawn from our experience in order to illustrate the need for such a technique and the benefits that can be derived from it.

Crude Split Optimization in a Preheat Train of a Refinery

In any refinery, the crude oil is initially heated by passing it through an interconnected set of heat exchangers called the crude preheat train before being fractionated. In a crude preheat train, typically the crude is split into one or more parallel streams, each of which is heated by passing it through a train of heat exchangers before being merged and sent to a furnace for further heating. The process streams that are used for heating the crude are the various product and pump-around streams from a downstream atmospheric or vacuum crude distillation column.

In order to maximize energy recovery from these process streams, the optimal flows of the crude splits through the different parallel heat exchanger trains should be determined online, say every few hours. For determining the optimal flows, the total inlet flow of crude and all hot process streams along with their inlet temperatures have to be specified. Moreover, details of all heat exchangers, such as heat exchanger areas, and overall heat transfer coefficients, also have to be specified.

Generally, in a crude preheat train, all the stream flows, as well as all intermediate temperatures, are measured. Thus there are more measurements than those required for performing the optimization. It is possible to ignore some of the measurements and use only the measurements of inlet flows and inlet temperatures of all streams for determining the optimal crude split flows. However, since all measurements contain errors, any optimization exercise carried out using such measurements will not necessarily result in the predicted gains.

In order to overcome this, steady-state reconciliation and gross error detection is applied to measured data to eliminate measurements containing gross errors and obtain reconciled estimates of all stream flows and temperatures which satisfy the flow and enthalpy balances of the crude preheat train. As part of the reconciliation, the overall heat transfer coefficients of all exchangers are also estimated.

These estimated heat transfer coefficients will more correctly reflect the actual current performance of the heat exchangers than their original design values. The reconciled estimates of inlet flows and temperatures of all streams, and the estimated overall heat transfer coefficients of all exchangers, are used to determine the optimal values of the crude splits which are then implemented in the plant. Use of reconciled estimates in the optimization is likely to result in actual energy recovery from the process being close to the predicted optimal values.

It should be noted that the time periods selected for reconciliation and optimization are selected based on the time constants of the system. Since a change in the crude split flows has an effect on the downstream crude distillation column and hence affects all the distillate streams which are used for pre-heating the crude, it takes two hours for all the stream flows and temperatures to reach a new steady state after a new set of optimal crude splits values are implemented. The process is operated at this steady state for an additional two hours after which the optimization of the crude split flows is repeated. The measurements made during the preceding two hours of steady-state operation are averaged and used as data for the reconciliation problem. This example is described in greater detail in the concluding chapter on industrial case studies.

Minimizing Water Consumption in Mineral Beneficiation Circuits

In a mineral beneficiation circuit, crushed ore is washed with water along with other additives in an interconnected network of classifiers or flotation cells in order to liberate the particles containing the minerals from the gangue material. In order to minimize the water consumption for a desired concentration of the beneficiated ore, the performance of the flotation cells has to be simulated for different flow conditions. The simulation model in turn requires data on the feed characteristics as well as on parameters such as pulp densities.

Generally, the flows of the feed stream and pure water streams are measured. Using samples drawn from different streams, the concentrations of different minerals in each stream and their pulp densities are also measured in the laboratory. These measurements contain errors and are also not consistent with the flow and component balances of the mineral beneficiation circuit. Steady-state reconciliation and gross error detection can be applied to the measurements in order to obtain reconciled estimates of all stream flows, pulp densities and mineral concentrations such that they satisfy the material balances. The reconciled estimates are used in the detailed simulation of the flotation cells and to determine the minimal amount of water to be added. In one such exercise, it was possible to reduce water consumption by five percent.

DATA RECONCILIATION PROBLEM FORMULATION

As stated in the preceding sections, data reconciliation improves the accuracy of process data by adjusting the measured values so that they satisfy the process constraints. The amount of adjustment made to the measurements is minimized since the random errors in the measurements are expected to be small. In the general case, not all variables of the process are measured due to economic or technical limitations.

The estimates of unmeasured variables as well as model parameters are also obtained as part of the reconciliation problem. The estimation of unmeasured values based on the reconciled measured values is also known as *data coaptation.* In general, data reconciliation can be formulated by the following constrained weighted *least-squares optimization* problem.

$$\underset{x_i, u_j}{\text{Min}} \sum_{i=1}^{n} w_i (y_i - x_i)^2 \qquad (1\text{-}1)$$

subject to

$$g_k(x_i, u_j) = 0 \qquad k = 1, \ldots, m \qquad (1\text{-}2)$$

The objective function 1-1 defines the total weighted sum square of adjustments made to measurements, where w_i are the weights, y_i is the measurement and x_i is the reconciled estimate for variable *i,* and u_j are the estimates of unmeasured variables. Equation 1-2 defines the set of model constraints. The weights w_i are chosen depending on the accuracy of different measurements.

The model constraints are generally material and energy balances, but could include inequality relations imposed by feasibility of process operations. The deterministic natural laws of conservation of mass or energy are typically used as constraints for data reconciliation because they are usually known. Empirical or other types of equations involving many unmeasured parameters are not recommended, since they are at best known only approximately. Forcing the measured variables to obey inexact relations can cause inaccurate data reconciliation solution and incorrect gross error diagnosis.

Any mass or energy conservation law can be expressed in the following general form [1]:

$$\text{input} - \text{output} + \text{generation} - \text{consumption} - \text{accumulation} = 0 \qquad (1\text{-}3)$$

The quantity for which the above equation is written could be the overall material flow, the flow of individual components, or the flow of energy. If there is no accumulation of any of these quantities, then these constraints are algebraic in character and define a steady-state operation.

For a dynamic process, however, the accumulation terms cannot be neglected and the constraints are differential equations. For most process units, there is no generation or depletion of material. In the case of reactors, though, the generation or depletion of individual components due to reaction should be taken into account.

For some simple units such as splitters, there is no change either in the composition or temperature of streams. For such units, the component and energy balances reduce to a simple form such as

$$x_i = x_j \tag{1-4}$$

where the variable x_i represents either the temperature or composition of stream *i*. The above equation is also useful when two or more sensors are used to measure the same variable, say flow rate or temperature of a stream.

The type of constraints that are imposed in reconciliation depend on the scope of the reconciliation problem and the type of process units. Furthermore, the complexity of the solution techniques used depends strongly on the constraints imposed. For example, if we are interested in reconciling only the flow rates of all streams, then the material balances constraints are linear in the flow variables and a *linear data reconciliation problem* results. On the other hand, if we wish to reconcile composition, temperature or pressure measurements along with flows, then a *nonlinear data reconciliation problem* occurs.

An issue to be addressed is the kind of constraints that we can legitimately impose in a data reconciliation application. Since data reconciliation forces the estimates of all variables to satisfy the imposed constraints, this issue assumes great importance. Usually, material and energy balance constraints are included because they are valid physical laws. It should be noted, however, that these equations are generally written assuming that there is no loss of material or energy from the process unit to the environment. While this may be valid for material flow, significant losses in energy may occur for example from improperly insulated heat exchangers. In such cases, it is better not to impose the energy balances or alternatively include an unknown loss term in the balance equation that can be estimated as part of the reconciliation.

Other than material and energy conservation constraints, a model of a process unit can contain equations involving the unit parameters. For example, a heat exchanger model can include a rating equation relating the heat duty to the overall heat transfer coefficient, the exchanger area available for heat transfer, and the stream flows and temperatures. Equation 1-5 describes this relationship.

$$Q - UA\Delta T_{ln} = 0 \tag{1-5}$$

where Q is the heat duty, U is the overall heat transfer coefficient, A is the exchanger area, and ΔT_{ln} is the logarithmic mean temperature difference.

Should this equation be included as a constraint when applying data reconciliation to processes involving heat exchangers? Generally, since the overall heat transfer coefficient is unknown and has to be estimated from the measured data, this equation may be included and U estimated as part of the reconciliation problem. If there is no prior information about U, however, and no feasibility restrictions on it, then inclusion of this constraint does not provide any additional information and estimates of all other variables will be the same regardless of whether this constraint is included or not. Thus, the data reconciliation problem can as well be solved without this constraint and U can subsequently be estimated by the above equation using the reconciled values of flows and temperatures.

On the other hand, if U has to be within specified bounds or if there is a good estimate for U from a previous reconciliation exercise (as in the crude preheat train example discussed in the previous section, where the estimates of U from the reconciliation solution of the most recent time period can be used as good *a priori* estimates), then the constraint should be included along with the additional information about U as part of the reconciliation problem. The overall heat transfer coefficient can also be related to the physical properties of the streams, their flows, temperatures and the heat exchanger characteristics using correlations. It is not advisable to use such an equation in the reconciliation model since the correlations themselves can be quite erroneous and forcing the flows and temperatures to fit this equation may increase the inaccuracy of the estimates.

Another important question is whether to perform reconciliation using a *steady-state* or a *dynamic model* of the process. Practically, a process is never truly at a steady state. However, a plant is normally operated for several hours or days in a region around a nominal steady-state operating point. For applications such as online optimization (as in the case of a crude split optimization example) where reconciliation is performed once every few hours, it is appropriate to employ steady-state reconciliation on measurements averaged over the time period of interest.

During transient conditions (such as during a changeover to a new crude type in a refinery) when the departure from steady state is significant, steady-state reconciliation should not be applied because it will result in large adjustments to measured values. Measurements taken during such transient periods can be reconciled, if necessary, using a dynamic model of the process. Similarly for process control applications where reconciliation needs to be performed every few minutes, dynamic data reconciliation is appropriate.

Data reconciliation is based on the assumption that only random errors are present in the measurements which follow a normal (Gaussian) distribution. If a gross error due to a measurement bias is present in some measurement or if a significant process leak is present which has not been accounted for in the model constraints, then the reconciled data may be very inaccurate. It is therefore necessary to identify and remove such gross errors. This is known as the *gross error detection problem.*

Gross errors can be detected based on the extent to which the measurements violate the constraints or on the magnitude of the adjustments made to measurements in a preliminary data reconciliation. Although gross error detection techniques were developed primarily to improve the accuracy of reconciled estimates, they are also useful in identifying measuring instruments that need to be replaced or recalibrated.

EXAMPLES OF SIMPLE RECONCILIATION PROBLEMS

In order to obtain a good understanding of the issues and underlying assumptions in data reconciliation, some of the simplest possible cases are introduced here. We assume a process operating at a steady state, constrained by a set of linear equations.

Systems With All Measured Variables

Let us first consider the simplest data reconciliation problem: the reconciliation of the stream flows of a process. Initially, all flow rates are assumed to be directly measured. The flow measurements contain unknown random errors. For that reason, the material input and output of every process unit and of the overall process do not balance. The aim of reconciliation is to make minor adjustments to the measurements in order to make them consistent with the material balances. The adjusted measurements, which are referred to as estimates, are expected to be more accurate than the measurements. Although the problem considered here is simple, it does have important industrial applications in accurate accounting for the material flows as, for example, in a lube blending plant, in the steam and water distribution subsystem of a plant, or in a complete refinery.

Example 1-1

Let us consider a simple process of a heat exchanger with a bypass as shown in Figure 1-2. Let us also ignore the energy flows of this process and focus only on the mass flows. It is assumed that the flows of all six streams of this process are measured and that these measurements con-

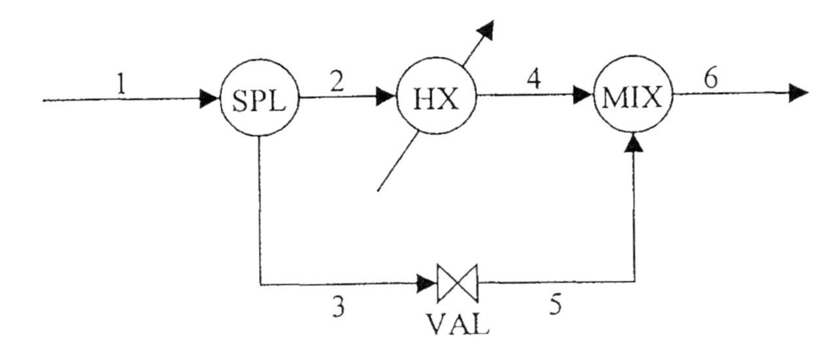

Figure 1-2. Heat exchanger system with bypass.

tain random errors. If we denote the true value of the flow of stream i by the variable x_i and the corresponding measured value by y_i, then we can relate them by the following equations

$$y_i = x_i + \varepsilon_i \qquad i = 1 \ldots 6 \tag{1-6}$$

where ε_i is the random error in measurement y_i.

The flow balances around the splitter, exchanger, valve, and mixer can be written as

$$x_1 - x_2 - x_3 = 0 \tag{1-7a}$$

$$x_2 - x_4 = 0 \tag{1-7b}$$

$$x_3 - x_5 = 0 \tag{1-7c}$$

$$x_4 + x_5 - x_6 = 0 \tag{1-7d}$$

The measured values (given in Table 1-1) do not satisfy the above equations, since they contain random errors. It is desired to derive estimates of the flows that satisfy the above flow balances. Intuitively, we can impose the condition that the differences between the measured and estimated flows, also referred to as *adjustments,* should be as small as possible. As a first choice, we can represent this objective as

$$\min_{x_i} \sum_{i=1}^{6} (y_i - x_i)^2 \tag{1-8}$$

The above function is the familiar least-squares criterion used in regression. Since it is immaterial whether the adjustments are positive or negative, the square of the adjustment is minimized. Although other types of criteria may be used such as minimizing the sum of absolute adjustment, they do not have a statistical basis and also make the solution of the problem more difficult.

The least-squares criterion is acceptable, if all measurements are equally accurate. The adjustment made to one measurement is given the same importance as any other. In practice, however, it is likely that some measurements are more accurate than others depending on the instrument being used and the process environment under which it operates. In order to account for this, we can use a weighted least-squares objective as a more general criterion, given by

$$\min_{x_i} \sum_{i=1}^{6} w_i (y_i - x_i)^2 \tag{1-9}$$

where the weights w_i are chosen to reflect the accuracy of the respective measurements. More accurate measurements are given larger weights in order to force their adjustments to be as small as possible. Generally, it is assumed that the error variances for all the measurements are known and that the weights are chosen to be the inverse of these variances.

The reconciliation problem is thus a constrained optimization problem with the objective function given by Equation 1-9 and the constraints given by Equations 1-7a through 1-7d. The solution of this optimization problem can be obtained analytically for flow reconciliation. Table 1-1 shows the true, measured, and reconciled flows for the process of Figure 1-2. The reconciled flows shown in column four of this table are obtained by assuming that all measurements are equally accurate (weights are all equal). It can be easily verified that while the measured values do not satisfy the flow balances, Equations 1-7a through 1-7d, the reconciled flows satisfy them.

Table 1-1
Flow Reconciliation for a Completely Measured Process

Stream Number	True Flow Values	Measured Flow Values	Reconciled Flow Values
1	100	101.91	100.22
2	64	64.45	64.50
3	36	34.65	35.72
4	64	64.20	64.50
5	36	36.44	35.72
6	100	98.88	100.22

Systems With Unmeasured Variables

In the previous example, we have assumed that all variables are measured. However, usually only a subset of the variables are measured. The presence of unmeasured variables not only complicates the problem solution, but also introduces new questions such as whether an unmeasured variable can be estimated, or whether a measured variable can be reconciled as illustrated by the following example.

Example 1-2

Let us consider the flow reconciliation problem of the simple process shown in Figure 1-2. However, we will not assume that all the flows are measured as before. Instead, we will assume that only selective flows are measured and in each case discuss the issues and problems involved in partially measured systems.

Case 1. *Flows of streams 1, 2, 5, and 6 are measured, while the other two stream flows are unmeasured.*

The objective in this case is to not only reconcile the measured flows, but also to estimate all the unmeasured flows as part of the reconciliation problem. As in Equation 1-6, we relate the measured and true stream flows,

$$y_i = x_i + \varepsilon_i \quad i = 1, 2, 5, 6 \tag{1-10}$$

The constraints are still given by Equation 1-7. It should be noted that the constraints involve both measured and unmeasured flow variables. The objective function is the weighted sum of squares of adjustments made to measured variables, and is given by

$$\min_{x_1, x_2, x_5, x_6} \quad w_1(y_1 - x_1)^2 + w_2(y_2 - x_2)^2 + w_5(y_5 - x_5)^2 + w_6(y_6 - x_6)^2 \tag{1-11}$$

Since the unmeasured variables are present only in the constraint set, the simplest strategy for solving the problem is to eliminate them from the constraints. This will not affect the objective function since it does not involve unmeasured variables. Variable x_3 can be eliminated by combining Equations 1-7a and 1-7c, while variable x_4 can be eliminated by combining Equations 1-7b and 1-7d. Thus, we obtain a reduced set of constraints which involves only measured variables.

$$x_1 - x_2 - x_5 = 0 \tag{1-12a}$$

$$x_2 + x_5 - x_6 = 0 \tag{1-12b}$$

The reduced data reconciliation problem is now to minimize 1-11 subject to the constraints of Equations 1-12a and 1-12b. It can be observed that this reduced problem involving the variables x_1, x_2, x_5, and x_6 is similar to the completely measured case, and an analytical solution can be used to obtain the reconciled values of the measured variables. Using the same measured values for x_1, x_2, x_5, and x_6 as given in Table 1-1, and assuming all measurements to be equally accurate, the reconciled values which are obtained are shown in Table 1-2 in the column under Case 1. Once the reconciled values for the measured variables are obtained, the estimates of the unmeasured variables can be calculated using the original constraints.

Thus the estimate of x_4 is equal to that of x_2, and the estimate of x_3 is equal to that of x_5. These values are also indicated in Table 1-2. By comparing with the results of Table 1-1, it can be observed that since there are fewer measured variables in this case, the estimates of some variables are less accurate than those derived for the completely measured system. The central idea that is gained from this case is that the reconciliation problem can be split or decomposed into subproblems—the first being a reduced reconciliation problem involving only measured variables, followed by an estimation or coaptation problem for calculating the estimates of unmeasured variables.

Table 1-2
Flow Reconciliation of Partially Measured Process

	Reconciled Flow Values		
Stream	Case 1—Streams 3 and 4 unmeasured	Case 2—Streams 3, 4, 5, 6 unmeasured	Case 3—Streams 2, 3, 4, 5 unmeasured
1	100.49	101.91	100.39
2	64.25	64.45	—
3	36.24	37.46	—
4	64.25	64.45	—
5	36.24	37.46	—
6	100.49	101.91	100.39

Case 2. *Only flows of streams 1 and 2 are measured.*

In this case, only Equations 1-7a and 1-7b contain measured variables and are useful in the reconciliation problem. The objective function is set up as before to minimize the adjustment made to measured variables and is given by

$$\min_{x_1, x_2} w_1(y_1 - x_1)^2 + w_2(y_2 - x_2)^2 \tag{1-13}$$

As in Case 1, we try to eliminate the unmeasured variables from the constraints 1-7a and 1-7b. Our attempt to produce an equation involving only measured variables by suitably combining the original constraints ends in failure. Thus, the reconciliation problem we obtain is to minimize 1-13 without any constraints. It is immediately obvious that the best estimates of x_1 and x_2 are given by their respective measured values which results in the least adjustment of zero for 1-13. The estimates of the unmeasured variables can now be calculated using the constraints. The estimate of x_6 is equal to x_1, the estimate of x_4 is equal to x_2, and the estimates of x_3 and x_5 are both equal to the difference between x_1 and x_2. These values are all given in Table 1-2 under Case 2.

Two important observations can be made in this case. First, no adjustment is made to the two measured variables x_1, and x_2. This is due to the fact that there is no additional information in the form of constraints that relate only the measured variables which can be exploited for adjusting their measurements. Such measured variables are also known as *nonredundant* variables. Second, a unique estimate for every unmeasured variable is obtained using the constraints and estimates of measured variables.

These unmeasured variables are also known as *observable.* A formal definition of the concepts of observability and redundancy is given in Chapter 3. It is sufficient at present to note that while the partially measured process in Case 1 gives a redundant and observable system, Case 2 gives rise to a nonredundant, observable system.

Case 3. *Only flows of streams 1 and 6 are measured.*

The reduced reconciliation problem we obtain for this case is

$$\min_{x_1, x_6} w_1(y_1 - x_1)^2 + w_6(y_6 - x_6)^2 \tag{1-14}$$

such that:

$$x_1 - x_6 = 0 \tag{1-15}$$

Equation 1-15 is obtained by adding all the constraints 1-7a through 1-7d. Assuming that the measurements of x_1 and x_6 are equally accurate, their reconciled values obtained are given in Table 1-2 under Case 3. We now attempt to calculate the estimates of the remaining four variables. We will not be successful, however, in obtaining unique estimates for these variables. In other words, there are many solutions—in fact, an infinite number—which can satisfy the constraints.

For example, one possible solution is to take the estimates of x_3 and x_5 to be both equal to that of x_1, and the estimates of x_2 and x_4 to be equal to zero. Alternatively, we can choose the estimates of x_2 and x_4 to be equal to that of x_1, while the estimates of x_3 and x_5 are chosen to be zero.

Without additional information, there is no way of determining which of these myriad possible solutions is more accurate. The variables x_2, x_3, x_4, and x_5 are denoted as *unobservable* in this case. An interesting feature of this case is that though there are some unmeasured variables which cannot be uniquely estimated, reconciliation of the variables x_1 and x_6 can still be performed utilizing the available measurements. Therefore, Case 3 is a redundant, unobservable system.

System Containing Gross Errors

In all the cases considered in Example 1, the measurements did not contain any systematic error or bias. In such cases, data reconciliation does reduce the error in measurements. We will now examine the case when one of the measurements contains a systematic bias or gross error and demonstrate the need to perform gross error detection along with data reconciliation.

Example 1-3

We reconsider the flow process shown in Figure 1-2 for which the true stream flows are as given in Table 1-1. We will assume that all flows are measured with measurements as given in Table 1-1, except that the measurement of stream 2 contains a positive bias of 4 units, so that its measured value is 68.45 instead of 64.45. As before, we reconcile these measurements and obtain estimates which are shown in Table 1-3, in column 2, when all the measurements are used.

A comparison of these estimates with those listed in Table 1-1, clearly shows that the accuracy of the estimates has decreased due to the presence of the gross error. Furthermore, although only the flow measurement of stream 2 contains a gross error, the accuracy of all the flow estimates has decreased. This is known as a *smearing effect* and it occurs due to reconciliation which exploits the spatial constraint relations between different variables.

In order for data reconciliation to be effective, it is therefore necessary to identify those measurements containing gross errors and either eliminate them or make appropriate compensation. The last column of Table 1-3 shows the reconciled estimates obtained when the flow measurement of stream 2 is discarded and not used in the reconciliation process. Clearly, the accuracy of the reconciled estimates has improved considerably, even though the redundancy has decreased by discarding the measurement.

Table 1-3
Flow Reconciliation When Stream 2 Flow Measurement Contains a Gross Error

	Reconciled Flow Values	
Stream	**All measurements used**	**Stream 2 measurement eliminated**
1	100.89	100.23
2	65.83	64.53
3	35.05	35.71
4	65.83	64.53
5	35.05	35.71
6	100.89	100.23

Thus far, we have not considered the important question of how to identify the measurement containing a gross error based only on the knowledge of the measured values and constraint relations between variables. There are several ways of tackling this problem and in this example we illustrate one approach. Given a set of measurements, we can initially reconcile them assuming that there are no gross errors in the data. In the flow process example considered here, the reconciled estimates obtained under this assumption have already been shown in the second column of Table 1-3. From these reconciled estimates we can compute the differences between the measured and reconciled values (measurement adjustments) for all measured variables, and these are shown in Table 1-4.

Table 1-4
Measurement Adjustments for Flow Process

Stream	Measurement adjustments
1	1.02
2	2.62
3	–0.40
4	–1.63
5	1.39
6	–2.01

If the constraints are linear as in this example, the expected variance of the adjustments can be analytically derived which will be a function of the constraint matrix and the measurement error variances. For the flow process example considered here, it can be shown that the standard deviation of measurement adjustments for every variable is 0.8165. A simple statistical test can be applied to determine if the computed measurement adjustments fall within a confidence interval, say within a $\pm 2\sigma$ interval. In this example, the $\pm 2\sigma$ interval (95% confidence interval) is [-1.6 1.6].

From Table 1-4, we can observe that the measurement adjustments for the flows of streams 2, 4, and 6 fall outside this interval and as a first cut the measurements of these streams can be suspected of containing a gross error. Among these the measurement adjustment of stream 2 has the largest magnitude and can be identified to contain a gross error. After discarding the measurement of stream 2, we can again reconcile the data and compute the measurement adjustments to examine if any more gross errors are present.

The procedure used above is a sequential procedure for gross error detection and makes use of the statistical test known as the *measurement test.* A variety of statistical tests and methods for identifying one or more gross errors have been developed and are described in Chapters 7 and 8. Although in this example we have only considered a gross error in measurements, it is possible for a gross error to be present in the constraints due to an unaccounted leak or loss of material. Some of the methods described in Chapters 7 and 8 can also be used to identify such gross errors. The example also clearly demonstrates that data reconciliation and gross error detection have to be applied together for obtaining accurate estimates.

BENEFITS FROM DATA RECONCILIATION AND GROSS ERROR DETECTION

Development of a data reconciliation and gross error detection package for a system and its practical implementation is a difficult and costly task and cannot be justified without its benefits for a particular industrial application. The justification for data reconciliation and gross error detection may come from the many important applications for improving process performance shown in Figure 1-1 which requires accurate data for achieving expected benefits as outlined below:

1. A direct application of data reconciliation is in evaluating process yields or in assessing consumption of utilities in different process units. Reconciled values provide more accurate estimates as compared to the use of raw measurements. For example, refinery-wide material balance reconciliation aids in a better estimate of overall refinery yields. Similarly, a plant-wide energy audit using reconciled flows and temperatures helps in a better identification of energy inefficient processes and equipment.
2. Applications such as simulation and optimization of existing process equipment rely on a model of the equipment. These models usually contain parameters which have to be estimated from plant data. This is also known as *model tuning,* for which accurate data is essential. The use of erroneous measurements in model tuning can give rise to incorrect model parameters which can nullify the benefits achievable through optimization. There are two possible ways in which data reconciliation can be used for such applications which we illustrate using a simple example.

Let us consider the problem of optimizing the performance of an existing distillation column. From the operating data, measurements of flows, temperatures and compositions of all inlet and outlet streams of the column can be obtained. One possible way is to reconcile these measurements using only overall material and energy balances around the column. The reconciled data can now be used along with a detailed tray to tray model of the column in order to estimate parameters such as tray efficiencies. The tuned model can then be used to optimize the performance of the column.

Alternatively, a simultaneous data reconciliation and parameter estimation can be performed using the detailed tray-to-tray model of the column. In this case, if measurements of tray temperatures and/or compositions are available, they can also be used and reconciled as part of the problem. Obviously, the second approach leads to a significant increase in effort and computation time. This approach is also referred to as *rigorous on-line modeling* and has been incorporated in many commercial steady-state simulators.

3. Data reconciliation can be very useful in scheduling maintenance of process equipment. Reconciled data can be used to accurately estimate key performance parameters of process equipment. For example, heat transfer coefficient of heat exchangers or the level of catalyst activity in reactors can be estimated and used to determine whether the heat exchanger should be cleaned or whether the catalyst should be replaced/regenerated, respectively.
4. Many advanced control strategies such as model-based control or inferential control require accurate estimates of controlled variables. Dynamic data reconciliation techniques can be used to derive accurate estimates for better process control.
5. Gross error detection not only improves the estimation accuracy of data reconciliation procedures but is also useful in *identifying instrumentation problems* which require special maintenance and correction. Incipient detection of gross errors can reduce maintenance costs and provide a smoother plant operation. These methods can also be extended to detect faulty equipment.

A BRIEF HISTORY OF DATA RECONCILIATION AND GROSS ERROR DETECTION

The problem of data reconciliation was first introduced in 1961 and during the past four decades more than 200 research publications in the two areas of data reconciliation and gross error detection have appeared. Our purpose in this section is to trace some of the significant contributions that spurred developments in these two areas.

Interestingly, the problem of data reconciliation was first posed by Kuehn and Davidson [2] who were then working in the systems engineering division of IBM Corporation. They derived the analytical solution for a linear material balance problem for the case when all variables are measured. In a series of papers between 1968 and 1976 [3, 4] several important ideas in data reconciliation and the optimal selection of measurements, particularly in linear processes, were introduced. These included the treatment of unmeasured variables, and the decomposition of the reconciliation and coaptation problems using a graph-theoretic approach.

The key concepts of *observability* and *redundancy* were also introduced in these papers. The classic paper by Mah et al. in 1976 [5] also treated the general linear data reconciliation problem including estimation of unmeasured variables. The interrelationship between linear algebraic and graph theoretic approaches were brought out in this paper. More importantly, the paper clearly demonstrated through simulation of a refinery process that data reconciliation does substantially improve accuracy especially when sufficient redundancy exists in the measurements. The problem of detecting gross errors caused by measurement biases and process leaks was also tackled in this work.

The next major contribution was the concept of a projection matrix introduced by Crowe et al. [6]. These authors decomposed the reconciliation and coaptation problems by using a projection matrix to eliminate the unmeasured variables. This approach is more general and can be used even if some of the unmeasured variables are unobservable. The use of the QR factorization in obtaining the projection matrix and in the solution of unmeasured variables was proposed by Swartz [7] and more recently by Sanchez and Romagnoli [8].

Data reconciliation for nonlinear processes was first addressed by Knepper and Gorman [9] who used the iterative technique proposed by Britt and Luecke [10] for parameter estimation in nonlinear regression. Their approach has some limitations as compared to the approach of successive linearization and use of projection matrix to solve the linearized subproblem proposed by Pai and Fisher [11]. In general, to solve the nonlinear data reconciliation problem which involves bounds and other inequality constraints, a constrained nonlinear optimization method has to be used. Tjoa and Biegler [12] made use of *successive quadratic pro-*

gramming (SQP) for solving a combined data reconciliation and gross error detection problem, as did Ravikumar et al. [13].

In parallel, methods for steady-state data reconciliation were being developed in the mineral processing area. One of the earliest applications of data reconciliation to a mineral processing circuit was published by Wiegel [14]. A representative sample of publications in this field are by Hodouin and Everell [15], Simpson et al. [16], and Heraud et al. [17], among others. A survey of computer packages for material balancing in mineral processing industries was published by Reid et al. [18].

The problem of data reconciliation in dynamic processes has received attention only recently, although it was first tackled using an extended Kalman filter by Stanley and Mah [19] who used a simple random walk model to describe the process dynamics. Almasy [20] used steady-state reconciliation techniques for dynamic balancing of a linear time invariant dynamic model of the process by considering the equivalent discrete input-output formulation. For a linear dynamic system, the optimal estimates are obtained using a Kalman filter which, however, cannot handle inequality constraints.

Dynamic data reconciliation has only recently been extended to nonlinear, constrained problems. Liebman et al. [21] have transformed the system of differential-algebraic equations describing a dynamic model into a standard *nonlinear program* (NLP) and reconciled the data using constrained nonlinear optimization methods. As compared to steady-state reconciliation which is increasingly being applied to industrial processes, it may take a few more years of development before dynamic data reconciliation is also ready and commercially available for industrial applications.

Within a few years after Kuehn and Davidson's paper on data reconciliation appeared, the problem of identifying gross errors in data and its importance in data reconciliation was pointed out by Ripps [22]. Ripps also proposed the procedure of measurement *elimination* as a technique for identifying the measurement containing a bias. This has now become one of the standard strategies in multiple gross error identification. Although statistical tests for gross error detection were proposed by Reilly and Carpani [23] as early as 1963, they did not attract much attention since they were presented in a conference paper. The *global test* and *measurement test* were proposed by Almasy and Sztano [24] in 1975 and the *nodal* or *constraint test* by Mah et al. [5] a year later.

More than a decade later, the *generalized likelihood ratio (GLR) test* was proposed by Narasimhan and Mah [25], the *Bayesian test* by Tamhane et al.

[26], and more recently the *principal component test* by Tong and Crowe [27]. Although, strategies for identifying the location of one or more gross errors were developed by Mah et al. [5] and Romagnoli and Stephanopoulos [28], a variety of serial elimination strategies using one or more statistical tests for multiple gross error identification were developed by Serth and Heenan [29] and Rosenberg et al. [30]. More importantly, they also compared the performance of these strategies through simulation for determining the best among them.

The method of simulation and its use in evaluating performance of gross error detection tests and strategies was first clearly explained by Jordache et al. [31]. Different measures for evaluating the performance of gross error detection strategies were also introduced in the above three papers. A different strategy called the serial compensation strategy for multiple gross error identification was proposed by Narasimhan and Mah [25]. Simultaneous strategies for multiple gross error detection have also been proposed by Rosenberg et al. [30] and more recently by Rollins and Davis [32]. The investigation for determining the best gross error detection method and for improving its performance is still being pursued.

Applications of data reconciliation to single process units either in the laboratory or in an operating plant were reported by Murthy [33], Madron et al. [34], Wang and Stephanopoulos [35], Crowe [36], Sheel and Crowe [37], among others who applied it to reactors, and by MacDonald and Howat [38] to a nonequilibrium isothermal flash unit. Reconciliation of flows in industrial processes were reported by Mah et al. [5] and Serth and Heenan [29], though it is not clear whether these were implemented in actual practice. Applications of data reconciliation to actual industrial processes were reported by Ravikumar et al.[13] and many other papers mentioned in Chapter 11. Development of commercial software for industrial applications of data reconciliation and gross error detection began in the late 1980s.

Excellent reviews of data reconciliation and gross error detection have been written at regular intervals by Hlavacek [39], Mah [40], Tamhane and Mah [41], Mah [42], and recently by Crowe [43]. The book by Mah [44] contains a chapter on this topic as does the book by Bodington [45]. Currently the only book wholly devoted to this area is by Madron [46], which has been revised and expanded recently by Ververka and Madron [47].

SCOPE AND ORGANIZATION OF THE BOOK

This book provides a summarized analysis of the various approaches to data reconciliation and gross error detection. Certain criteria for selecting various techniques and guidelines for their practical implementation are also indicated.

In Chapter 1, we have presented the need for data conditioning in process monitoring. Various signal processing and error reduction techniques were briefly mentioned. Data reconciliation which provides a model based error analysis and correction was introduced and illustrated by a simple example. Major concepts in data reconciliation, such as redundancy and observability, were also defined.

Chapter 2 introduces the statistical characterization of measurement errors and various univariate error reduction techniques. Data filtering, which is widely used for data conditioning, is described in more detail. Various filtering techniques are presented and compared.

Proceeding from Chapter 3 onward, the material is presented in increasing level of complexity. Chapter 3 describes the problem of steady-state linear reconciliation. Both theoretical and computational issues related to the linear data reconciliation are elucidated. Decomposition techniques for linear models with both measured and unmeasured variables are described here. Observability and redundancy are important issues for this case. Variable classification techniques related to the observability and redundancy concepts are therefore presented. Both graph-theoretical and matrix-based approaches are described.

Chapter 4 deals with steady-state data reconciliation for bilinear systems. Bilinear constraints, such as component material balances and certain heat balance equations occur frequently in many industrial reconciliation applications. Bilinear equations contain terms that are products of two random variables. Specialized reconciliation solution methods have been proposed for bilinear constraints. This chapter presents some of them along with their associated benefits and shortcomings.

Chapter 5 treats nonlinear data reconciliation. Nonlinear models are often used to accurately describe most chemical processes. Various techniques used for solving the nonlinear reconciliation problem are discussed. Some are based on successive linearization, while others are derived from general nonlinear programming techniques. The most efficient and widely used solution methods are presented in this chapter. Decomposition techniques for nonlinear problems are also analyzed. Inequality constraints such as bounds on variables are often imposed with nonlinear models in order to obtain a feasible solution. The treatment of inequality constraints is finally analyzed in this chapter.

In the previous chapters only steady-state processes are considered. Data reconciliation techniques for dynamic systems are discussed in

Chapter 6. The reconciliation problem for a linear dynamic process becomes a state estimation problem which can be solved via Kalman filtering methods. General optimization techniques have to be used for dynamic nonlinear problems which are described as part of nonlinear dynamic data reconciliation techniques.

While data reconciliation attempts to eliminate inaccuracies caused by random errors in measurements, gross error detection deals with the identification and removal of systematic biases in measurements and leaks. Chapter 7 introduces the issues involved in gross error detection and describes the basic statistical tests that can be used to detect gross errors. The underlying assumptions, characteristics and relative advantages and disadvantages of various statistical tests are also discussed. Interaction between gross error detection and data reconciliation is also highlighted.

In any industrial application using process data, it is very important to identify all gross errors, so they can be removed or appropriately accounted for. None of the statistical tests described in the previous chapter provides satisfactory gross error identification for all practical scenarios and more complex strategies are required. Chapter 8 describes some of the most successful such strategies. The applicability of these methods to nonlinear processes is further discussed. Finally, the effect of bounds or inequality constraints on gross error detection is analyzed.

The previous two chapters describe gross error detection methods for steady-state processes. Chapter 9 treats the gross error identification for dynamic systems. The dynamic feature of a process introduces new issues such as combining information from measurements collected over a period of time and on-line implementation.

The efficacy of data reconciliation and gross error detection depends significantly upon location of measured variables. Recent attempts to optimally design the sensor network for maximizing accuracy of data reconciliation solution and at minimum cost are described in Chapter 10.

Several large scale industrial applications and existent software systems for data reconciliation and gross error detection are discussed in Chapter 11. Various aspects such as the context of the industrial application, the problems associated with each type of application and the methods used to solve them are analyzed in this last chapter.

SUMMARY

- Measurement errors occur frequently in process instrumentation. Some errors are small and random (random errors), others are large and systematic (gross errors).
- Data validation and data filtering are used to reduce the errors in process data. Filtered data, however, usually do not satisfy the plant model.
- Data reconciliation exploits redundancy in process data in order to determine the necessary measurement adjustments used to create a set of data consistent with the plant model.
- No data reconciliation is possible without data redundancy (more measurements are available than the minimum needed to solve the simulation problem).
- Data reconciliation solution obtained from data with gross errors is not reliable because a large error spreads over other variables, causing unreasonable data adjustments.
- Data reconciliation and gross error detection are closely interrelated. They need to be implemented together in order to obtain a reliable data reconciliation.
- Statistical tests are useful tools for gross error detection.
- Location of instrumentation is important for both data reconciliation and gross error detection. An optimal sensor placement can be predetermined.
- Unmeasured variables and model parameters can be estimated by data reconciliation, providing that enough measured data is available in order to make them observable.
- Only natural material and energy conservation laws are acceptable for plant models used in data reconciliation. Correlations or approximate relations among process variables are not recommended, since they introduce additional sources of error.
- Data reconciliation can be applied to both steady-state and dynamic processes.

REFERENCES

1. Reklaitis, G. V. *Introduction to Material and Energy Balances.* New York: John Wiley & Sons, 1983
2. Kuehn, D. R., and Davidson, H. "Computer Control. II. Mathematics of Control." *Chem. Eng. Progress* 57 (1961) :44–47.
3. Vaclavek, V. "Studies on System Engineering. I. On the Application of the Calculus of Observations in Calculations of Chemical Engineering Balances." *Coll. Czech. Chem. Commun.* 34 (1968): 3653.
4. Vaclavek, V., and Loucka, M. "Selection of Measurements Necessary to Achieve Multicomponent Mass Balances in Chemical Plant." *Chem. Eng. Sci.* 31 (1976): 1199–1205.
5. Mah, R.S.H., G. M. Stanley, and D.W. Downing. "Reconciliation and Rectification of Process Flow and Inventory Data." *Ind. & Eng. Chem. Proc. Des. Dev.* 15 (1976): 175–183.
6. Crowe, C. M., Y.A.G. Campos, and A. Hrymak. "Reconciliation of Process Flow Rates by Matrix Projection. I: Linear Case." *AIChE Journal* 29 (1983): 881–888.
7. Swartz, C.L.E. "Data Reconciliation for Generalized Flowsheet Applications."American Chemical Society National Meeting, Dallas, Tex. (1989).
8. Sanchez, M., and J. Romagnoli. "Use of Orthogonal Transformations in Data Classification-Reconciliation." *Computers Chem. Engng.* 20 (1996): 483–493.
9. Knepper, J. C., and J. W. Gorman. "Statistical Analysis of Constrained Data Sets." *AIChE Journal* 26 (1980): 260–264.
10. Britt, H. I., and R. H. Luecke. "The Estimation of Parameters in Nonlinear Implicit Models." *Technometrics* 15 (1973): 233–247.
11. Pai, C.C.D., and G. D. Fisher. "Application of Broyden's Method to Reconciliation of Nonlinearly Constrained Data." *AIChE Journal* 34 (1988): 873–876.
12. Tjoa, I. B., and L. T. Biegler. "Simultaneous Strategies for Data Reconciliation and Gross Error Detection of Nonlinear Systems." *Computers Chem. Engng.* 15 (1991): 679–690.
13. Ravikumar, V., S. R. Singh, M. O. Garg, and S. Narasimhan. "RAGE—A Software Tool for Data Reconciliation and Gross Error Detection," in *Foundations of Computer-Aided Process Operations* (edited by D.W.T. Rippin, J. C. Hale, and J. F. Davis). Amsterdam: CACHE/Elsevier, 1994, 429–436.

14. Wiegel, R. I. "Advances in Mineral Processing Material Balances." *Canad. Metall. Q.* 11 (1972): 413–424.

15. Hodouin, D., and M. D. Everell. "A Hierarchical Procedure for Adjustment and Material Balancing of Mineral Processes Data." *Int. J. Miner. Proc.* 7 (1980): 91–116.

16. Simpson, D. E., V. R. Voller, and M. G. Everett. "An Efficient Algorithm for Mineral Processing Data Adjustment." *Int. J. Miner. Proc.* 31 (1991): 73–96.

17. Heraud, N., D. Maquin, and J. Ragot. "Multilinear Balance Equilibration: Application to a Complex Metallurgical Process." *Min. Metall. Proc.* 11 (1991): 197–204.

18. Reid K. J, K. A. Smith, V. R. Voller, and M. Cross. "A Survey of Material Balance Computer Packages in the Mineral Industry," in *17th Applications of Computers and Operations Research in the Mineral Industry* (edited by T. B. Johnson and R. J. Barnes). New York: AIME, 1982.

19. Stanley, G. M., and R.S.H. Mah. "Estimation of Flows and Temperatures in Process Networks." *AIChE Journal* 23 (1977): 642–650.

20. Almasy, G. A. "Principles of Dynamic Balancing." *AIChE Journal* 36 (1991): 1321–1330.

21. Liebman, M. J., T. F. Edgar, and L. S. Lasdon. "Efficient Data Reconciliation and Estimation for Dynamic Processes Using Nonlinear Programming Techniques." *Computers Chem. Engng.* 16 (1992): 963–986.

22. Ripps, D. L. "Adjustment of Experimental Data." *Chem. Eng. Progr. Symp. Ser. No. 55* 61 (1965): 8–13.

23. Reilly, P. M., and R. E. Carpani. "Application of Statistical Theory to Adjustment of Material Balances," presented at the 13th Can. Chem. Eng. Conf., Montreal, Quebec, 1963.

24. Almasy, G. A, and T. Sztano. "Checking and Correction of Measurements on the Basis of Linear System Model." *Prob. Control Inform. Theory* 4 (1975): 57–69.

25. Narasimhan, S., and R.S.H. Mah. "Generalized Likelihood Ratio Method for Gross Error Identification." *AIChE Journal* 33 (1987): 1514–1521.

26. Tamhane. A. C., C. Jordache, and R.S.H. Mah. "A Bayesian Approach to Gross Error Detection in Chemical Process Data. Part I: Model Development." *Chemometrics and Intel. Lab. Sys.* 4 (1988): 33.

27. Tong, H., and C. M. Crowe. "Detection of Gross Errors in Data Reconciliation by Principal Component Analysis." *AIChE Journal* 41 (1995): 1712–1722.

28. Romagnoli, J. A., and G. Stephanopoulos. "Rectification of Process Measurement Data in the Presence of Gross Errors." *Chem. Eng. Sci.* 36 (1981): 1849–1863.

29. Serth, R.W., and W. A. Heenan. "Gross Error Detecting and Data Reconciliation in Steam-Metering Systems." *AIChE Journal* 30 (1986): 743–747.

30. Rosenberg, J., R.S.H. Mah, and C. Jordache. "Evaluation of Schemes for Detecting and Identification of Gross Errors in Process Data." *Ind. & Eng. Chem. Proc. Des. Dev.* 26 (1987): 555–564.

31. Jordache, C., R.S.H. Mah, and A. C. Tamhane. "Performance Studies of the Measurement Test for Detecting of Gross Error in Process Data." *AIChE Journal* 31 (1985): 1187–1201.

32. Rollins, D. K., and J. F. Davis. "Unbiased Estimation Technique for Identification of Gross Errors." *AIChE Journal* 38 (1992): 563–571.

33. Murthy, A.K.S. "Material Balance around a Chemical Reactor, II." *Ind. Eng. Chem. Process. Des. Dev.* 13 (1974): 347.

34. Madron, F., V. Veverka, and V. Vanacek. "Statistical Analysis of Material Balance of a Chemical Reactor." *AIChE Journal* 23 (1977): 482–486.

35. Wang, N. S. and G. Stephanopoulos. "Application of Macroscopic Balances to the Identification of Gross Measurement Errors." *Biotechnol. Bioeng.* 25 (1983): 2177–2208.

36. Crowe, C. M. "Reconciliation of Process Flow Rates by Matrix Projection, II. The Nonlinear Case." *AIChE Journal* 32 (1986): 616–623.

37. Sheel, J. P., and C. M. Crowe. "Simulation and Optimization of an Existing Ethylbenzene Dehydrogenation Reactor." *Can. J. Chem. Eng.* 47 (1969): 183–187.

38. MacDonald, R. J., and C. S. Howat. "Data Reconciliation and Parameter Estimation in Plant Performance Analysis." *AIChE Journal* 34 (1988): 1–8.

39. Hlavacek, V. "Analysis of a Complex Plant—Steady State and Transient Behavior I—Plant Data Estimation and Adjustment." *Computers Chem. Engng.* 1 (1977): 75–81.

40. Mah, R.S.H. "Design and Analysis of Performance Monitoring Systems," in *Chemical Process Control II* (edited by D. E. Seborg and T. F. Edgar). New York: Engineering Foundation, 1982.

41. Tamhane, A. C., and R.S.H. Mah "Data Reconciliation and Gross Error Detection in Chemical Process Networks." *Technometrics* 27 (1985): 409–422.

42. Mah, R.S.H. "Data Screening," in *Foundations of Computer-Aided Process Operations* (edited by G. V. Reklaitis and H. D. Spriggs). Amsterdam: CACHE/Elsevier, 1987, 67–94.
43. Crowe, C. M., "Data Reconciliation—Progress and Challenges." *J. Proc. Cont.* 6 (1996): 89–98.
44. Mah, R.S.H. *Chemical Process Structures and Information Flows.* Boston: Butterworths, 1990.
45. Bodington, C. E. *Planning, Scheduling and Control Integration in Process Industries.* New York: McGraw-Hill, 1995.
46. Madron, F. *Process Plant Performance: Measurement and Data Processing for Optimization and Retrofits.* Chichester, West Sussex, England: Ellis Horwood Limited Co., 1992.
47. Veverka, V. V. and Madron, F. *Material and Energy Balancing in Process Industries: From Microscopic Balances to Large Plants.* Amsterdam, The Netherlands: Elsevier, 1997.

2

Measurement Errors and Error Reduction Techniques

CLASSIFICATION OF MEASUREMENT ERRORS

As mentioned in Chapter 1, there are many sources for instrument errors which determine a measurement error in virtually all measured process data. Some of the measurement errors are random and small (*random errors*), while others are systematic and large (*gross errors*). Some authors, such as Madron [1] and Liebman et al. [2], prefer to define a separate class called *systematic errors* which is distinguished from the gross error category. In their classification, systematic errors are consistent measurement biases, while the gross error class includes large measurement-related errors (biases and outliers), instrument complete failure or process-related errors such as process leaks.

In this text, in order to simplify the notation and terminology we classify all instrument and process errors in two categories: **random errors** and **gross errors** (as defined above). Any significant systematic bias is included in the gross error category.

Random Errors

It is generally observed that if the measurement of a process variable is repeated under identical conditions, the same value is not obtained. This is due to the presence of random errors in measurements. Random errors cannot be either predicted nor accurately explained. We choose to model the effect of random errors on measurements as additive contributions.

Thus, the relation between the measured value, true value and random error in the measurement of a variable *i* is expressed by Equation 1-6. In this chapter, unless otherwise required, we drop the subscript *i* and rewrite Equation 1-6 as

$$y = x + \varepsilon \tag{2-1}$$

where y is the measured value, x is the true value and ε is the random error. The random error usually oscillates around zero. Its characteristics can be described using statistical properties of random variables which are described in Appendix C. Its *mean* or *expected value* is therefore given by,

$$E(\varepsilon) = 0 \tag{2-2}$$

and its *variance*

$$\mathrm{var}(\varepsilon) = E[\varepsilon^2] = \sigma^2 \tag{2-3}$$

where σ is the standard deviation of the measurement error. *Standard deviation* is a measure of the measurement precision. The smaller the standard deviation, the more precise is the measurement and the higher the probability that the random error will be close to zero.

If the random errors in the measurements of two different variables *i* and *j* are also considered statistically independent, then they have zero correlation, that is,

$$\mathrm{cov}(\varepsilon_i, \varepsilon_j) = E(\varepsilon_i \varepsilon_j) = 0 \tag{2-4}$$

Although statistical independence does not always represent reality, this assumption is widely used in data reconciliation literature because it offers a simpler mathematical description of the measurement errors. Measurements obtained from two different instruments can be correlated if they share a common source of error (for example, a change in the ambient conditions affecting accuracy in a group of measuring devices). This type of correlation is known as *spatial correlation.* The degree of association between errors ε_i and ε_j is expressed by means of a *correlation coefficient,* r_{ij}:

$$r_{ij} = \frac{\mathrm{cov}(\varepsilon_i, \varepsilon_j)}{\sigma_i \ \sigma_j} \tag{2-5}$$

Equation 2-5 can be used to estimate $cov(\varepsilon_i,\varepsilon_j)$, because r_{ij} can be obtained from statistical analysis of a set of repeated measurements [3].

Another type of correlation occurs when the same source of the error persists for a number of measurement periods. In that case, the measurements errors of the same variable at different time instants are serially correlated. *Serial correlation* is also produced by delay time in control operations, due to unit capacity or inertia. For instance, if there is a delay of k time periods, an output-measured value $y_{out}(t)$ can be correlated with an input value at time t-k, say $y_{in}(t-k)$. Kao et al. [4] showed that neglecting serial correlations can be significant for gross error detection and suggested remedies to account for serially correlated data. A summarized statistical description of the serially correlated data, based on time series analysis, can be found in Mah [5].

Estimating the standard deviation. The standard deviation of a measurement error plays an important role in data reconciliation and various other error reduction techniques. Since the true standard deviation is never known, an estimate of the standard deviation can be obtained by using a sample standard deviation, according to the following formula [3]:

$$s = \frac{1}{N-1}\left[\sum_{i=1}^{N}(y_i - \bar{y})\right]^{1/2} \qquad (2\text{-}6)$$

where s is the estimated value of standard deviation, y_i is the *i*th observation and $\bar{y}$ is the arithmetic average of *N* observations of the same variable. This formula provides an unbiased estimate of the standard deviation. Sample size *N* is important for the reliability of the estimate. The more observations, the more reliable the estimate. Madron [1] indicates that a minimum of 15 observations (for a steady process variable) should be used.

An important requirement for estimating the standard deviation of a measurement error from a sample of measurements using the above equation is that all the measurements of the variable should be drawn from the same statistical population. Practically, this implies that if we use a sample of *N* measurements of a variable made at successive time instants for estimating the standard deviation of the measurement error, then it is implicitly assumed that during this time interval the true value of the variable has not changed. Moreover, it is also assumed that the measurement errors at different time instants all have the same standard deviation. Alternative ways to estimate the standard deviations (in fact, the entire covariance matrix of measurement errors) when the true values are not constant and when gross errors also occur are described in Chapter 3.

A complete mathematical description and statistical treatment of the random errors requires a *probability density function* (see Appendix C). This implies knowledge or assumption concerning the distribution type. The usual assumption in data reconciliation literature is that process data follow a normal distribution. Madron [1] summarizes the main reasons for selecting the normal distribution as follows:

1. It was found that the normal distribution approximates well the behavior of measurements in natural sciences, particularly within the range mean $\pm 3\sigma$.
2. An error is often the sum of a large number of single, elementary errors. According to the central limit theorem, under certain generally acceptable conditions, the distribution of such a sum approaches the normal distribution (for a large number of elementary errors).
3. The theory of normal model error is well developed and is easy to treat mathematically. The values of probability density and distribution functions for standard normal distribution are available in tabulated form in any statistical textbook which facilitates solving of practical problems.

One immediate practical use of the probability density function for the normal distribution is in estimating the standard deviation of a function of random variables. This problem is important for estimating standard deviation for a *secondary random variable* which is calculated based on some other directly measured variables, denoted as *primary variables.* It is assumed that probability properties (mean value, standard deviation) of the directly measured variables are known. For example, a flow rate F can be estimated by using pressure measurements in an orifice gauge according to the formula:

$$F = k(p_0 \Delta p / T)^{1/2} \qquad (2\text{-}7)$$

where k is the orifice gauge constant, Δp is the pressure difference on the orifice, p_0 is the inlet orifice pressure and T is the fluid temperature [1].

The mean value of a function $z = f(x)$ of random variables x is defined as:

$$\mu_z = E[f(x)] = \int_{-\infty}^{+\infty} \dots \int_{-\infty}^{+\infty} f(x)p(x)\, dx_1 \dots dx_m \qquad (2-8)$$

where p(x) is the m-variate probability density function of the vector of random variables x. If f(x) is a linear function, i.e.,

$$f(x) = \sum_i c_i x_i \qquad (2-9)$$

then the mean value of f(x) is also linear:

$$\mu_z = \sum_i c_i \mu_i \qquad (2-10)$$

The variance of a function f(x) of random variables is defined as

$$\sigma_z^2 = \int_{-\infty}^{+\infty} \dots \int_{-\infty}^{+\infty} \left[f(x) - \mu_z \right]^2 p(x)\, dx_1 \dots dx_m \qquad (2-11)$$

If f(x) is linear and the primary errors are uncorrelated, 2-11 reduces to

$$\sigma_z^2 = \sum_i c_i^2 \sigma_i^2 \qquad (2-12)$$

If f(x) is a nonlinear function—such as Equation 2-7 above—the problem becomes more complex and a solution can be obtained by either integrating the Equation 2-11, or by linearization of f(x) (Taylor expansion) which enables using Equation 2-12 for an approximate solution [1]. The latter approach gives rise to the following general approximation formula:

$$\sigma_z^2 = \sum_{i=1}^{m} \left[\frac{\partial f(x_1, \dots, x_m)}{\partial x_i} \right]^2 \sigma_{x_i}^2 \qquad (2-13)$$

A practical aspect of the estimation problem of the standard deviation for linear functions deals with computing an *overall accuracy for a measurement system.* For example, let us assume that three devices contribute to produce a measured value: a sensor, a transmitter, and a recorder. Each component has its own error and standard deviation. The overall error and standard deviation can be obtained by a linear combination of each component error and standard deviation. An overall standard deviation is obtained by using the Equation 2-12. Zalkind and Shinskey [6] used similar analytical derivations as Madron [1] and provide examples of estimating instrument error and standard deviation by combining component information.

An extension of these type of calculations is given in Nair and Jordache [7]. They applied the linear combination rule to estimate the *effective standard deviation* of a measurement system based on the information obtained from the accuracy and precision of the measurement system. The *accuracy* of the system is a measure of the agreement between the instrument reading and the true value. This information is provided by the instrument vendor and it is usually estimated by the linear combination rule applied to measuring and processing components as presented above. The *precision* of a measurement is a measure of the agreement of several repeated readings of the same measurement. The sample standard deviation estimated by Equation 2-6 is a measure for the repeatability of a measurement in steady conditions. An overall (effective) standard deviation can be estimated as

$$s_e^2 = s_a^2 + s_p^2 \tag{2-14}$$

where s_a is the instrument accuracy (usually given as a percentage of the instrument range) and s_p is the precision of the instrument (the repeatability standard deviation).

Gross Errors

A detailed definition of the gross errors was given in Chapter 1 and at the beginning of this chapter. Usually gross errors are associated with sensor faults. Figure 2-1, reproduced from Dunia et al. [8], illustrates graphically the most common types of instrument faults: bias, complete failure, drifting, and precision degradation.

If a gross error exists in a measured value, the measurement equation Equation 2-1 changes to:

$$y = x + \varepsilon + \delta \tag{2-15}$$

where δ is the magnitude of the gross error. Note that process leaks, which are also categorized as gross errors, cannot be modeled by Equation 2-15. They represent model errors and therefore affect the constraint equations as shown in Chapter 7.

Gross errors significantly affect the accuracy of any industrial application using process data. They have to be detected and removed. Some of

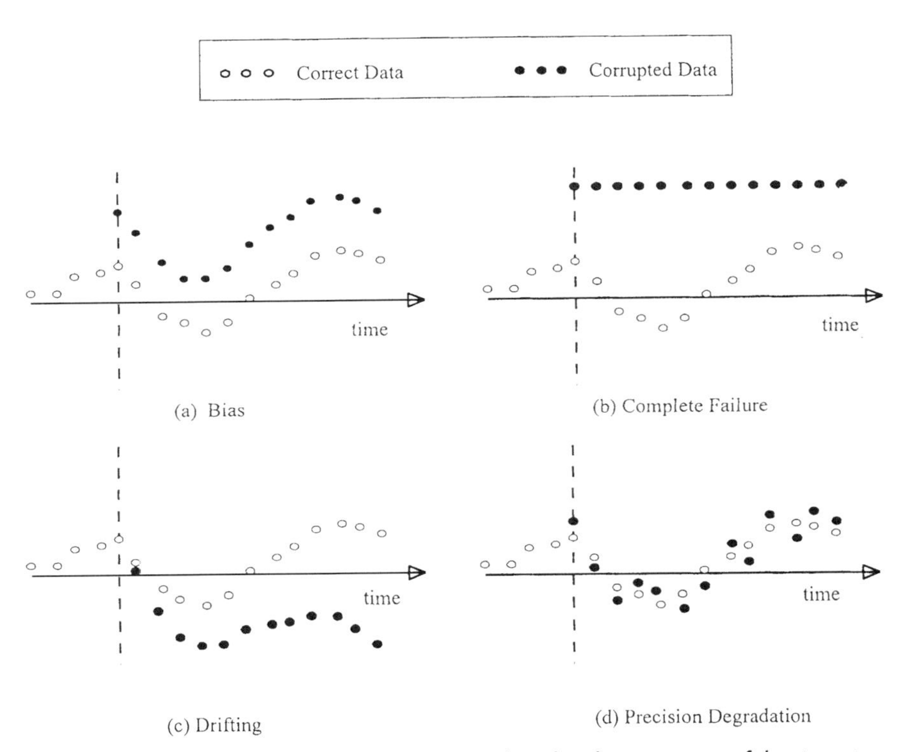

Figure 2-1. Instrument types of faults. *Reproduced with permission of the American Institute of Chemical Engineers. Copyright ©1996 AIChE. All rights reserved.*

them, such as occasional outliers (spikes), can be detected by using special filtering techniques or statistical quality control (also known as *statistical process control*). Other types might be more difficult to detect without a physical model. Data reconciliation is the appropriate tool in most cases.

ERROR REDUCTION METHODS

Analog and digital filters have been widely used to reduce random errors (high-frequency noise) in process values. Inadequate sampling frequency converts a high frequency signal into an artificial low-frequency signal. This phenomenon is known as *signal aliasing.* Analog filters are used to prefilter process data before sampling and prevent aliasing. Digital filters are used afterward to further attenuate high-frequency noise. Seborg et al. [9] provide a summarized presentation of both analog and digital filters for process data used in control applications. This text includes a review of various digital filters, which are very helpful tools for data conditioning before data reconciliation.

Various classical digital filters have been designed. Each filter type has its own advantages, as well as related shortcomings. Some are able to significantly reduce noise but they introduce a sizable delay in the filtered response. There are other filtering procedures that do not add a long delay but do not produce satisfactory noise removal. Other types of filters give both satisfactory noise removal and time delay in some cases, but perform poorly for measurement having variable frequencies or noise associated with a fast dynamic in the process variables. Overshooting/undershooting is a common problem with the last case.

In general, a trade-off between the amount of noise attenuation and the time delay in the filtered results is required in order to achieve the best performance for any type of filter. This can be accomplished by tuning the filter parameters which, unfortunately, is not an easy task. The random noise is often combined with instrument bias, slow drifts, or fast process changes, or other disturbances such as a cycling in the feedback control loops. To distinguish between a true process change and random noise, the dynamics of the process must be well known or a diagnostic system, such as an expert system, should be used. In the absence of such information, it is advisable to avoid excessive filtering. Too much filtering tends to mask significant changes in the process variables.

A brief description and analysis of the most widely used classical digital filters is given below. The discussion is restricted to *data filtering,* which means noise removal in the most recent measurements. Data filtering is to be distinguished from *data smoothing* which deals with past data. The former estimates the current value based on the current and past measurements and it is of primary concern in process control. The latter estimates the value of the central point from past and recent measurements (values from both sides of the central point) and it is mainly used for fault diagnosis and steady-state process optimization. Many authors, however, do not distinguish between the two terms and use the data "smoothing" term for data filtering as well.

An *integral of absolute errors* (IAE) similar to that of Kim and Lee [10] will be used to compare various filtering techniques in this text. Here the IAE is the summation of the absolute difference of the filtered values and the corresponding true values over a specified number of time steps. Note that Kim and Lee used smoothed values instead of true values, but using simulated true values gives a cleaner measure for the amount of filtering and delay. The lower the IAE, the more filtering (with reduced delay) is obtained.

Exponential Filters

This filter is by far the most commonly used in industrial applications. It is a discrete-time filter, equivalent to the first order lag in a continuous system, and is a standard filter incorporated in many DCS systems. It is also known as *first-order filter* in control area. It can be analytically described by the following equation:

$$y_k = \theta x_k = (1 - \theta) y_{k-1} \qquad (2\text{-}16)$$

where x_k = raw (unfiltered) measurement at time t_k
y_k = filtered value at time t_k
θ = filter parameter.

The exponential filter requires filter initialization: at $t_k = 0$, $y_0 = x_0$. The filter parameter θ is a tuning parameter with a range $0 < \theta \leq 1$. If θ is close to zero, significant filtering is obtained, while for θ close to 1, very little filtering is done. Note that the exponential filter is of the *infinite impulse response* (IIR) type, which means that the effect of any input signal is felt forever but with diminishing effects.

Exercise 2-1. *Derive Formula 2-16 from the first-order differential equation used in control literature for the first-order lag,*

$$\tau_f \frac{dy(t)}{dt} + y(t) = x(t)$$

where τ_f = *filter time constant, in units of time. Discuss the relationship between the filter time constant* τ_f *and the filter parameter* θ.

Hint: *Express the derivative* dy(t)/dt *as an approximation by a backward step:*

$$\frac{dy(t)}{dt} \approx \frac{y_k - y_{k-1}}{\Delta t}$$

where Δt = *time interval between two consecutive samples* $(t_k - t_{k-1})$ *and* $y(t) = y_k$; $x(t) = x_k$.

The exponential filter has several advantages. The effect of an impulse (spike) input x_k is immediately reduced to θx_k. It is computationally efficient and is easy to use and tune for steady state or slow dynamic signals (single parameter tuning). It does not overshoot and ultimately approaches the proper steady-state value. For these reasons, the exponential filter is used in many control systems. It has a problem, however, in that significant measurement noise attenuation is accompanied by relatively large delay in the filtered signal.

Example 2-1

Figure 2-2 illustrates the filtering provided by the exponential filter for two different parameters ($\theta = 0.2$ and $\theta = 0.4$) to a data set presented in Table 2-1. This particular data set contains two step signals, steady-state noise, and a spike (outlier). The true values are plotted with the dark solid line. Raw values were simulated by adding random errors (generated with a selected standard deviation) to the true values. A spike was also simulated in one data point.

The *integral absolute error* (IAE) defined above was included in the chart for performance comparison. As shown by the individual filter plots, the exponential filter with lower filter parameter ($\theta = 0.2$) indeed does more filtering in steady-state situations than the filter with $\theta = 0.4$.

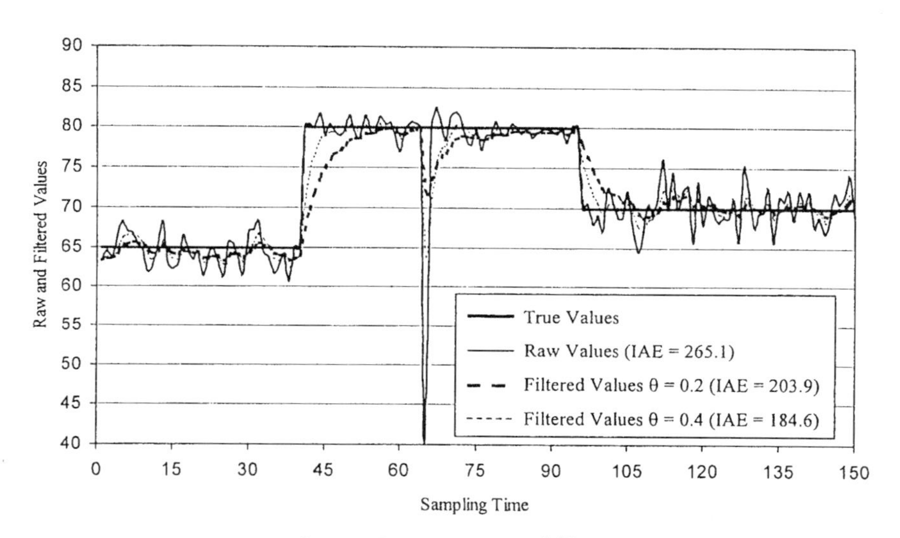

Figure 2-2. Exponential filters.

However, the overall IAE for $\theta = 0.2$ is higher than the IAE for $\theta = 0.4$ because of the increased delay after step changes and the spike. Tuning the exponential filters for noisy data accompanied by frequent step changes and occasional spikes is a challenging task.

Table 2-1. True Values and Raw Data for Example 2-1

Time	True Value	Raw Value
1	65.0000	63.2684
2	65.0000	64.4329
3	65.0000	63.7492
4	65.0000	66.3014
5	65.0000	68.1928
6	65.0000	67.0725
7	65.0000	66.7491
8	65.0000	65.2475
9	65.0000	64.5384
10	65.0000	61.8974
11	65.0000	62.4045
12	65.0000	65.3847
13	65.0000	68.1843
14	65.0000	62.9013
15	65.0000	62.1988
16	65.0000	62.9564
17	65.0000	66.4241
18	65.0000	64.6219
19	65.0000	63.4319
20	65.0000	64.2522
21	65.0000	61.3372
22	65.0000	63.7754
23	65.0000	64.8156
24	65.0000	63.1959
25	65.0000	61.2669
26	65.0000	65.6751
27	65.0000	65.3788
28	65.0000	63.5500
29	65.0000	61.5714
30	65.0000	66.9907
31	65.0000	67.1368
32	65.0000	68.2234
33	65.0000	64.3802
34	65.0000	63.7198
35	65.0000	61.6801
36	65.0000	62.6957
37	65.0000	63.8664
38	65.0000	60.5902
39	65.0000	65.0443
40	65.0000	65.0000
41	80.0000	80.0000
42	80.0000	80.3490
43	80.0000	80.0741
44	80.0000	81.7151

Table 2-1. True Values and Raw Data for Example 2-1 *(continued)*

Time	True Value	Raw Value
45	80.0000	79.0523
46	80.0000	80.3484
47	80.0000	79.5040
48	80.0000	79.0081
49	80.0000	79.7818
50	80.0000	81.4910
51	80.0000	79.2872
52	80.0000	78.6627
53	80.0000	81.4937
54	80.0000	79.6349
55	80.0000	79.3214
56	80.0000	81.3281
57	80.0000	80.4372
58	80.0000	80.5982
59	80.0000	78.7022
60	80.0000	77.0366
61	80.0000	79.5797
62	80.0000	80.6583
63	80.0000	80.4631
64	80.0000	80.0000
65	80.0000	40.0000
66	80.0000	80.0000
67	80.0000	82.4981
68	80.0000	80.6631
69	80.0000	78.4428
70	80.0000	81.2815
71	80.0000	81.9524
72	80.0000	80.9163
73	80.0000	78.6781
74	80.0000	77.8396
75	80.0000	79.6754
76	80.0000	77.2718
77	80.0000	78.9643
78	80.0000	79.2137
79	80.0000	80.7974
80	80.0000	79.1659
81	80.0000	79.1004
82	80.0000	80.4222
83	80.0000	80.1734
84	80.0000	79.2181
85	80.0000	79.8151
86	80.0000	78.7353
87	80.0000	79.1694
88	80.0000	79.6649
89	80.0000	79.8545
90	80.0000	79.5434
91	80.0000	79.2140
92	80.0000	79.8671
93	80.0000	78.1465
94	80.0000	80.3220
95	80.0000	80.0000

(table continued on next page)

Table 2-1. True Values and Raw Data for Example 2-1 *(continued)*

Time	True Value	Raw Value
95	80.0000	80.0000
96	70.0000	70.0000
97	70.0000	70.6065
98	70.0000	67.9913
99	70.0000	68.9857
100	70.0000	67.1895
101	70.0000	72.4049
102	70.0000	71.0063
103	70.0000	68.7471
104	70.0000	68.7806
105	70.0000	72.0691
106	70.0000	67.9624
107	70.0000	64.4622
108	70.0000	66.7040
109	70.0000	70.2901
110	70.0000	69.5325
111	70.0000	70.9931
112	70.0000	76.0272
113	70.0000	69.8086
114	70.0000	74.4137
115	70.0000	69.8983
116	70.0000	72.3646
117	70.0000	72.6424
118	70.0000	65.6905
119	70.0000	73.2806
120	70.0000	67.8898
121	70.0000	71.5254
122	70.0000	69.5290
123	70.0000	68.2254
124	70.0000	68.8187
125	70.0000	68.8263
126	70.0000	71.0346
127	70.0000	65.7383
128	70.0000	74.9540
129	70.0000	72.4687
130	70.0000	67.6565
131	70.0000	69.2903
132	70.0000	69.5289
133	70.0000	72.3914
134	70.0000	65.6847
135	70.0000	71.2073
136	70.0000	70.7821
137	70.0000	71.9797
138	70.0000	68.4319
139	70.0000	72.0655
140	70.0000	69.9292
141	70.0000	71.9856
142	70.0000	67.6774
143	70.0000	68.8004
144	70.0000	66.7914
145	70.0000	68.7419
146	70.0000	71.8966
147	70.0000	70.5520
148	70.0000	71.8154
149	70.0000	73.9484
150	70.0000	69.0318

Various modifications have been proposed to enhance the performance of the exponential filter:

1. Rhinehart [11] developed a method for *automatic tuning of the first-order filter.* The method assumes that the sampling period is small in comparison to the time for real process changes and that the noise is a random error which follows a normal distribution with zero mean. Instead of specifying the filter parameter θ or the filter time constant τ_f, the user needs to specify the desired confidence interval for the filter value (usually 95%). The method adjusts the filter time constant to minimize the time lag while maintaining the desired accuracy.
2. The *double exponential filter,* or second-order filter, is equivalent to two first-order filters in series where the second one filters the output from the first exponential filter. This type of filter was used by Tham and Parr [12] for signal reconstruction when an outlier is detected by a validation test. The classical derivation of this filter, coming from time series analysis is given in Seborg et al. [9].

> ***Exercise 2-2.*** *Derive the regular double filtering formula starting with Equation 2-16 and use α and β as parameters for the first and the second filter, respectively. For simplification, consider $\alpha = \beta$. Apply the obtained formula to calculate the filtered values and the IAE's for the data given in Table 2-1.*

3. The *nonlinear exponential filter* is another variation of the exponential filter (Weber, [13]). This filter heavily filters the noise, while reducing the delay. The nonlinear filter uses a design noise band, determined as a multiple R of standard deviations. The form of the filter is as Equation 2-16, except that the filter parameter θ is defined as

$$\theta = \min\left[1, \frac{x_k - y_{k-1}}{R\sigma}\right] \tag{2-17}$$

where: R = tuning parameter
σ = standard deviation of the measurement error

The relationship between the nonlinear exponential filter and the exponential filter is easy to obtain. Equation 2-16 can be written as

$$y_k = y_{k-1} + \theta(x_k - y_{k-1}) = y_{k-1} + \theta\Delta x \tag{2-18}$$

Then

$$\theta = \min\left[1, \frac{|\Delta x|}{R\sigma}\right] \tag{2-19}$$

i.e.,

$$\theta = \frac{|\Delta x|}{R\sigma} \quad \text{if } |\Delta x| < R\sigma \quad \text{and} \tag{2-20}$$

$$\theta = 1 \qquad \text{if } |\Delta x| > R\sigma \tag{2-21}$$

Since the true standard deviation is never known, a sample standard deviation can be used as an estimate. The filter parameter θ is then used in Equation 2-16 (as before, $0 < \theta \leq 1$). Typically, R lies between 3 and 5. If R is less than 3, little noise reduction is achieved. On the other hand, if R is greater than 5, significant delay results with a marginal improvement in the noise filtering.

The nonlinear filter acts like an exponential filter with a filter parameter θ that varies depending on the magnitude of the difference between filtered and raw measurements. The filter parameter θ is low for signals close to the previous filtered value and high for signals far from the previous filter value. Measurements far from the previous filtered value have $\theta = 1$ (no filtering at all outside the design noise band). Therefore, delay is eliminated for situations when there is a rapid significant measurement change. Since the nonlinear exponential filter is tuned for the frequency of a particular noise level, it performs optimally for signals whose noise level is steady. It is not recommended for filtering signals with spikes since such signals are not sufficiently filtered or not filtered at all.

Example 2-2

The filtering performance of the nonlinear exponential filter (with $\sigma = 1$ and $R = 4$) for the same data set as in previous examples, is shown in Figure 2-3. As noticed from the overall IAE for the filtered values, the performance of this filter is higher than that of the simple exponential filters presented in Example 2-1. The only problem with this filter is that it does not filter out spikes at all. Therefore, this filter type is not appropriate for signals with significant outliers.

> ***Exercise 2-3. Reverse nonlinear exponential filter.*** *Modify the definition of the filter parameter θ for the nonlinear exponential filter so it filters more data outside a chosen noise band and less (or no filtering at all) inside the noise band. Using data in Table 2-1, calculate filter values and IAE and plot the results as in Example 2-2. Describe a situation where the reverse nonlinear exponential filter can be useful. For what kind of signal is this filter type the most inappropriate?*

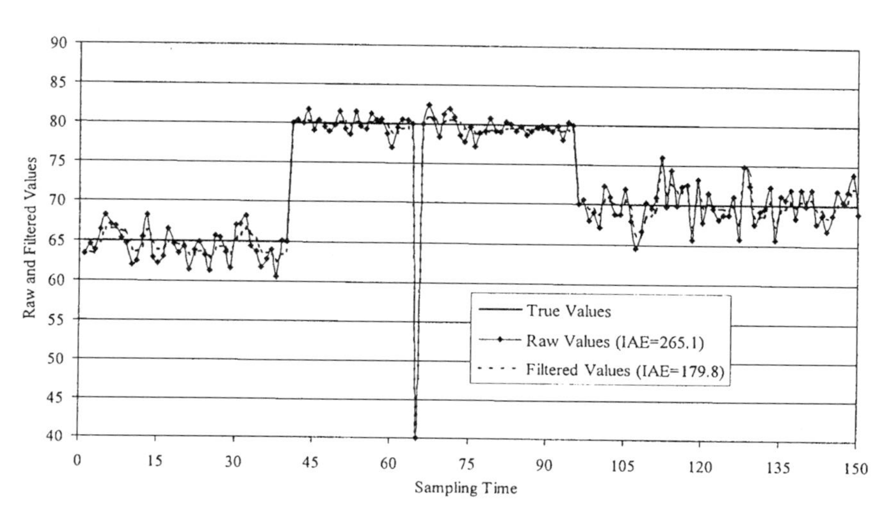

Figure 2-3. Nonlinear exponential filter.

Moving Average Filters

This is another common class of filters. The general analytical expression is:

$$y_k = \sum_{i=k-N+1}^{k} w_i x_i \tag{2-22}$$

where N = number of data points in filter
w_i = weight for the measurement x_i

In the absence of past data for the previous *N* time points, a history initialization is required:

$x_i = x_0, i = 1, \ldots, N-1; k = N$

The moving average is a *finite impulse response* (FIR) filter which means that the effect of any input lasts only for *N* steps. For the common form, all input data is given equal weight, i.e., $w_i = 1/N$. The equal weight moving average cancels out periodic noise. Like in the case of the exponential filter, the moving average is easy to tune for steady-state or quasi-steady-state signals, requiring only the adjustment of the number of input values used to calculate the average. Furthermore, as with the exponential filter, the moving average does not overshoot and reaches correct steady state after a step change. The moving average is also easy to implement and fast to compute, although it requires more storage and calculation than the exponential filter. Moreover, being an FIR filter, the moving average requires special initialization as shown above.

The moving average is most effective when estimating the center point rather than the current value. It is particularly useful for estimating a fixed value or a linear trend.

Example 2-3

Figure 2-4 illustrates the filtering provided by the moving average filter with equal weights for two different parameters ($N = 10$ and $N = 20$) to the data set presented in Table 2-1. As noticed from the individual filter plots, the moving average filter with higher number of data points ($N = 20$) does more filtering in steady-state situations than the filter with $N = 10$. However, the overall IAE for N = 20 is much higher than the IAE for N = 10, because of the increased delay after step changes and the spike. The previous history persists longer in the filtered values for the case with a larger number of data points. For this reason, the moving average filter with equal weights is not recommended for signals with step changes or spikes.

For dynamic data, a better performance can be obtained by using a moving average with unequal weights. As in the case with equal weights, the summation of all w_i weights over $i = 1, \ldots, N$ must be equal to 1. One possible set of such weights can be provided by the following formula [14]:

$$w_i = \frac{1}{\left[\dfrac{e^r - 1}{r} - 1\right]} \int_{(i-1)/N}^{1/N} (e^{rx} - 1)\,dx \tag{2-23}$$

where the weights w_i are exponentially increasing with i and satisfy the summation condition. Two tuning parameters are involved: the number of data points N and an exponent r. Usually $0 < r \leq 10$. For a fixed number of points N, a higher exponent r results in a higher weight for the more recent measurement, i.e., less filtering. A lower value for the exponent r provides more filtering but also adds more delay.

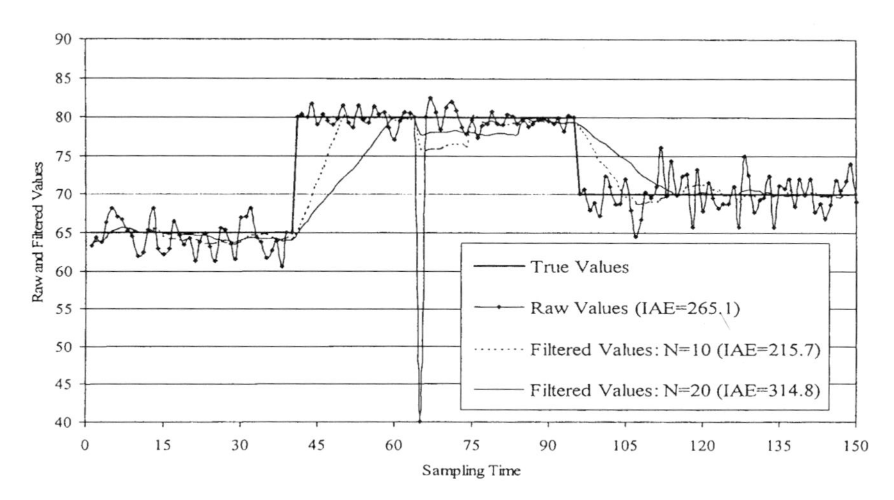

Figure 2-4. Moving average filters.

Exercise 2-4. *Prove that the summation over N of w_i's given by Equation 2-23 is equal to 1.*

Exercise 2-5. *Repeat the filter calculation for the data presented in Table 2-1 with the moving average filter with unequal weights as given by Equation 2-23. Choose N=20 and r=4. Compare the results with the results in Figure 2-4. Is there any incentive to use the more complex filter weights given by Equation 2-23 rather than a simple moving average filter with equal weights?*

Note that the filter with unequal exponential weights described above by Equation 2-22 and 2-23 is to be distinguished from the *exponentially weighted moving average (EWMA) filter* which is often used in statistical process control area. The EWMA filter is analytically described as [3]:

$$y_k = \lambda \bar{y}_k + (1-\lambda) y_{k-1} \qquad (2\text{-}24)$$

where $\bar{y}_k$ = sample mean (moving average with equal weights) at time t_k
y_k = filtered value at time t_k
λ = filter parameter ; $0 < \lambda < 1$

Initially, y_0 is taken as the control target μ_0 ($y_0 = \mu_0$). This filter can also be expressed as a weighted average of past sample means, i.e.,

$$y_k = \sum_{j=0}^{k-1} \lambda (1-\lambda)^j \bar{y}_{k-j} + (1-\lambda)^k \mu_0 \qquad k = 1, 2, 3, \ldots, \qquad (2\text{-}25)$$

Equation 2-25 indicates that the weights assigned to the sample means decrease geometrically with age. For that reason, this filter is sometimes referred to as the *geometric moving average filter* [3]. Some authors [8, 15], prefer to use the current measurement, x_k, instead of the sample mean $\bar{y}_k$. A recommended range for λ is $0.05 < \lambda < 0.5$ [3]. MacGregor [15] indicates that a common choice is $\lambda = 0.2$.

Polynomial Filters

Polynomial filters can be derived from the least-squares filters, which have been designed for data smoothing. While least-squares polynomials are widely used in data smoothing, they are also suitable for data filtering [10, 14]. The general form of the polynomial filter is shown by the following equation:

$$y_k = b_m t_k^m + b_{m-1} t_k^{m-1} + \ldots + b_2 t_k^2 + b_1 t_k + b_0 \quad (2\text{-}26)$$

where: t_k = current time
m = filter order (a nonzero, positive integer)
$b_0, \ldots, b_m$ = filter parameters chosen such that:

$$\min_{b_0,\ldots b_m} \sum_{i=1}^{N} (x_i - y_i)^2 \quad (2\text{-}27)$$

where: N = number of time steps (data points) included in the filter.

Polynomial filters are FIR type filters since they use a limited history of inputs. They provide good noise reduction, while following the overall trend of the measurement data. The amount of filtering and the delay depend on the number of data points used and the order of polynomial.

The disadvantage with Equation 2-26 for the polynomial filter is that the filter parameters $b_0, \ldots, b_m$ vary with each time step and must be recalculated by solving the least squares problem at each time step. Because of excessive computations, initial commercial applications were generally limited to first order filters.

An alternative form of the polynomial filter using time invariant filter parameters can be derived for measurement signals sampled at uniform intervals. Uniform sampling intervals are typical with digital sample data control systems. In this form, the polynomial filter becomes:

$$y_N = c_N x_N + c_{N-1} x_{N-1} + \ldots + c_1 x_1 \quad (2\text{-}28)$$

where $x_1, \ldots, x_N$ = unfiltered signal values at time $t_1, \ldots, t_N$
$c_1, \ldots, c_N$ = time invariant filter factors
$t_N - t_{N-1} = \ldots = t_2 - t_1$, uniform sampling interval

The important characteristic of this polynomial filter form is that the filter factors $c_1, \ldots, c_N$ are not functions of the time step as are the $b_0, \ldots, b_m$ parameters in the conventional formulation shown in Equation 2-26. Once the polynomial order *m*, and the number of points *N* are selected, the $c_1, \ldots, c_N$ filter factors are constants. For a particular set *m* and *N*, the filter factors $c_1, \ldots, c_N$ can be calculated by a procedure described in Exercise 2-6.

This form of the polynomial filter was developed and applied to industrial processes in the early 1970s and was recently published in first-order form [10, 14].

Exercise 2-6. *Prove Equation 2-28 for uniform sampling intervals and find an expression for calculating the filter factors* $c_1, \ldots, c_N$. ***Hint:*** *Derive the least-squares objective function 2-27 with respect to* $b_0, \ldots, b_m$ *and find first a regression equation for the vector* $\mathbf{b}$*, which has a general form* $\mathbf{b} = (\mathbf{P}^T\mathbf{P})^{-1}\mathbf{P}^T\mathbf{x}$ *where* $\mathbf{x}$ *is the vector of unfiltered signal values and* $\mathbf{P}$ *is some regression matrix (see [16]). Next, use Equation 2-26 written as* $y_k=[t_k^0\ t_k^1\ t_k^2 \ldots t_k^m]\mathbf{b}$*, and by replacing vector* $\mathbf{b}$ *with the previous regression, and taking* $k=N$ *(for the last or current data point) and assuming that* $t_N=N\Delta t$ *where* Δt *is a constant sampling interval, get an expression for* y_N *similar to Equation 2-28.*

When using a large number of points in a polynomial filter, it may be more convenient to use an analytical representation of the filter factors [14]. These forms of the filters can be derived by recognizing that the filter factors can be represented analytically as

$$c_i = a_m i^m + a_{m-1} i^{m-1} + \cdots + a_1 i^1 + a_0 \qquad (2\text{-}29)$$

where: $a_0, \ldots, a_m$ = filter factor coefficients

Substituting c_N factors from Equation 2-29 into Equation 2-28 gives a filter form in a_i's which greatly reduces data storage requirements. For example, a polynomial filter using 100 points in the form of Equation 2-28 requires storage of 100 *c* factors. The same filter in the form of Equation 2-29 for the *c* factors requires storage of only two *a* coefficients. The filter factor coefficients $a_0, \ldots, a_m$ can be determined by a least-squares solution of the non-square system of equations formed by writing Equation 2-29 repeatedly, for i=1,2, . . ., N.

For a given number of filter data points, the higher the polynomial order the more closely the filtered response follows the measurement data. The high-frequency noise is not removed, however. A low-order polynomial is usually preferred for filtering, although the lower the order, the larger the delay. Typically, a first or second order is used for most polynomial filters in process control systems. For a selected polynomial order, the only tuning parameter is the number of data points *N*. A high number of points gives a smoother output but also more delay. Overshooting is another negative behavior of polynomial filters. It occurs when there is a fast rate of change in a signal value and affects the result even after the signal becomes stabilized. The larger the number of points used by the filter, the more overshooting occurs.

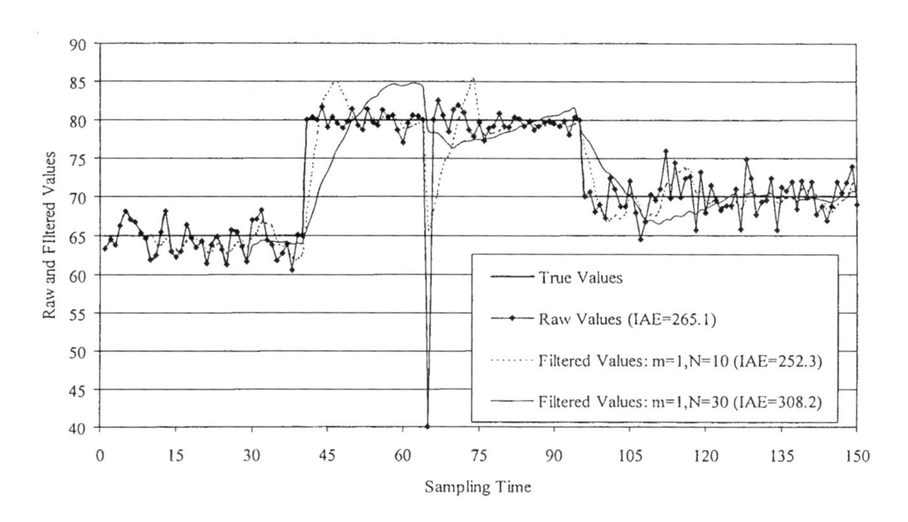

Figure 2-5. First-order polynomial filters.

Example 2-4

Figure 2-5 illustrates the filtering performance provided by a first-order polynomial filter with two different number of data points ($N = 10$ and $N = 30$) to the data set presented in Table 2-1. As shown by the two individual filter plots, the polynomial filter with higher number of data points ($N = 30$) does more filtering in steady-state situations than the filter with $N = 10$, but it takes longer to reach steady state after a step change. It also does more overshooting after a step change or a spike than the filter with the lower number of data points. Tuning the polynomial filter is relatively easier for steady-state signals but, as with the other filter types, it is more cumbersome for signals with step changes or spikes.

Hybrid Filters

As seen in the filter examples presented above, none of the individual filters behave in a satisfactory way for unsteady-state signals. The performance of classical digital filters can be enhanced by creating hybrids that combine the features of different filters. The simplest hybrid one can build is an arithmetic average of the filtered values from two types of filters. To create a better filter, the two participating individual filters values should have opposite features. For example, if one filter is able to significantly reduce noise but introduces a long delay, the other filter needs to create a much shorter delay although it might follow the noisy data too closely. For that reason, the most appropriate combinations for such hybrids are: polynomial/moving average, polynomial/exponential and exponential/nonlinear exponential.

A more complex hybrid that eliminates overshooting but follows the process dynamics is presented in Clinkscales and Jordache [14]. To achieve this behavior, their filter first detects the type of the change in the process value by analyzing the trend. A modified Shewhart test used in statistical process control [3] has been used to detect a change in the state of a variable as follows:

$$Z_i = \frac{X_i - X_a}{\sigma} \tag{2-30}$$

where X_i is the current (ith) raw value, X_a is a long-term average for the most recent steady-state situation and σ is the steady-state standard deviation of the measured variable X. If $|Z_i| > SCL$, where SCL is a selected Shewhart control limit, a significant change in process data is detected. This test is much simpler than the usual CUSUM tests used in statistical process control. It is easier to implement, and, with proper tuning, enables a faster state-change detection than the CUSUM tests. For this reason, the Shewhart test is often used in association with CUSUM tests in statistical process control [17].

Five major signal types can be determined with their algorithm: *steady-state, step change, ramp, impulse (spike),* and *undetermined transient state.* The filter is forced to follow more closely the process dynamics in the case of a true process change and to do more filtering in the case of the random noise (steady state) or spikes. To eliminate overshooting/undershooting, the filter value is limited to the maximum/minimum short term unequally weighted moving average (more weight on the most recent measurement).

A similar approach has been used by Tham and Parr [12]. They also apply statistical tests to determine if there is a trend in the data. They classify the trends in three categories: (a) a distinct trend; (b) a trend that, due to noise, is not immediately discernible; and (c) no trend. Additional tests are applied to detect outliers for each type of trend. If an outlier is found, a signal reconstruction formula is used to estimate a value which is used to replace the outlier.

Exercise 2-7. *Create hybrids (simple arithmetic averages) for polynomial/exponential and polynomial/moving average using data from Table 2-1. Use $\theta = 0.2$ for the exponential filter, $N=10$ for the moving average and $N=20$ for the first-order polynomial filter. For the first-order polynomial filter and $N = 20$, the filter factor coefficients are as follow: $a_0 = -0.1$ and $a_1=0.01429$; the filter factors c_i can be calculated with Equation 2-29. Compare the results with those obtained with the individual filters.*

There are many other error reduction techniques, but analyzing all of them is beyond the scope of this book. Statistical process control represent an area of special interest for data validation and data conditioning in connection with process control applications [3, 15, 17, 18]. Stanley [19] provided an excellent review of almost all known error reduction methods, including data reconciliation. The focus of this textbook, however, is data reconciliation which exploits redundancy in process data to more accurately adjust the process values and detect gross errors.

SUMMARY

- Random errors can be naturally described by a normal probability distribution which is suitable to most measurements associated with physical sciences. Other distribution types can also be used.
- Standard deviations of secondary random variables can be estimated from the standard deviation of the primary random variables.
- Individual filtering techniques can be used for error reduction in process measurements, but they are not easy to tune. Some reduce significantly the errors, but with large delay. Others have less delay, but overshoot/undershoot after a true step process change.
- Hybrid filters perform better than individual digital filters, but for best performance they need to be able to recognize the type of process signal.

REFERENCES

1. Madron, F. *Process Plant Performance: Measurement and Data Processing for Optimization and Retrofits.* Chichester, West Sussex, England: Ellis Horwood Limited Co., 1992.
2. Liebman, M. J., T. F. Edgar, and L. S. Lasdon. "Efficient Data Reconciliation and Estimation for Dynamic Processes Using Nonlinear Programming Techniques." *Computers Chem. Engng.* 16 (no. 10/11, 1992): 963–986.
3. Wadsworth, H. M. *Handbook of Statistical Methods for Engineers and Scientists.* New York: McGraw-Hill, 1990.
4. Kao, C. S., A. C. Tamhane, and R.S.H. Mah. "Gross Error Detection in Serially Correlated Process Data." *Ind. & Eng. Chem. Research* 29 (no. 6, 1990): 1004–1012.
5. Mah, R.S.H. *Chemical Process Structures and Information Flows.* Boston: Butterworths, 1990.
6. Zalkind, C. S., and F. G. Shinskey. "Statistical Methods for Computing Over-All System Accuracy." *ISA Journal* (Oct. 1963): 63–66.
7. Nair, P., and C. Jordache "Rigorous Data Reconciliation is Key to Optimal Operations." *Control for the Process Industries,* Vol. IV, no. 10, pp. 118–123. Chicago: Putman, 1991.
8. Dunia, R., S. J. Qin, T. F. Edgar, and T. J. McAvoy. "Identification of Faulty Sensors Using Principal Component Analysis." *AIChE Journal* 42 (no. 10, 1996): 2797–2812.
9. Seborg, D. E., T. F. Edgar, and D. A. Mellichamp. *Process Dynamics and Control.* New York: John Wiley & Sons, 1989.
10. Kim Y. H., and J. M. Lee. "Improve Process Measurements with a Least Squares Filter." *Hydrocarbon Processing* (Aug. 1992): 143–146.
11. Rhinehart, R. R. "Method for Automatic Adaptation of the Time Constant for a First-Order Filter." *Ind. & Eng. Chem. Research* 30 (no. 1, 1991): 275–277.
12. Tham, M. T., and A. Parr. "Succeed at Online Validation and Reconstruction of Data." *Chem. Eng. Progress* (May 1994): 46–56.
13. Weber, R. "Measurement Smoothing with a Nonlinear Exponential Filter." *AIChE Journal* 26 (no. 1, 1980): 132–134.
14. Clinkscales, T. A., and C. Jordache. "Hybrid, Digital Filtering Techniques for Process Data Noise Attenuation with Reduced Delay," presented at the AIChE Spring National Meeting, Atlanta, Ga., April 1994.

15. MacGregor, J. F. "On-line Statistical Process Control." *Chem. Engng. Progress* (Oct. 1988): 21–31.

16. Montgomery, D. C., and E. A. Peck. *Introduction to Linear Regression Analysis.* New York: John Wiley & Sons, 1982.

17. Lucas, J. M. "Combined Shewhart—CUSUM Quality Control Schemes," *Journal of Quality Technology* 14 (no. 2, 1982): 51–59.

18. Rinehart, R. R. "A CUSUM Type On-line Filter." *Process Control and Quality* (Amsterdam: Elsevier) (no. 2, 1992): 169–176.

19. Stanley, G. M. "Avoiding the Chattering Rule: Filtering and Other Techniques for Ignoring Noisy Data but Noticing Instrument Problems," presented at Gensym Diagnostics Working Group, Woodlands, Tex., Oct. 1, 1992.

3

Linear Steady-State Data Reconciliation

Linear data reconciliation for steady-state systems has already been introduced in Chapter 1. The examples analyzed in Chapter 1 are instances of a linear data reconciliation problem. The general formulation and solution of linear data reconciliation problems is discussed in this chapter. Vector notation is used in this and subsequent chapters because it provides a compact representation and allows powerful concepts from linear algebra and matrix theory to be exploited. Appendix A provides an introduction to some basic concepts of vectors and matrices.

LINEAR SYSTEMS WITH ALL VARIABLES MEASURED

As shown in Chapter 1, the simplest data reconciliation problem involves a linear model with all variables directly measured. We also assume that the measurements do not contain any systematic biases.

General Formulation and Solution

The model for the measurements described by Equation 1-6 can be written as

$$\mathbf{y} = \mathbf{x} + \varepsilon \tag{3-1}$$

where $\mathbf{y}$ is a vector of n measurements, $\mathbf{x}$ is the corresponding vector of true values of the measured variables and ε is the vector of unknown random errors. Although in Equation 3-1 we have assumed that the measurements and variables have one to one correspondence, it does not impose any limitation on the applicability of the method. Other forms of the measurement model in which the variables are assumed to be indirectly measured can be converted to the above model using appropriate transformations. These issues are discussed in Chapter 7 along with gross error detection strategies.

The constraints described by Equations 1-7a through 1-7d can be represented in general by

$$\mathbf{Ax} = \mathbf{0} \tag{3-2}$$

where $\mathbf{A}$ is a matrix of dimension $m \times n$, and $\mathbf{0}$ is a $m \times 1$ vector whose elements are all zero. Each row of Equation 3-2 corresponds to a constraint. It can be easily verified that for a flow reconciliation problem, the elements of each row of matrix $\mathbf{A}$ are either +1, −1 or 0, depending on whether the corresponding stream flow is input, output or, respectively, not associated with the process unit for which the flow balance is written. In general, if some of the variables are known exactly, the RHS of Equation 3-2 is a constant nonzero vector, $\mathbf{c}$.

The objective function, Equation 1-9, can be represented in general by

$$\underset{x}{\text{Min}} \ (\mathbf{y} - \mathbf{x})^T \mathbf{W} (\mathbf{y} - \mathbf{x}) \tag{3-3}$$

The $n \times n$ matrix $\mathbf{W}$ is usually a diagonal matrix, the diagonal elements representing the weights as in Equation 1-5. However, in general, it can also contain nonzero off-diagonal elements. The interpretation of the elements of $\mathbf{W}$ in terms of the statistical properties of the errors ε is discussed in the next section.

The analytical solution to the above problem can be obtained using the method of Lagrange multipliers [1, 2]

$$\hat{\mathbf{x}} = \mathbf{y} - \mathbf{W}^{-1}\mathbf{A}^T(\mathbf{A}\mathbf{W}^{-1}\mathbf{A}^T)^{-1}\mathbf{A}\mathbf{y} \tag{3-4}$$

where we have denoted the solution for the estimates using the notation $\hat{\mathbf{x}}$. In deriving the above solution it is assumed that the matrix $\mathbf{A}$ is of full row rank, which implies that there are no linearly dependent constraints in Equation 3-2. If the RHS of Equation 3-2 is not identically zero, but a known constant vector $\mathbf{c}$, then the estimates are obtained by replacing the vector $\mathbf{Ay}$ in Equation 3-4 by $\mathbf{Ay}-\mathbf{c}$.

> ***Exercise 3-1.*** *Using the method of Lagrange multipliers for minimizing 3-3 subject to equality constraints, Equation 3-2, obtain the solution for the reconciled estimates given by Equation 3-4.*

The estimates given by Equation 3-4 satisfy the constraints. In practice, there may be bounds on the variables. Since these have not been included in the reconciliation problem, the estimates obtained using Equation 3-4 may not satisfy these bounds and in some cases negative values for the estimates may be obtained which are physically meaningless. The only way to obtain physically meaningful results in such cases is to impose bounds on variables as additional constraints. But this complicates the problem significantly and an analytical solution can no longer be obtained. The solution of the bounded data reconciliation problem is discussed in Chapter 5.

> ***Exercise 3-2.*** *Prove that the estimates given by Equation 3-4 satisfy the constraints of Equation 3-2.*

Statistical Basis of Data Reconciliation

So far we have described the formulation of the data reconciliation problem from a purely intuitive viewpoint, especially with regard to the selection of the objective function weights to be used for different measurements. The data reconciliation problem can also be explained using a statistical theoretical basis, which not only helps in understanding this subject better, but also provides useful quantitative information about the improvement in the accuracy of the data obtained through reconciliation and the statistical properties of the resulting estimates. These can be used to identify grossly incorrect data or to design sensor networks as described in Chapter 10.

The statistical basis for data reconciliation arises from the properties that are assumed for the random errors in the measurements. Generally, as mentioned in Chapter 2, it is assumed that the random errors follow a multivariate normal distribution with zero mean, and a known variance-covariance matrix Σ. However, it should be kept in mind that sometimes the primary measured signal is transformed into the final indicated variable of interest. If the transformation is nonlinear such as Equation 2-7, then the error in the indicated variable need not be normally distributed.

As indicated in Chapter 2, only the linearized form can be approximated by a normal distribution. Thus, if possible, the variables **x** in the measurement model of Equation 3-1 should represent the primary measured variables, and relationships between the primary measured variable and the variables of interest should be included as constraints. In case the constraints are nonlinear, then a nonlinear data reconciliation technique as described in Chapter 5 can be used to solve the problem.

The matrix Σ contains information about the accuracy of the measurements and the correlations between them. The diagonal element of Σ, σ_i^2 is the variance in measured variable *i*, and the off-diagonal element σ_{ij}^2 is the covariance of the errors in variables *i* and *j*. If the measured values are given by the vector **y**, then the most likely estimates for **x** are obtained by maximizing the likelihood function of the multivariate normal distribution:

$$\underset{x}{\text{Max}} \frac{1}{(2\pi)^{n/2} |\Sigma|^{n/2}} \exp\{-0.5(\mathbf{y}-\mathbf{x})^T \Sigma^{-1} (\mathbf{y}-\mathbf{x})\} \tag{3-5}$$

where |Σ| is the determinant of Σ. The above maximum likelihood estimation problem is equivalent to minimizing the function

$$\underset{x}{\text{Min}} \ (\mathbf{y}-\mathbf{x})^T \Sigma^{-1} (\mathbf{y}-\mathbf{x}) \tag{3-6}$$

The estimates are also required to satisfy the constraints, Equation 3-1. Comparing Equations 3-6 and 3-3, we note that the formulation of the data reconciliation problem from a statistical viewpoint, simply requires that the weight matrix **W** be chosen to be the inverse of the covariance matrix Σ. This choice is also reasonable, if we consider the matrix Σ to be diagonal. In this case, Equation 3-6 becomes

$$\underset{x}{\text{Min}} \sum_{i=1}^{n} (y_i - x_i)^2 / \sigma_i^2 \tag{3-7}$$

where σ_i is the standard deviation of the error in measurement *i*. Equation 3-7 shows that the weight factor for each measurement is inversely proportional to the standard deviation of its error. Since a higher value of standard deviation implies that the measurement is less accurate, the above choice gives larger weights to more accurate measurements. Another advantage of using Equation 3-7 is that it is dimensionless since the standard deviation of a measurement error has the same units as the measurement. The estimates can now be obtained using Equation 3-4 by replacing **W** with Σ^{-1}.

It is also now possible to derive the statistical properties of the estimates obtained through data reconciliation. Consider the case when all the variables are measured. The estimates are given by

$$\hat{\mathbf{x}} = \mathbf{y} - \Sigma\mathbf{A}^T(\mathbf{A}\Sigma\mathbf{A}^T)^{-1}\mathbf{A}\mathbf{y} = [\mathbf{I}-\Sigma\mathbf{A}^T(\mathbf{A}\Sigma\mathbf{A}^T)\mathbf{A}]\mathbf{y} = \mathbf{B}\mathbf{y} \quad (3\text{-}8)$$

Equation 3-8 shows that the estimates are obtained using a linear transformation of the measurements. The estimates, therefore, are also normally distributed, with expected value and covariance matrix given by

$$E[\hat{\mathbf{x}}] = \mathbf{B}E(\mathbf{y}) = \mathbf{B}\mathbf{x} = \mathbf{x} \quad (3\text{-}9)$$

$$Cov[\hat{\mathbf{x}}] = E\{(\mathbf{B}\mathbf{y})(\mathbf{B}\mathbf{y})^T\} = \mathbf{B}\Sigma\mathbf{B}^T \quad (3\text{-}10)$$

Equation 3-9 implies that the estimates are unbiased, which is a property of maximum likelihood estimates for the linear systems. Equation 3-10 gives a measure of the accuracy of the estimates. In the case where some of the variables are unmeasured, it is possible to derive similar properties. These statistical properties are exploited to identify measurements with gross errors as well as to design sensor networks.

LINEAR SYSTEMS WITH BOTH MEASURED AND UNMEASURED VARIABLES

For partially measured systems, the reconciliation problem is usually solved by decomposing it into two subproblems [3, 4]. In the first subproblem, the redundant measured variables are reconciled, followed by a coaptation problem in which the observable unmeasured variables are estimated. This strategy is more efficient than an attempt to estimate all the variables simultaneously. The general formulation and solution of the reconciliation problem for partially measured systems is now described.

Let the number of unmeasured variables be p. The variables are classified into two sets, the vector **x** of measured variables and the vector **u** of unmeasured variables. The measurement model is still given by Equation 3-1 and the objective function by 3-6. However, the constraints have to be recast in terms of both the *measured* and *unmeasured* variables. Equation 3-2 is written as

$$\mathbf{A}_x\mathbf{x} + \mathbf{A}_u\mathbf{u} = 0 \tag{3-11}$$

where the columns of $\mathbf{A}_x$ correspond to the measured variables and those of $\mathbf{A}_u$ correspond to the unmeasured variables. Matrices $\mathbf{A}_x$ and $\mathbf{A}_u$ are of dimensions $m \times n$ and $m \times p$, respectively.

The unmeasured variables $\mathbf{u}$ have to be eliminated from Equation 3-11 using suitable linear combinations of the constraints. This is equivalent to premultiplying the constraints by a matrix $\mathbf{P}$, also known as a *projection matrix* [4]. The matrix $\mathbf{P}$ should satisfy the property

$$\mathbf{P}\mathbf{A}_u = \mathbf{0} \tag{3-12}$$

Premultiplying Equation 3-11 by matrix $\mathbf{P}$, we get the reduced set of constraints involving only measured variables as

$$\mathbf{P}\mathbf{A}_x\mathbf{x} = \mathbf{0} \tag{3-13}$$

The number of columns of $\mathbf{P}$ should clearly be equal to the number of constraints, m. As many independent rows as possible are constructed for $\mathbf{P}$ which satisfy property 3-12. The number of such rows, t, is linked to the observability of the unmeasured variables. If all the unmeasured variables are observable as in Cases 1 and 2 of Example 1-2, then t is equal to $m - p$. This can be easily inferred by noting that t is equal to the number of constraints in the reduced constraints of Equation 3-13. If all p unmeasured variables can be uniquely estimated, then this requires p of the constraint equations. Thus, only the remaining $m - p$ constraints are available for reconciling the measured variables. It can also be proved that for all the unmeasured variables to be observable, the p columns of $\mathbf{A}_u$ should be independent.

> ***Exercise 3-3.*** *Prove that if the columns of matrix $\mathbf{A}_u$ are linearly independent, then unique estimates for the variables $\mathbf{u}$ exist.*

> ***Exercise 3-4.*** *Solve the reconciliation problem for the case of linear constraints with constant terms:*
> $\boldsymbol{A_x x} + \boldsymbol{A_u u} = \mathbf{c}$, *when the columns of* $\boldsymbol{A_u}$ *may or may not be linearly independent.*

If not all unmeasured variables are observable, then t is equal to m-s, where s is the number of independent columns of $\mathbf{A}_u$. The interpretation of this result can be done as follows. Since the unmeasured variables cannot be uniquely estimated, the estimates of a few of the unmeasured variables have to be additionally specified in order to uniquely solve for the remaining unmeasured variables. If $p - s$ is the number of unmeasured variables whose estimates have to be additionally specified, then to solve for the other s unmeasured variables requires s of the constraint equations, resulting in $m - s$ remaining constraints for reconciliation.

A comparison with Case 3 of the Example 1-2 shows that $m = 4$, $n = 6$, and $p = 4$. However, only three of the columns of $\mathbf{A}_u$ are independent, and an estimate of one of the unmeasured among x_2 to x_5 have to be additionally specified in order to estimate all the other variables. Thus, $p - s = 1$ or $s = 3$, and the number of constraints in the reduced set, m–s is equal to 1 as observed from Equation 1-15. Note that the number of constraints in the reduced set is also known as *degrees of redundancy.*

The reduced data reconciliation problem is to minimize 3-6 subject to the constraints, Equation 3-13. Since the constraints are similar to Equation 3-2, the reconciled values for **x** can be obtained using Equation 3-8, with the matrix **A** being replaced by the reduced matrix $\mathbf{PA}_x$.

$$\hat{\mathbf{x}} = \mathbf{y} - \Sigma(\mathbf{PA}_x)^T[(\mathbf{PA}_x)\Sigma(\mathbf{PA}_x)^T]^{-1}(\mathbf{PA}_x)\mathbf{y} \tag{3-14}$$

Using Equation 3-14, we can now substitute for **x** in Equation 3-12 and obtain the estimates $\hat{\mathbf{u}}$ for the variables **u**, provided all the variables are observable (or the columns of $\mathbf{A}_u$ are independent). Since $\mathbf{A}_u$ is a $m \times p$ matrix with $p < m$, a least-squares approximate solution can be used. From the theory of generalized inverse [5], the least-squares solution is given by

$$\hat{\mathbf{u}} = -(\mathbf{A}_u^T \mathbf{A}_u)^{-1} (\mathbf{A}_x \hat{\mathbf{x}}) \tag{3-15}$$

The general solution for $\hat{\mathbf{u}}$ for when all the variables are not observable is developed in the next section. The decomposition strategy described above is also useful for data reconciliation of processes with nonlinear constraints as described in Chapter 5. The only additional issue to be discussed is the construction of the projection matrix **P**, which follows.

The Construction of a Projection Matrix

There are several different matrix methods for the construction of the projection matrix. One such method is given by Crowe [4]. However, probably the most efficient method is to use the QR factorization [5] of the matrix $\mathbf{A}_u$. Such a method was first applied to data reconciliation by Swartz [6] and recently utilized by Sanchez and Romagnoli [7] to decompose and solve linear and bilinear data reconciliation problems.

Consider the case when the columns of the $m \times p$ matrix $\mathbf{A}_u$ are linearly independent. Then it is possible to factorize $\mathbf{A}_u$ as

$$\mathbf{A}_u = \mathbf{QR\Pi}_u = \begin{bmatrix} \mathbf{Q}_1 & \mathbf{Q}_2 \end{bmatrix} \begin{bmatrix} \mathbf{R}_1 \\ \mathbf{0} \end{bmatrix} \mathbf{\Pi}_u \tag{3-16}$$

where Π_u is a permutation matrix (that is, the columns of Π_u are the permuted columns of the identity matrix), $\mathbf{R}_1$ is a nonsingular $p \times p$ upper triangular matrix, and **Q** is a $m \times m$ orthogonal matrix, that is,

$$\mathbf{Q}^T\mathbf{Q} = \mathbf{I} \tag{3-17}$$

In essence, the columns of **Q** form a basis for the m-dimensional space, while the matrix $\mathbf{R}_1$ represents the p columns of $\mathbf{A}_u$ in terms of the first p basis vectors, $\mathbf{Q}_1$. Since **Q** is orthogonal, the matrix $\mathbf{Q}_2$ has the property

$$\mathbf{Q}_2^T\mathbf{A}_u = \mathbf{Q}_2^T\begin{bmatrix} \mathbf{Q}_1 & \mathbf{Q}_2 \end{bmatrix} \begin{bmatrix} \mathbf{R}_1 \\ \mathbf{0} \end{bmatrix} \Pi_u = \begin{bmatrix} \mathbf{0} & \mathbf{I} \end{bmatrix} \begin{bmatrix} \mathbf{R}_1 \\ \mathbf{0} \end{bmatrix} \Pi_u = \mathbf{0} \tag{3-18}$$

From Equation 3-18, it is clear that the matrix $\mathbf{Q}_2^T$ is the desired projection matrix **P**.

The QR factorization is also useful in estimating the unmeasured variables easily. Using the QR factorization, Equation 3-11 can be written as

$$\mathbf{A}_x\mathbf{x} + \mathbf{QR}\Pi_u\mathbf{u} = \mathbf{0} \tag{3-19}$$

where $\Pi_u\mathbf{u}$ is a reordered vector $\mathbf{u}$. Premultiplying Equation 3-19 by $\mathbf{Q}^T$ we get

$$\mathbf{Q}^T\mathbf{A}_x\mathbf{x} + \mathbf{R}\Pi_u\mathbf{u} = \mathbf{0} \tag{3-20}$$

or, rearranging

$$\mathbf{R}\Pi_u\mathbf{u} = -\mathbf{Q}^T\mathbf{A}_x\mathbf{x} \tag{3-21}$$

Using Equation 3-16 for $\mathbf{R}$ in Equation 3-21 we get

$$\begin{bmatrix}\mathbf{R}_1\\ \mathbf{0}\end{bmatrix}\Pi_u\mathbf{u} = -\begin{bmatrix}\mathbf{Q}_1^T\\ \mathbf{Q}_2^T\end{bmatrix}\mathbf{A}_x\mathbf{x} \tag{3-22}$$

or,

$$\mathbf{R}_1\Pi_u\mathbf{u} = -\mathbf{Q}_1^T\mathbf{A}_x\mathbf{x} \tag{3-23}$$

Since $\mathbf{R}_1$ is a $p \times p$ upper triangular matrix, Equation 3-23 can be easily solved by backward substitution to give the estimates of $\mathbf{u}$. The solution can be formally expressed as:

$$\Pi_u\mathbf{u} = -\mathbf{R}_1^{-1}\mathbf{Q}_1^T\mathbf{A}_x\mathbf{x} \tag{3-24}$$

By substituting for the estimates of $\mathbf{x}$ (obtained using $\mathbf{Q}_2^T$ for $\mathbf{P}$ in Equation 3-14) in the above equation, we obtain the estimates for $\mathbf{u}$ (since $\Pi_u\mathbf{u}$ is a reordered form of the original vector $\mathbf{u}$).

In the case when only s of the columns of $\mathbf{A}_u$ are independent, then the QR factorization takes the form

$$\mathbf{A}_u = \mathbf{QR} = [\mathbf{Q}_1 \quad \mathbf{Q}_2]\begin{bmatrix}\mathbf{R}_1 & \mathbf{R}_2\\ \mathbf{0} & \mathbf{0}\end{bmatrix}\mathbf{\Pi}_u \tag{3-25}$$

where $\mathbf{R}_1$ now is a $s \times s$ nonsingular upper triangular matrix, and $\mathbf{R}_2$ is a $s \times (p - s)$ matrix. The projection matrix is still given by $\mathbf{Q}_2^T$. In the same way, the unmeasured variables can be partitioned into two subsets of s and $p - s$ variables.

$$\Pi_u\mathbf{u} = \begin{bmatrix}\mathbf{u}_s\\ \mathbf{u}_{p-s}\end{bmatrix} \tag{3-26}$$

In order to use the QR factorization for estimating the unmeasured variables we substitute for **R** in Equation 3-21 using Equation 3-25 and for $\Pi_u \mathbf{u}$ using Equation 3-26 and obtain

$$\begin{bmatrix} \mathbf{R}_1 & \mathbf{R}_2 \\ \mathbf{0} & \mathbf{0} \end{bmatrix} \begin{bmatrix} \mathbf{u}_s \\ \mathbf{u}_{p-s} \end{bmatrix} = -\begin{bmatrix} \mathbf{Q}_1^T \\ \mathbf{Q}_2^T \end{bmatrix} \mathbf{A}_x \mathbf{x} \tag{3-27}$$

The upper part of the matrix Equation 3-27 involves only the unmeasured variables:

$$\mathbf{R}_1 \mathbf{u}_s + \mathbf{R}_2 \mathbf{u}_{p-s} = -\mathbf{Q}_1^T \mathbf{A}_x \mathbf{x} \tag{3-28}$$

which, since $\mathbf{R}_1$ is nonsingular, gives the solution

$$\mathbf{u}_s = -\mathbf{R}_1^{-1}\left(\mathbf{Q}_1^T \mathbf{A}_x \mathbf{x} + \mathbf{R}_2 \mathbf{u}_{p-s}\right) \tag{3-29}$$

Equation 3-29 indicates that the solution for the first s (reordered) unmeasured variables can be obtained only if estimates of the remaining p − s unmeasured variables are specified. This is also consistent with the fact that not all unmeasured variables are observable. The QR factorization described here is also useful in identifying which of the unmeasured variables are unobservable as described in the next section.

Example 3-1

We illustrate the construction of the projection matrix by QR factorization and its utility in determining observable and unobservable variables by using the flow reconciliation problem used in Case 3 of Example 1-2 where flows of streams 1 and 6 are the only variables measured. From the constraints Equations 1-7a through 1-7d for this process, we can obtain the matrices corresponding to measured and unmeasured variables, and these are given by

$$\mathbf{A}_x = \begin{bmatrix} 1 & 0 \\ 0 & 0 \\ 0 & 0 \\ 0 & -1 \end{bmatrix} \qquad \mathbf{A}_u = \begin{bmatrix} -1 & -1 & 0 & 0 \\ 1 & 0 & -1 & 0 \\ 0 & 1 & 0 & -1 \\ 0 & 0 & 1 & 1 \end{bmatrix}$$

The QR factorization of matrix A_u gives

$$\mathbf{Q} = \begin{bmatrix} -0.7071 & -0.4082 & -0.2887 & 0.5 \\ 0.7071 & -0.4082 & -0.2887 & 0.5 \\ 0 & 0.8165 & -0.2887 & 0.5 \\ 0 & 0 & 0.8660 & 0.5 \end{bmatrix}$$

$$\mathbf{R} = \begin{bmatrix} 1.4142 & 0.7071 & -0.7071 & 0 \\ 0 & 1.2247 & 0.4082 & 0.8165 \\ 0 & 0 & 1.1547 & 1.1547 \\ 0 & 0 & 0 & 0 \end{bmatrix}$$

From the matrix **R**, it can be inferred that $s = 3$, and that the submatrix corresponding to the first three columns and first three rows is $\mathbf{R}_1$. The projection matrix is the transpose of the last column of **Q**. The reduced constraint matrix is given by

$$\mathbf{Q}_2^T \mathbf{A}_x = [0.5 - 0.5]$$

The reduced constraint matrix can be seen to be equivalent to Equation 1-15 which was obtained using simple algebraic manipulation.

Observability and Redundancy

In Chapter 1, we introduced the concepts of observability and redundancy without formally defining them. In this section, we define these terms clearly and discuss different techniques for variable classification.

The concepts of observability and redundancy are intimately linked with the solvability and estimability of variables. In medium and large scale process plants, there are hundreds of variables, and, for technical and economic reasons it is not possible to measure all of them. It is thus important to know for a given process and a set of measured variables, which of the unmeasured variables can be estimated. The concept of observability deals with this issue.

It is also useful to know whether a measured variable can be estimated even if its sensor fails for some reason. Redundancy deals with this question. Observability and redundancy analysis can be exploited for adding new measuring instruments or for altering the choice of the set of variables to be measured. It can also play a useful role in efficient decomposition and solution of the data reconciliation problem.

Definition of Observability: A variable is said to be observable if it can be estimated by using the measurements and steady-state process constraints.

Definition of Redundancy: A measured variable is said to be redundant if it is observable even when its measurement is removed.

From the above definition of observability, it is obvious that a measured variable is observable, since its measurement provides an estimate of the variable. However, an unmeasured variable is observable only if it can be indirectly estimated by exploiting process constraint relationships and measurements in other variables. Measured variables are redundant if they can also be estimated indirectly through other measurements and constraints.

The observability and redundancy of variables depend both on the measurement structure (also called the *sensor network*) as well as on the nature of the constraints. We have already seen how the measurement selection affects the observability of flow variables in the three cases for the example shown in Chapter 1. A systematic approach is necessary for determining which of the unmeasured variables are observable. There are broadly two approaches that have been followed for solving this problem. One class of methods is based on the use of linear algebra and matrix theory, while the other uses principles of graph theory. Both approaches are discussed here since they provide valuable insights.

Matrix Decomposition Methods

Observability and redundancy classification of variables can be carried out as part of the solution of the data reconciliation problem [6, 7]. We first describe how unobservable variables can be identified during the construction of the projection matrix.

Unobservable variables are present only when the columns of $\mathbf{A}_u$ are not linearly independent. In such cases, the QR factorization of $\mathbf{A}_u$ has the form shown in Equation 3-25, which can be used to rearrange the constraints in the form of Equation 3-27. The solution for unmeasured variables given by Equation 3-29 can be written as

$$\mathbf{u}_s = -\mathbf{R}_1^{-1}\mathbf{Q}_1^T\mathbf{A}_x\mathbf{x} - \mathbf{R}_u\mathbf{u}_{p-s} \tag{3-30}$$

where

$$\mathbf{R}_u = \mathbf{R}_1^{-1}\mathbf{R}_2 \tag{3-31}$$

The matrix $\mathbf{R}_u$ contains all the necessary information to classify unmeasured variables. If a row of $\mathbf{R}_u$ has no nonzero element, then the corresponding unmeasured variable in the LHS of Equation 3-30 can be estimated purely from the estimates of $\mathbf{x}$ and is therefore observable. If on the other hand, a row of $\mathbf{R}_u$ contains a nonzero element, then the corresponding unmeasured variable on the LHS of Equation 3-30 is unobservable since it depends on the estimates chosen for the p–s unmeasured variables on the RHS of this equation. All the p–s unmeasured variables in the RHS of Equation 3-30 are also unobservable, since their estimates have to be specified.

Redundant measured variables can be identified either by looking at their reconciled estimates or by considering the reduced constraint matrix $\mathbf{Q}_2^T\mathbf{A}_x$. A nonredundant measured variable will not be adjusted since it is not possible to estimate this variable indirectly through other variables. Hence, its reconciled value will be identical to its measured value. Corresponding to this variable, the elements in the column of matrix $\mathbf{Q}_2^T\mathbf{A}_x$ will all be zero.

Example 3-2

In order to classify the measured and unmeasured variables of Example 3-1, we can make use of the QR factorization already computed in the example. The matrix $\mathbf{R}_u$, which is useful for classifying unmeasured variables, can be computed as

$$\mathbf{R}_u = \mathbf{R}_1^{-1}\mathbf{R}_2 = \begin{bmatrix} 0.3333 \\ 0.3333 \\ 1 \end{bmatrix}$$

Since all the rows of $\mathbf{R}_u$ contain a nonzero element, this implies that all unmeasured variables are unobservable. In order to classify the measured variables, we make use of matrix $\mathbf{Q}_2^T\mathbf{A}_x$ computed in Example 3-1. Since both the columns of this matrix contain a nonzero element, we infer that both measurements (flows of streams 1 and 6) are redundant. These results can be counter-checked with the results of Example 1-2, Case 3.

Observability and redundancy classification using the projection matrix was also used by Crowe [8]. Crowe applied theoretical rules and derived algorithms to classify flows and concentrations in material balance data reconciliation. The procedure allows the inclusion of chemical reactions, flow splitters and pure energy flows. Fewer classification rules, however, are required by the QR projection matrix approach described above. Matrix methods for observability and redundancy classification in bilinear processes were developed by Ragot et al. [9].

Graph Theoretic Method

The use of graph theoretic concepts for observability and redundancy classification when only overall flows are considered have been developed by Vaclavek [10] and Mah et al. [11]. Later, Vaclavek and Loucka [12] extended their classification ideas to multicomponent systems, while Stanley and Mah [13] developed classification algorithms for energy systems. Kretsovalis and Mah [14, 15, 16] developed graph theoretic algorithms for classifying flows, temperatures and composition variables in general processes, while Meyer et al. [17] developed a simpler algorithm applicable to bilinear processes. In this section, we focus only on overall flows of the process.

In order to use graph theory, the process under consideration should be represented as a process graph. The process graph can be simply derived from the flowsheet of the process by adding an extra node denoted as the environment node and connecting all process feeds and products to it. Thus, for the process of Figure 1-2 in Chapter 1, the process graph is shown in Figure 3-1, where the direction of the streams are not indicated since they are irrelevant for the present analysis. The following simple yet powerful result is obtained from graph theory for identifying unobservable flows:

An unmeasured flow is unobservable if and only if it forms part of some cycle consisting solely of unmeasured flow streams of the process graph.

As an example, consider Case 3 of Example 1-2 for which the unmeasured flows of streams 2 to 5 were shown to be unobservable. The process graph for this case is shown again for convenience in Figure 3-2 with the measured streams marked by a cross. It can be easily observed from Figure 3-2 that these streams form a cycle. On the contrary, for Case 2, the unmeasured flows of streams 3 through 6 do not form any cycle among them and are therefore observable. This can be verified from the process graph for this case shown in Figure 3-3 in which the measured edges are marked.

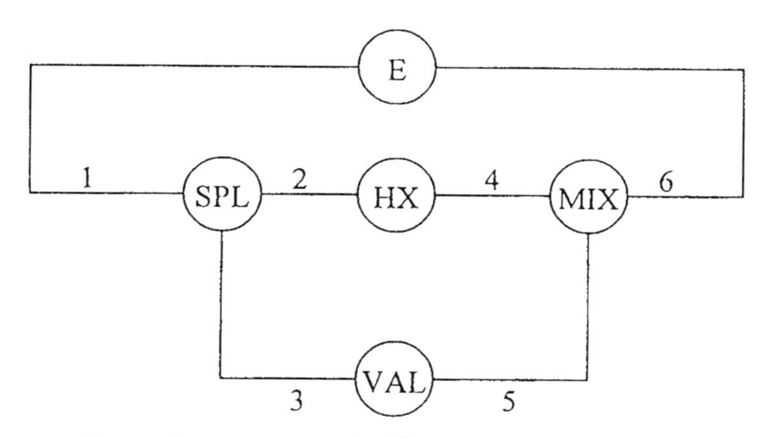

Figure 3-1. Process graph of heat exchanger with bypass.

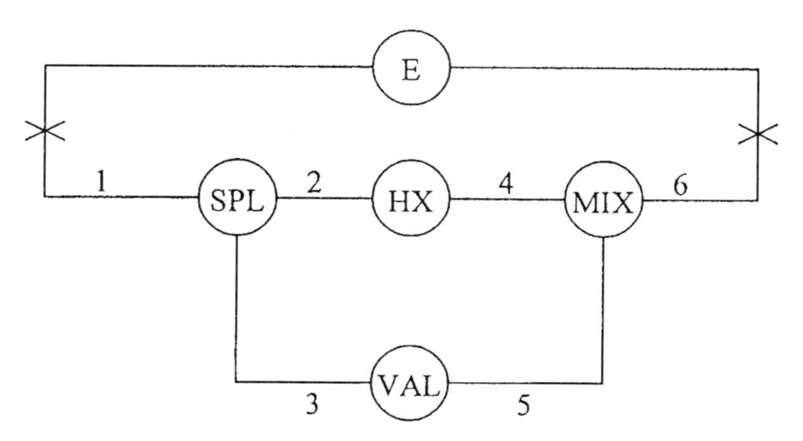

Figure 3-2. Heat exchanger process graph with unobservable values.

Redundant measured variables can also be identified by using the following simple procedure. We merge every pair of nodes which are linked by an unmeasured stream, obtaining in the process a reduced graph which contains only measured streams. All the measured streams of this reduced graph are redundant and will be reconciled. Any measured stream that gets eliminated as a result of the merging process is nonredundant. For example, we apply the merging process to Figure 3-2.

The reduced graphs obtained after merging, in sequence, nodes linked by streams 2, 3, 4, and 5 are shown in Figures 3-4a to 3-4c, respectively.

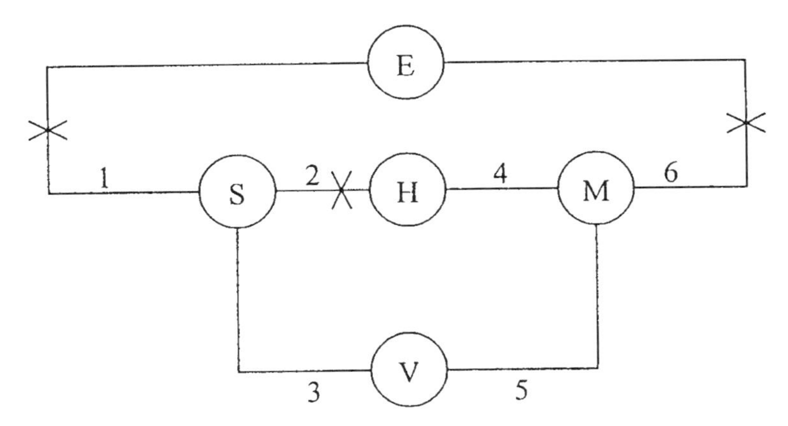

Figure 3-3. Heat exchanger process graph with observable variables.

The final reduced graph of Figure 3-4c contains the measured edges 1 and 6, which implies that they are redundant and will be reconciled. This can be compared with the results of Case 3 of Example 1-2 which shows that flows of streams 1 and 6 are present in the reduced data reconciliation problem. It can also be observed that the reduced data reconciliation problem can be obtained by writing the constraints based on the reduced process graph of Figure 3-4c. Thus, for flow reconciliation of processes containing unmeasured variables, the reduced data reconciliation problem can be formulated using a reduced graph instead of using a projection matrix technique.

Example 3-3

This example is used to illustrate the presence of observable/unobservable unmeasured variables and redundant/nonredundant measured variables coexisting in the same process. The process graph for this example is drawn from Mah [11] and is shown in Figure 3-5. The measured flows of this process are indicated in the figure. For classifying the unmeasured variables easily, the measured edges from Figure 3-5 can be deleted resulting in the graph shown in Figure 3-6a. From this figure it is observed that streams 8, 11, and 14 form a cycle and are therefore unobservable. The remaining unmeasured flows are observable. In order to identify the redundant measurements, the nodes linked by unmeasured edges are merged, resulting in the reduced graph shown in Figure 3-6b. All the measured flows present in Figure 3-6b are redundant, but the measured flow of edge 1 is nonredundant since it is eliminated during the merging process.

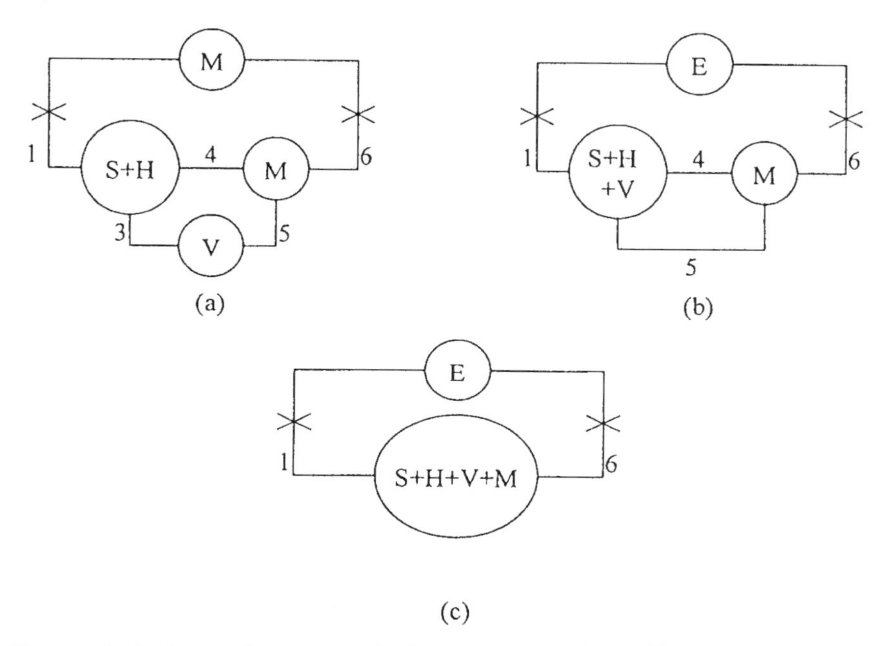

Figure 3-4. Heat exchanger graph after merging unobservable streams (a) stream 2, (b) stream 3, (c) streams 4 and 5. *Reprinted with permission from [11]. Copyright ©1976 American Chemical Society.*

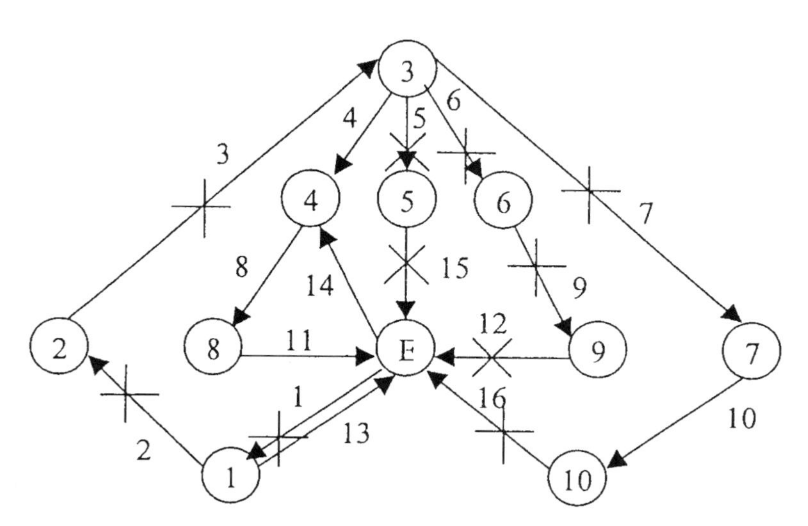

Figure 3-5. Process graph of a refinery subsection. *Reprinted with permission from [11]. Copyright ©1976 American Chemical Society.*

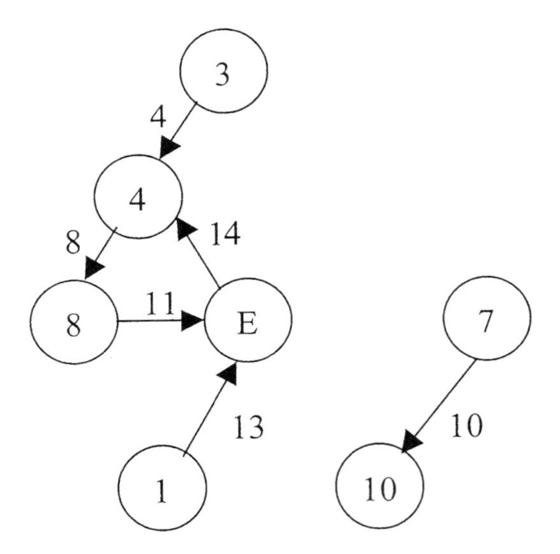

Figure 3-6a. Subgraph of unmeasured variables of refinery process. *Reprinted with permission from [11]. Copyright ©1976 American Chemical Society.*

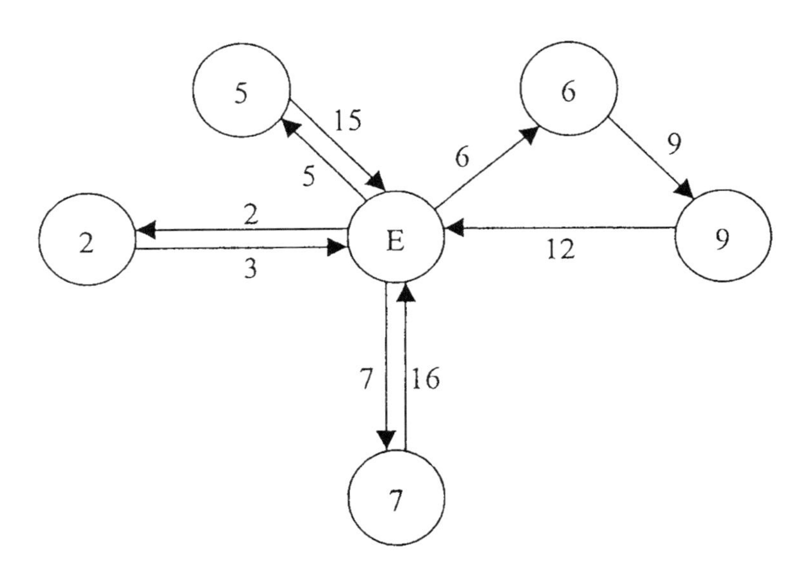

Figure 3-6b. Reconciliation subgraph of refinery process. *Reprinted with permission from [11]. Copyright ©1976 American Chemical Society.*

Other Classification Methods

Besides the two major variable classification methods previously described, a different approach was developed by Romagnoli and Stephanopoulos [18, 19]. They used an output set assignment of the mass and energy balance equations. This approach has been more recently revised and used in a data reconciliation computer package (PLADAT [20]). The most important merit of this approach is a classification of process constraints which enables building a reduced set of constraints for data reconciliation (redundant subset).

If such a redundant set exists, the reconciliation problem can be decomposed into a *redundant subproblem* (solving the reconciliation problem with the redundant set of constraints) and a *coaptation subproblem* (solving for the observable unmeasured variables). But a QR decomposition approach for both variable classification and solving of the data reconciliation problem is more straightforward. Recently, Sanchez and Romagnoli [7] also used it for reconciliation problems involving linear and bilinear constraints.

ESTIMATING MEASUREMENT ERROR COVARIANCE MATRIX

Hitherto, we have assumed that the measurement error covariance matrix Σ is completely known. One possible method of obtaining the error variances and covariances is from the error characteristics of different components (such as sensor, transmitter, recorder, etc.) as explained in the preceding chapter. In order to use this approach, we need information about the standard deviations of the errors committed by the different components, as well the transformations used in processing and transmitting the data.

It is generally difficult to obtain this information, although the *Instrument Engineers Handbook* by Liptak [21] is a good source for such data. It should be noted that the data given in this handbook is under ideal laboratory conditions and may not be valid under actual process conditions. If nonlinear transformations are involved in processing the raw measured data, then the standard deviation of the measurement error can be computed only by using linear approximations of the transforming functions as in Equation 2-13. Bagajewicz [22] has shown that the measurement error obtained in this manner can be considered to be normally distributed, if the range of the measuring instrument is large compared to the standard deviation of the measurement error.

An alternative way of estimating the covariance matrix is from a sample of measurements made in a time window. If $\mathbf{y}_i$, i = 1...N is the vector of measurements made (typically at successive sampling times), then an estimate of the covariance matrix can be obtained analogous to Equation 2-6 as

$$\hat{\Sigma} = \frac{1}{N-1}\sum_{i=1}^{N}(\mathbf{y}_i - \bar{\mathbf{y}})(\mathbf{y}_i - \bar{\mathbf{y}})^T \tag{3-32}$$

where $\bar{\mathbf{y}}$ is the sample mean given by

$$\bar{\mathbf{y}} = \frac{1}{N}\sum_{i=1}^{N}\mathbf{y}_i \tag{3-33}$$

The above method of estimating the covariance matrix is known as the *direct method.* An important requirement for estimating Σ using Equation 3-32 is that the true values of all variables should be constant during the time interval in which the above measurements are made or, in other words, the process should truly be in a steady state. In actual practice, the true values change continuously and the estimates obtained will be poor if these changes are comparable in magnitude to the measurement errors. On the other hand, if the measurements contain a gross error, the estimate of the covariance matrix is not affected provided the magnitude of the gross error is constant in this time interval.

Almasy and Mah [23] first proposed an *indirect method* of estimating the covariance matrix when the true values of the process are undergoing constant changes. Their method exploits the constraint model given by Equation 3-2. For this purpose, we define the constraint residuals **r** as

$$\mathbf{r} = \mathbf{A}\mathbf{y} \tag{3-34}$$

Using Equations 3-1 and 3-2 and the assumption that the measurement errors follow a normal distribution with **0** mean and covariance matrix Σ, we can prove that **r** follows a normal distribution with **0** and covariance matrix **V** given by

$$\mathbf{V} = \mathbf{A}\Sigma\mathbf{A}^T \tag{3-35}$$

It can be observed from Equation 3-34 that the constraint residuals do not depend on the true values of the variables and do not need any information concerning their behavior. From a sample of measurements, we can obtain an estimate of the covariance matrix of constraints residuals as

$$\hat{\mathbf{V}} = \frac{1}{N-1}\sum_{i=1}^{N}(\mathbf{Ay}_i)(\mathbf{Ay}_i)^T \tag{3-36}$$

Note that in obtaining an estimate of $\mathbf{V}$ using Equation 3-36 we do not make use of an estimate for the mean of the constraint residuals from the sample of measurements, since we know its true mean to be $\mathbf{0}$. The estimate of $\mathbf{V}$ can be used in Equation 3-35 to back-calculate and estimate for Σ.

We first note that the matrices $\mathbf{V}$ and Σ are square symmetric matrices of dimensions m and n, respectively. The use of Equation 3-35 to estimate Σ from $\mathbf{V}$, therefore, implies that we have to solve for $n(n + 1)/2$ parameters from $m(m + 1)/2$ equations. Since n is greater than m, several possible solutions for $\hat{\Sigma}$ can be obtained. In order to obtain a unique solution, Almasy and Mah [23] suggested that the sum square of off-diagonal elements of Σ be minimized subject to satisfying Equation 3-35. This is based on the argument that, in practice, Σ is usually diagonal or diagonally dominant. An analytical solution can be obtained for this problem as follows:

Let the vector $\mathbf{d}$ consist of the diagonal elements and the vector $\mathbf{t}$ consist of the off-diagonal elements of Σ (which can be formed by placing the columns one below the other, considering only the elements below the diagonal). We can similarly arrange the diagonal and off-diagonal elements of $\hat{\mathbf{V}}$ as a vector $\mathbf{p}$ and rewrite Equation 3-35 as

$$\mathbf{p} = \mathbf{Md} + \mathbf{Nt} \tag{3-37}$$

where the matrices $\mathbf{M}$ and $\mathbf{N}$ are mapping matrices that can be constructed from the elements of the constraint matrix $\mathbf{A}$ [24]. The solution for the diagonal and off-diagonal elements of Σ are given by

$$\mathbf{d} = \mathbf{SM}^T\mathbf{Rp} \tag{3-38}$$

$$\mathbf{t} = \mathbf{N}^T(\mathbf{R} - \mathbf{RMSM}^T\mathbf{R})\mathbf{p} \tag{3-39}$$

where

$$\mathbf{R} = (\mathbf{N}\mathbf{N}^T)^{-1} \tag{3-40}$$

$$\mathbf{S} = (\mathbf{M}^T\mathbf{R}\mathbf{M})^{-1} \tag{3-41}$$

Keller et al. [24] extended the above indirect method for obtaining an estimate of Σ which has only a few nonzero elements in specified locations.

In order to obtain a good estimate of Σ using the indirect method, the following two conditions have to be met:

(1) The true values of process variables corresponding to each measurement should satisfy the constraint Equation 3-2.
(2) The measurements should not contain any gross errors. If a gross error exists in any measurement, then the constraint residuals will not have zero mean.

Typically, when the true values of process variables undergo changes, we cannot ignore accumulation terms in the material and energy conservation constraints and the first condition may not be met. In such cases, it has to be questioned whether the indirect method offers an advantage over the direct method. In order to tackle this problem, Almasy and Mah [23] recommend that each $\mathbf{y}_i$ used in Equation 3-36 to obtain an estimate of $\mathbf{V}$ should be the average of the measurements made within a time interval in which the process is operating around a nominal steady state. A set of N such time periods should be chosen to obtain a sample of averaged measurement values to be used in Equation 3-36.

A justification for this recommendation can be given using the fact that, in practice, steady-state data reconciliation is applied to measurements averaged over a time interval in which the process operates around a nominal steady state (see industrial examples discussed in Chapter 1). Even if the true values in this time period are randomly fluctuating about the nominal steady-state values, we can expect the average of the true values to satisfy the steady-state conservation constraints as also assumed in data reconciliation.

In order to tackle the problem of gross errors in the indirect method, Chen et al. [25] proposed a robust method of estimation in which the different constraint residual vectors are given appropriate weights when computing the estimate of $\mathbf{V}$ using Equation 3-36. A small weight is assigned to a constraint residual vector if it is not consistent with the other vectors in the sample set. The estimation algorithm is iterative and described by Chen et al. [25]. This procedure is useful if only some of the measurement vectors have gross errors.

If a gross error of constant magnitude is present in all measurements, then the above procedure will not eliminate the problem. One practical solution is to choose the data from time periods that are widely separated in time so that they do not share common features such as the same gross error with the same magnitude present in all the samples. This can be done if a large historical database of operating data is available.

One can also choose to combine the estimates obtained from different methods in a judicious manner. The indirect estimation method, however, has not yet been extended to treat nonlinear constraints.

SIMULATION TECHNIQUE FOR EVALUATING DATA RECONCILIATION

We conclude this chapter by describing a simulation technique that can be used for evaluating the effectiveness of data reconciliation and to estimate the error reduction that can be achieved. For the purposes of simulation, the following input data has to be obtained:

(i) The process flowsheet, which indicates the number of process units, the process streams and their connectivity. The type of process unit need not be specified if simulation of only overall flow reconciliation is to be performed.
(ii) The "true" or "nominal" steady-state flow values of all streams. These true values must be consistent with the flow balances of the process. These true values are useful for judging the improvement in accuracy achieved through data reconciliation.
(iii) The set of measured flows of the process and the standard deviation of the error in each measurement. The standard deviation may be expressed as a fraction of the true value or specified as an absolute value.

At first, random errors are generated which follow a standard normal distribution with mean zero and given standard deviations. These are added to the true values to obtain the simulated "measurements." The constraint matrix, $\mathbf{A}$, is obtained based on the process connectivity information, and the submatrices, $\mathbf{A}_x$ and $\mathbf{A}_u$, are also obtained corresponding to the measured and unmeasured variables. The projection matrix is now computed using a QR factorization of $\mathbf{A}_u$. The estimates can now be computed using Equations 3-14 and 3-24 or 3-29. The error in the estimates can now be computed by comparison with the true values.

In order to obtain a statistically accurate estimate of the error reduction achievable through data reconciliation, it is necessary to perform several simulation trials with different random measurements generated in each trial. The reduction in error due to data reconciliation is computed and averaged over all the trials. Typically about 1,000–10,000 simulation trials are used to obtain this estimate.

Many software packages like MATLAB and mathematical libraries such as IMSL or HARWELL have pseudo-random number generators that can be used for simulation purposes. It should be noted, however, that it is implicitly assumed in the simulation that there are no model errors (that is, the true values of variables satisfy constraints) and that the measurement errors are normally distributed with known variances. In practice, since these assumptions may be violated, the error reduction that can be achieved will be less than that estimated through simulation. Ultimately, the benefits of data reconciliation should be evaluated in practice through actual improvement in process performance.

SUMMARY

- An analytical solution is available for linear data reconciliation with all variables measured.
- Unmeasured variables can be eliminated from the reconciliation model by a projection matrix. A reduced model is obtained, which can be used to reconcile the measured variables.
- Variable classification (as *redundant/nonredundant* measured variables and *observable/unobservable* unmeasured variables) can be performed using matrix methods, while solving the data reconciliation problem, or by a separate graph-theoretic algorithm.

REFERENCES

1. Kuehn, D. R., and H. Davidson. "Computer Control. II. Mathematics of Control." *Chem. Eng. Progress* 57 (1961): 44–47.
2. Seber, G.A.F. *Linear Regression Analysis.* New York: John Wiley & Sons, 1977.
3. Mah, R.S.H. *Chemical Process Structures and Information Flows.* Boston: Butterworths, 1990.
4. Crowe, C. M., Y.A.G. Campos, and A. Hrymak. "Reconciliation of Process Flow Rates by Matrix Projection. I: Linear Case." *AIChE Journal* 29 (1983): 881–888 .
5. Noble, B., and J. W. Daniel. *Applied Linear Algebra.* Englewood Cliffs, N.J.: Prentice-Hall, Inc., 1977.
6. Swartz, C.L.E. "Data Reconciliation for Generalized Flowsheet Applications," American Chemical Society National Meeting, Dallas, Tex., 1989.
7. Sanchez, M., and J. Romagnoli. "Use of Orthogonal Transformations in Data Classification-Reconciliation." *Computers Chem. Engng.* 20 (1996): 483–493.
8. Crowe, C. M. "Observability and Redundancy of Process Data for Steady State Reconciliation." *Chem. Eng. Sci.* 44 (1989): 2909–2917.
9. Ragot, J., D. Maquin, G. Bloch, and W. Gomolka. "Observability and Variables Classification in Bilinear Processes." *Benelux Quarterly J. Automatic Control—Journal A* 31 (1990): 17–23.
10. Vaclavek, V. "Studies on System Engineering III. Optimal Choice of the Balance Measurements in Complicated Chemical Engineering Systems." *Chem. Eng. Sci.* 24 (1969): 947–955.
11. Mah, R.S.H., G. M. Stanley, and D. W. Downing. "Reconciliation and Rectification of Process Flow and Inventory Data." *Ind. & Eng. Chem. Proc. Des. Dev.* 15 (1976): 175–183.
12. Vaclavek, V., and M. Loucka. "Selection of Measurements Necessary to Achieve Multicomponent Mass Balances in Chemical Plant." *Chem. Eng. Sci.* 31 (1976): 1199–1205.
13. Stanley, G. M., and R.S.H. Mah. "Observability and Redundancy Classification in Process Networks. Theorems and Algorithms." *Chem. Eng. Sci.* 36 (1981): 1941–1954.

14. Kretsovalis, A., and R.S.H. Mah. "Observability and Redundancy Classification in Multicomponent Process Networks." *AIChE Journal* 33 (1988): 70–82.

15. Kretsovalis, A., and R.S.H. Mah. "Observability and Redundancy Classification in Generalized Process Networks. I: Theorems." *Computers Chem. Engng.* 12 (1988): 671–688.

16. Kretsovalis, A., and R.S.H. Mah. "Observability and Redundancy Classification in Generalized Process Networks. II: Algorithms." *Computers Chem. Engng.* 12 (1988): 689–703.

17. Meyer, M., B. Koehret, and M. Enjalbert. "Data Reconciliation on Multicomponent Network Process." *Computers. Chem. Eng.* 17 (1993): 807–817.

18. Romagnoli, J., and G. Stephanopoulos. "On Rectification of Measurement Errors for Complex Chemical Plants." *Chem. Eng. Science* 35 (1980): 1067–1081.

19. Romagnoli, J., and G. Stephanopoulos. "A General Approach to Classify Operational Parameters and Rectify Measurement Errors for Complex Chemical Processes." *Comp. Appl. to Chem. Eng.* (1980): 53–174.

20. Sanchez, M., A. Bandoni, and J. Romagnoli. "PLADAT: A Package for Process Variable Classification and Plant Data Reconciliation." *Computers Chem. Engng. 616 (Suppl.)* (1992): S499–S506.

21. Liptak, B. G. *Instrument Engineers' Handbook—Process Measurement and Analysis,* 3rd ed. Oxford: Butterworth-Heineman, 1995.

22. Bagajewicz, M. "On the Probability Distribution and Reconciliation of Process Plant Data." *Comput. Chem. Eng.* 20 (1996): 813–819.

23. Almasy, G. A., and R.S.H. Mah. "Estimation of Measurement Error Variances from Process Data." *Ind. Eng. Chem. Process Des. Dev.* 23 (1984): 779–784.

24. Keller, J. Y., M. Zasadzinski, and M. Darouach. "Analytical Estimator of Measurement Error Variances in Data Reconciliation." *Computers Chem. Engng.* 16 (1992): 185–188.

25. Chen, J., A. Bandoni, and J. A. Romagnoli. "Robust Estimation of Measurement Error Variance/Covariance from Process Sampling Data." *Computers Chem. Engng.* 21 (1997): 593–600.

4

Steady-State Data Reconciliation for Bilinear Systems

BILINEAR SYSTEMS

In a chemical plant, the process streams contain several species or components. Besides stream flow rates, the compositions of some of the streams are also measured. Since composition analyzers are comparatively more expensive, on-line analyzers may not be used in many cases, and these measurements are obtained from a laboratory, which may also increase the errors in the reported data. Neither the overall flow balance nor the component balances are generally satisfied by the measurements. It is therefore necessary to reconcile both flow and composition measurements simultaneously.

The constraints of the data reconciliation problem are linear if we consider only overall flow balances. However, if we wish to simultaneously reconcile flow and composition measurements, then component balances also have to be included as constraints of the data reconciliation problem. These constraints contain component flow rate terms which are products of the flow rate and composition variables. Since these constraints are nonlinear, it is possible to obtain the solution using a nonlinear data reconciliation technique. It is also possible to solve the multicomponent data reconciliation problem more efficiently by exploiting the fact that the nonlinear terms in the constraints are at most products of two variables.

The term *bilinear data reconciliation* is used to refer to problems containing this specific form of constraints. The reasons for developing special techniques for solving bilinear data reconciliation problems are twofold. First, these techniques will be more efficient as compared to techniques used for solving nonlinear data reconciliation problems. This becomes especially important when plant-wide data reconciliation is performed. Second, a significant number of industrial applications of data reconciliation is for multicomponent systems.

An important example is the mineral beneficiation circuit where mineral concentration measurements and flows are reconciled. Other typical examples are reconciliation of flows and compositions around a single distillation column or a sequence of columns such as a chill-down train of a petrochemical complex. In several cases, reconciliation of flows and temperatures of energy flow subsystems are also bilinear problems if the specific enthalpy is only a function of temperature. A crude-preheat train of a refinery and a steam distribution network of a chemical process are important examples. It should be kept in mind, however, that these special techniques only solve the problem efficiently, but do not give any additional benefits.

In this chapter, we describe two methods that have been specifically developed for reconciling data of bilinear systems. While these methods are more efficient than nonlinear programming techniques, they have the disadvantage that, at present, neither of the methods can rigorously handle inequality constraints, such as simple bounds on variables. Thus, in certain cases, it is possible that these methods may give rise to negative estimates of flows and compositions.

DATA RECONCILIATION OF BILINEAR SYSTEMS

Example 4-1

In order to illustrate a typical bilinear data reconciliation problem, we consider a simple example of reconciling the flows and compositions of a binary distillation column as shown in Figure 4-1. We will assume that the flows and component mole fractions of feed, distillate, and bottom streams are measured. A typical set of measured values is shown in the last column of Table 4-1. The discrepancies in the material flows and normalization equations are shown in Table 4-2. It is observed from this table that the measured flows and compositions do not satisfy the material flows or normalization equations.

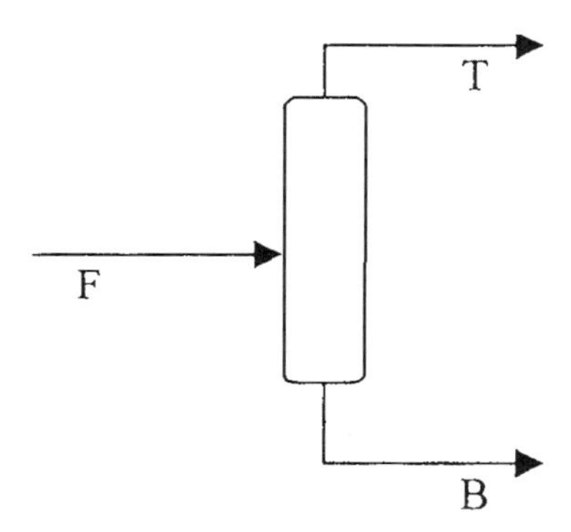

Figure 4-1. Binary distillation column.

Table 4-1
Operating Data of Binary Distillation Column

Stream	Variables	Measured Values
F	Flow	1095.47
	Component 1 %	48.22
	Component 2 %	51.70
D	Flow	478.40
	Component 1 %	94.10
	Component 2 %	5.01
B	Flow	488.23
	Component 1 %	1.97
	Component 2 %	97.48

Table 4-2
Constraint Balance Residuals before Reconciliation

Balance Type	Residuals
Overall Flow	–128.841
Component Balances	
1	–68.417
2	–66.406
Normalization Equations	
F	–0.089
D	–0.892
B	–0.551

The data reconciliation objective is formulated as in Equation 2-3 and is given by

$$\begin{aligned}\underset{F,x}{\text{Min}}\; J &= W_F(\tilde{F}-F)^2 + W_B(\tilde{B}-B)^2 + W_D(\tilde{D}-D)^2 \\ &+\sum_{k=1}^{2}\Big[W_{x_{Fk}}(\tilde{x}_{Fk}-x_{Fk})^2 + W_{x_{Bk}}(\tilde{x}_{Bk}-x_{Bk})^2 \\ &+W_{x_{Dk}}(\tilde{x}_{Dk}-x_{Dk})^2\Big] \qquad (4\text{-}1)\end{aligned}$$

where W's are the weights, and x_{jk}'s are the mole fractions of components. The first three terms in the above objective function are the weighted sum of squared adjustments made to stream flows and the other terms involve the adjustments made to the mole fraction measurements.

The reconciled estimates have to satisfy the material balances around the column. The different types of constraints that can be imposed are

(i) Overall flow balance around the column
(ii) Component flow balances for all the components
(iii) Normalization equations for mole fractions of each stream

All of the above constraints need not be imposed since they are not all independent. For a separator, such as the distillation column considered in this example, a complete set of independent constraints is the component flow balances and the normalization equations. The overall flow balance can be derived using these two types of equations and, thus, it need not be imposed. One common mistake is to assume that by imposing the overall flow balance and component flow balances, the reconciled mole fraction estimates will automatically satisfy the normalization constraint for all streams. This is not the case, however, as shown later through this example.

Thus, the constraints for this example are

$$Fx_{F1} - Bx_{B1} - Dx_{D1} = 0 \qquad (4\text{-}2)$$

$$Fx_{F2} - Bx_{B2} - Dx_{D2} = 0 \qquad (4\text{-}3)$$

$$x_{F1} + x_{F2} = 1 \qquad (4\text{-}4)$$

$$x_{B1} + x_{B2} = 1 \tag{4-5}$$

$$x_{D1} + x_{D2} = 1 \tag{4-6}$$

The component balance constraints, Equations 4-2 and 4-3, contain products of the flow rate and compositions which make the data reconciliation problem more difficult to solve as compared to the linear case considered in the preceding chapter. The objective function 4-1 along with the constraints Equations 4-2 through 4-6 can be treated as a nonlinear equality constrained optimization problem and can be solved using a constrained nonlinear optimization program. However, efficient methods to solve these types of problems have been developed. Using any of these methods, the reconciled data for this example may be obtained and are given in Table 4-3.

In Table 4-3, the second column shows the reconciled estimates when component flow balances and normalization equations are imposed while the third column gives the estimates when the overall flow and component flow balances are used.

Table 4-3
Reconciled Data of Binary Distillation Column

	Reconciled Values	
Variables	**With Normalization Constraints**	**Without Normalization Constraints**
Flow	1009.51	1009.48
Component 1 %	48.24	48.05
Component 2 %	51.76	51.54
Flow	502.22	503.14
Component 1 %	94.99	94.42
Component 2 %	5.01	5.01
Flow	507.29	506.35
Component 1 %	1.97	1.97
Component 2 %	98.03	97.77

The constraint imbalances after reconciliation for these two cases are given in Table 4-4. The results in this table clearly demonstrate the necessity of including normalization constraints in multicomponent data reconciliation problems.

Table 4-4
Constraint Balance Residuals After Reconciliation

	Residual Values	
Balance Type	**Normalization Constraints Imposed**	**Normalization Constraints Not Imposed**
Overall Flow	0.0000E+00	0.0000E+00
Component balances		
1	8.5453E-13	9.9349E-10
2	< 1.0E-13	4.8778E-10
Normalization Equations		
F	0.0000E+00	0.41
D	0.0000E+00	0.57
B	0.0000E+00	0.26

General Problem Formulation

The preceding example shows that multicomponent data reconciliation for a distillation column is a bilinear problem. In a similar manner, the data around a sequence of separation columns can be reconciled which also gives rise to a bilinear problem. In the mineral processing industry, a common application of bilinear data reconciliation is the reconciliation of flow and mineral compositions of a beneficiation circuit. We first present the general formulation for multicomponent data reconciliation of such typical processes.

Depending on the process and the subsystem that is considered, several different types of process units may be encountered. In chemical process industries, the different units where the flow or compositions of streams undergo changes may be classified as mixers, splitters, separators, and reactors. The type of constraints that can be imposed depends on the nature of the process unit. It is therefore important to have a clear understanding of the complete set of independent constraints that can be written for each unit and, hence, for the entire sub-process. Although, for each process unit different combinations of independent constraints can be written, usually the independent equations are chosen as described below.

Mixers

A mixer has two or more input streams and has one output stream as shown schematically in Figure 4-2a. If the streams are single phase then the constraints imposed for these units are

(i) Component flow balances:

$$\sum_{j=1}^{S} F_j x_{jk} - F_o x_{ok} = 0 \qquad k = 1 \ldots C$$

(ii) Normalization equations:

$$\sum_{k=1}^{C} x_{jk} = 1 \qquad j = 1 \ldots S$$

$$\sum_{k=1}^{C} x_{ok} = 1$$

Splitters

A splitter splits an input stream into two or more output streams as shown schematically in Figure 4-2b. The constraints that can be written for this unit are

(i) Component flow balances (equality of compositions):

$$x_{ik} - x_{jk} = 0 \qquad j = 1 \ldots S, \qquad k = 1 \ldots C$$

(ii) Overall flow balance:

$$F_i - \sum_{j=1}^{S} F_j = 0$$

(iii) Normalization equation for input stream:

$$\sum_{k=1}^{C} x_{ik} = 1$$

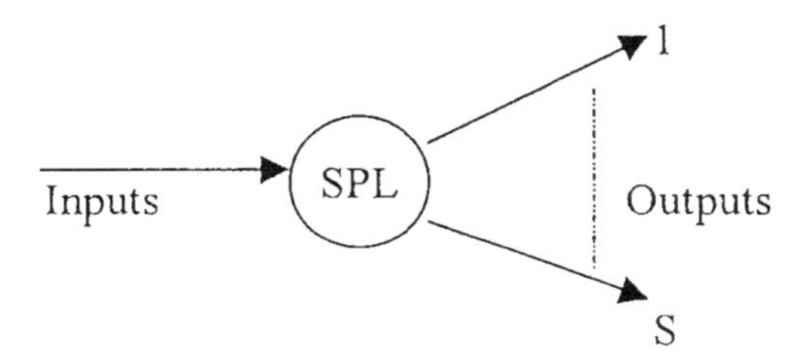

Figure 4-2b. Splitter unit.

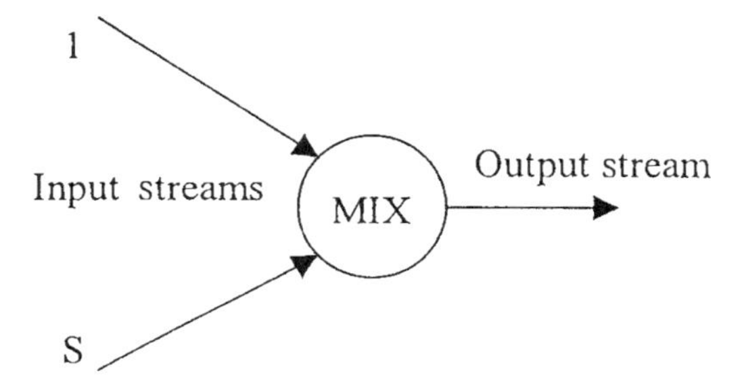

Figure 4-2a. Mixer unit.

All other constraints such as component balances and normalization constraints for output streams can be derived by appropriate combinations of above equations. It should also be kept in mind that if a splitter is a part of a subsystem, then the normalization equations should be written only for the input stream of the splitter and not for the output streams of the splitter.

> ***Exercise 4-1.*** *Show that for a splitter shown in Figure 4-2b, the number of independent equations is CS+2. Also show that the component balances and normalization equations of all output streams, can be derived using the above CS+2 equations imposed for a splitter.*

An alternative formulation which makes use of the definition of split fractions is sometimes more convenient. Let α_j be the ratio of the flow rate of outlet stream j to that of the inlet stream of the splitter. Then the following equations also constitute a complete nonredundant set of equations for the splitter.

(i) Component flow balances:

$$F_j x_{jk} - \alpha_j F_i x_{ik} = 0 \qquad j = 1 \ldots S;\ k = 1 \ldots C$$

(ii) Overall flow balance:

$$\sum_{j=1}^{S} \alpha_j = 1$$

(iii) Normalization equation for input stream:

$$\sum_{k=1}^{C} x_{ik} = 1$$

(iv) Split fraction definitions:

$$F_j - \alpha_j F_i = 0 \qquad j = 1 \ldots S$$

The use of split fraction variables introduces as many additional variables as the number of output streams and thus the number of independent equations that must be written for a splitter using split fraction variables is equal to CS+S+2. The use of split fraction variables also complicates the problem further, since the component balances are no longer bilinear but are trilinear (products of three variables).

Exercise 4-2. *Demonstrate that by eliminating the split fraction variables from the above alternative set of CS+S+2 equations, it is possible to obtain the CS+2 independent equations of the first formulation for a splitter.*

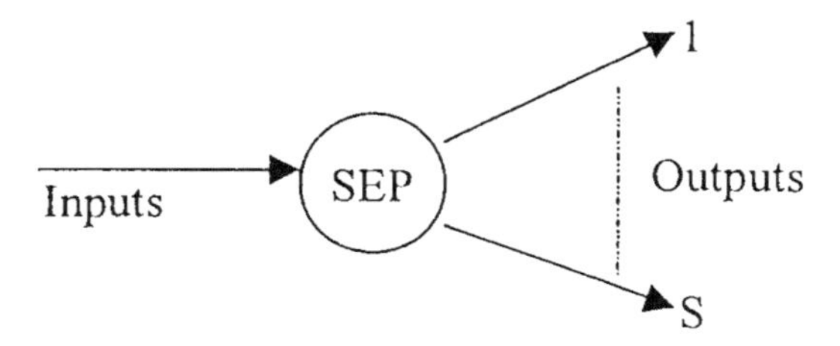

Figure 4-2c. Separator unit.

Separators

A separator, which is the inverse of a mixer, takes an input stream and separates it into two or more streams of different compositions as shown in Figure 4-2c. If all streams are single phase, the equations for this unit are similar to those of a mixer.

(i) Component flow balances:

$$F_i x_{ik} - \sum_{j=1}^{S} F_j x_{jk} = 0 \qquad k = 1 \ldots C$$

(ii) Normalization equations:

$$\sum_{k=1}^{C} x_{ik} = 1$$

$$\sum_{k=1}^{C} x_{jk} = 1 \qquad j = 1 \ldots S$$

Reactors

We consider a reactor with a single feed stream and a single product stream as shown in Figure 4-2d. Reactors with multiple feed or product streams can be modeled by using a mixer before the reactor and a separator after the reactor. Due to the reactions that occur, neither the overall molar flow nor the molar flows of components are conserved. There are two alternative choices of model equations for a reactor. In the first approach, we assume that independent reactions which occur in the reactor are specified. Let n_{kj} be the stoichiometric coefficient of component *k* in reaction *j,* and let the unknown extents of reaction be ξ_j, $j = 1 \ldots R$, where R is the number of independent reactions specified. Using the extents of reactions we can write the following equations

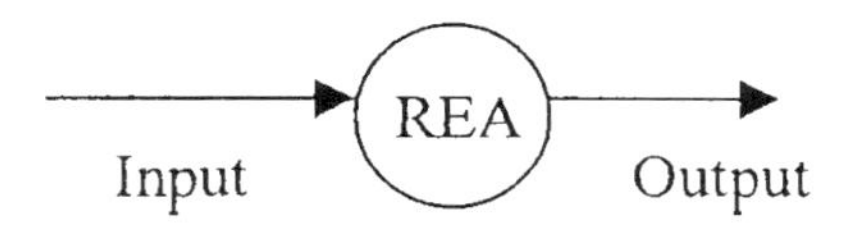

Figure 4-2d. Reactor unit.

(i) Component balances:

$$F_i x_{ik} + \sum_{j=1}^{R} \nu_{kj}\xi_j - F_o x_{ok} = 0 \quad k = 1 \ldots C$$

(ii) Normalization equations:

$$\sum_{k=1}^{C} x_{ik} = 1$$

$$\sum_{k=1}^{C} x_{ok} = 1$$

The alternative set of model equations is obtained by using the fact that each elemental species is conserved. If we denote the number of atoms of element *j* in component *k* by a_{jk} then the following equations can be written for a reactor

(i) Elemental balances:

$$\sum_{k=1}^{C} \left(a_{jk} x_{ik} F_i - a_{jk} x_{ok} F_o\right) = 0 \quad j = 1 \ldots E$$

(ii) Normalization equations:

$$\sum_{k=1}^{C} x_{ik} = 1$$

$$\sum_{k=1}^{C} x_{ok} = 1$$

As shown by Reklaitis [1], these sets of equations are equivalent and give identical results only if a complete set of independent reactions that can occur among the components present is specified. In the absence of any information regarding the reactions that occur, the elemental balance model can be used. However, if energy balances also have to be included as part of the reconciliation, then the extent of reaction model is convenient as shown subsequently.

The model equations of the various units can be classified as either process unit type or as stream-type relations. Overall flow and component balances are categorized as unit type equations whereas normalization equations and pulp density relations which relate variables of a stream are classified as stream type equations. This classification helps in easily writing down the constraints for a process or subsystem which has a combination of some of the above units. First, the unit-type equations of all process units which are part of the process are written followed by the stream-type equations corresponding to the process units.

Example 4-2

Consider a simple process of an organic synthetic juice plant by Meyer et al. [2] shown in Figure 4-3 which consists of a splitter and three separators. The following constraints are written for this process.

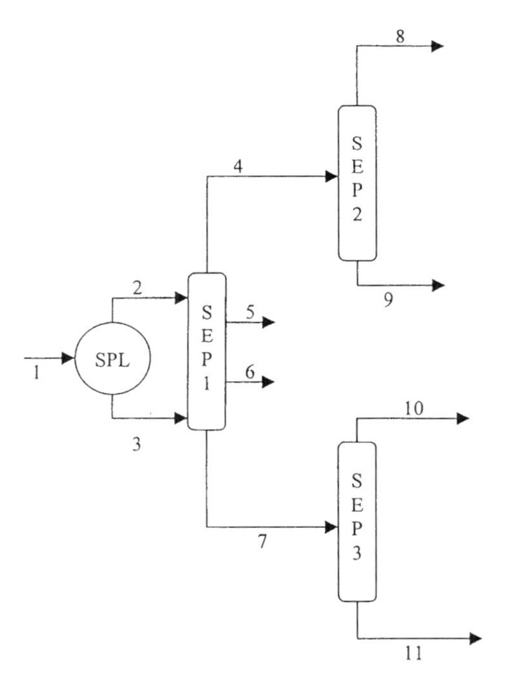

Figure 4-3. Synthetic organic juice process.

(i) Unit type: Component balances around each of the three separators and equality of compositions of streams 1 and 2 as well as streams 1 and 3 (splitter equations).
(ii) Stream type: Normalization equations for stream 1 and streams 4 to 11.

Normalization equations for streams 2 and 3 are not written since these are outlet streams from a splitter.

BILINEAR DATA RECONCILIATION SOLUTION TECHNIQUES

An examination of the equations for the different units described above shows that the only nonlinearity that is present occurs as products of two variables. The techniques that have been proposed to solve bilinear data reconciliation problems exploit this feature in an efficient manner. Although, several different techniques have been proposed especially in the mineral processing literature, we have chosen to describe only two of these methods that in our opinion are the most efficient among them. We only present an overall description of these methods and refer the reader to the appropriate references for details of the derivations.

Crowe's Projection Matrix Method

The first bilinear data reconciliation technique we describe is the method developed by Crowe [3]. In order to better understand this method, we initially consider a multicomponent process consisting of single phase mixers and separators only. Furthermore, we consider the case when all stream flows and compositions are measured. It can be verified from the model equations for mixers and separators that the constraints for such a process consists of component balances around every unit and normalization equations for every stream. If there are *m* units and *n* streams and *C* components, then the process constraints are

$$\sum_{j=1}^{S} a_{ij} F_j x_{jk} = 0 \qquad i = 1 \ldots m;\ k = 1 \ldots C \tag{4-7}$$

$$\sum_{k=1}^{C} x_{jk} = 1 \qquad j = 1 \ldots S \tag{4-8}$$

The objective is to determine estimates of all flows and compositions such that the total weighted sum square of adjustments made to flow and composition measurements is minimized. The objective function is given by

$$\underset{F_j,\, x_{jk}}{\text{Min}} \quad J = \sum_{j=1}^{S} (\tilde{F}_j - F_j)^2 W_{F_j} + \sum_{j=1}^{S} \sum_{k=1}^{C} (\tilde{x}_{jk} - x_{jk})^2 W_{x_{jk}} \tag{4-9}$$

The above formulation of the DR problem is in terms of flow rate and mole fraction variables. Alternatively, we can also formulate the problem in terms of overall flow and component flow variables, where the component flow N_{jk} of component *k* in stream *j* is defined as

$$N_{jk} = F_j x_{jk} \qquad k = 1 \ldots C \tag{4-10}$$

Using these variables, the component balances can be written as

$$\sum_{j=1}^{S} a_{ij} N_{jk} = 0 \quad i = 1 \ldots m;\ k = 1 \ldots C \tag{4-11}$$

The normalization equations can also be written as

$$F_j - \sum_{k=1}^{C} N_{jk} = 0 \quad j = 1 \ldots S \tag{4-12}$$

It can be observed from Equations 4-11 and 4-12 that the constraints are linear in the flow variables and this feature can be exploited in the solution procedure. Although the constraints are now in terms of flow variables, the objective function still contains mole fraction variables since these are the measured quantities. In order to overcome this problem, Crowe [3] proposed a modified objective of the DR problem which is to minimize the sum square of adjustments made to flows and component flow variables. In this case the modified objective function is

$$\underset{F_j,\, N_{jk}}{\text{Min}} \quad J = \sum_{j=1}^{S} (\tilde{F}_j - F_j)^2 W_{F_j} + \sum_{j=1}^{S} \sum_{k=1}^{C} (\tilde{N}_{jk} - N_{jk})^2 W_{N_{jk}} \tag{4-13}$$

Since the component flows are not the measured quantities, in the above objective function it is necessary to clarify the notion of the measured value of the component flow variables and the weight factors to be used for these variables. A component flow N_{jk} is taken to be a measured quantity if both the flow F_j and the composition x_{jk} are measured.

In the previous chapter, it was shown that the weight factor of a measured variable can be chosen to be the inverse of the variance of its measurement error. An estimate of the variance of the error in the product $\tilde{N}_{jk}$ is obtained by linearizing it in terms of the flow rate and composition measurements as

$$\tilde{N}_{jk} \approx x^*_{jk}(\tilde{F}_j - F^*_j) + F^*_j(\tilde{x}_{jk} - x^*_{jk}) + F^*_j x^*_{jk} \tag{4-14}$$

The variance $\sigma^2_{\tilde{N}jk}$ of the error in $\tilde{N}_{jk}$ can be obtained by applying the rule for linear sum of independent normally distributed variables

$$\sigma^2_{\tilde{N}_{jk}} = \left(x^*_{jk}\right)^2 \sigma^2_{\tilde{F}_j} + \left(F^*_j\right)^2 \sigma^2_{\tilde{x}_{jk}} \tag{4-15}$$

The weight $W_{N_{jk}}$ can be taken to be equal to the inverse of the variance $\sigma^2_{\tilde{N}jk}$.

The choice of a modified objective function for DR and the weight factors for the "measured" component flows can lead to larger adjustments being made to the measurements. However, the modified objective still indirectly attempts to minimize the total adjustment made to the measured variables.

The modified objective function 4-13 subject to the constraints Equations 4-11 and 4-12 gives rise to a linear DR problem in the flow variables. For the special case considered here, all variables are measured and we can immediately obtain the estimates of all flows using the analytical solution of Equation 3-4. From these estimates, the reconciled values of the mole fractions can be obtained as

$$\hat{x}_{jk} = \hat{N}_{jk} / \hat{F}_j \tag{4-16}$$

Example 4-3

Crowe's method is applied to reconcile the data of the binary distillation column discussed in Example 4-1. The measured flows and compositions are as given in Table 4-1. The true flows and compositions and the reconciled values obtained using Crowe's method are given in Table 4-5. In order to obtain the reconciled values, the measurement error variances in flows are taken as 5% of the true values and for the compositions they are taken as 1% of the true values. As compared to the reconciled values shown in Table 4-3 which are obtained using a nonlinear optimization technique, Crowe's method gives more accurate flow estimates at the expense of greater inaccuracy in the composition estimates. This is due to the fact that Crowe's method adjusts the component flows rather than the compositions.

Table 4-5
Reconciled Data of Binary Distillation Column Using Crowe's Method

Variables	True Values	Reconciled Values
Flow	1000	1002.7
Component 1 %	48.00	46.34
Component 2 %	52.00	50.33
Flow	494.624	496.1
Component 1 %	95.00	91.22
Component 2 %	5.00	5.28
Flow	505.376	506.6
Component 1 %	2.00	2.39
Component 2 %	98.00	94.43

Treatment of Unmeasured Variables

The presence of unmeasured flows or composition variables introduces subtle complications in Crowe's method. Depending on the measurements made, the streams can be classified into two categories:

(i) Streams with measured flows and some or all compositions unmeasured
(ii) Streams with unmeasured flows and some or all compositions unmeasured

A measured value for the component flow of a stream cannot be obtained if the corresponding composition variable is unmeasured or if the stream flow is unmeasured or both. Since there is one to one correspondence between composition variables and their component flows, it is appropriate to consider the component flow as unmeasured if the corresponding composition variable is unmeasured regardless of whether the stream flow is measured or not. However, if a stream flow is unmeasured, then treating all component flows of this stream as unmeasured will result in a loss of information of the measured compositions of this stream. In order to avoid this, Crowe's method classifies the stream flows and component flows into the following three categories:

(i) **Category I** consists of all measured stream flow variables and "measured" component flow variables. Thus, this category consists of measured variables only.
(ii) **Category II** consists of all component flows corresponding to measured compositions, but unmeasured stream flow. It also contains all the unmeasured stream flow variables. Thus, this category consists of a mixture of measured compositions and unmeasured stream flow variables.
(iii) **Category III** consists of all component flows corresponding to unmeasured compositions for which the stream flow may or may not be measured. Thus, this category consists of unmeasured variables only.

The flows and component flow variables in the different categories are denoted by superscripts I, II, and III. The objective function for the DR problem can now be formulated as

$$\underset{F^I, N_k^I, x_k^{II}}{\text{Min}} \ (\tilde{\mathbf{F}}^I - \mathbf{F}^I)^T \mathbf{W}_{F^I} (\tilde{\mathbf{F}}^I - \mathbf{F}^I) + \sum_{k=1}^{C} (\tilde{\mathbf{N}}_k^I - \mathbf{N}_k^I)^T \mathbf{W}_{N_k^I} (\tilde{\mathbf{N}}_k^I - \mathbf{N}_k^I)$$
$$+ \sum_{k=1}^{C} (\tilde{\mathbf{x}}_k^{II} - \mathbf{x}_k^{II})^T \mathbf{W}_{x_k^{II}} (\tilde{\mathbf{x}}_k^{II} - \mathbf{x}_k^{II}) \qquad (4\text{-}17)$$

The above objective function has been expressed compactly using vectors $\mathbf{F}$, $\mathbf{N}_k$, and $\mathbf{x}_k$, corresponding to overall flows, component k flows, and compositions of all streams in each category, respectively. The weight matrices $\mathbf{W}$ are diagonal matrices with the diagonal entries being the weights for the appropriate variables of all streams in each category.

The constraints of the DR problem are the material balances for each unit written as described earlier. These equations can be cast in terms of the variables in the three categories. For solving this problem, Crowe [3] proposed a two-stage decomposition strategy for eliminating unmeasured variables from the constraint equations. In the first stage, unmeasured component flows in Category III are eliminated by using a projection matrix. For this, the procedure used in linear DR can be followed because the constraints are linear in the component flows. In the second stage, the unmeasured flow variables in Category II are eliminated by using a second projection matrix. This requires some algebraic manipulation of the constraint equations which are described in [3].

The reduced DR problem still requires an iterative procedure to solve for reconciled compositions of Category II and component flows of Category I, starting with guesses of Category II flow variables. It can be verified that if estimates of Category II flows are given, then the reduced reconciliation problem becomes a linear DR problem which can be solved analytically. These reconciled estimates are used to back-calculate the unmeasured flows of Category II using a similar procedure as described in Chapter 3, which are used as starting guesses for the next iteration until convergence.

After the estimates of variables in Categories I and II are obtained, they can be used to back-calculate the estimates of unmeasured component flows of Category III as described in Chapter 3. Since Crowe's method directly gives the estimates of component flows in Categories I and III, the mole fraction estimates are obtained using Equation 4-16.

Example 4-4

We consider the mineral flotation process analyzed by Smith and Ichiyen [5] shown in Figure 4-4. The process consists of three flotation cells (separators) and a mixer, and eight streams each consisting of two minerals, copper and zinc, in addition to gangue material. The flow of stream 1 is taken to be unit mass (basis), while the other stream flows are unmeasured. The mineral concentrations of all streams except 8 are measured. These values are shown in the first row of Table 4-6. Based on this information the flow and component variables can be classified as

Category I—F_1, N_{11}, N_{12}

Category I—F_2, N_{21}, N_{22},...F_7, N_{71}, N_{72}

Category III—F_8, N_{81}, N_{82}

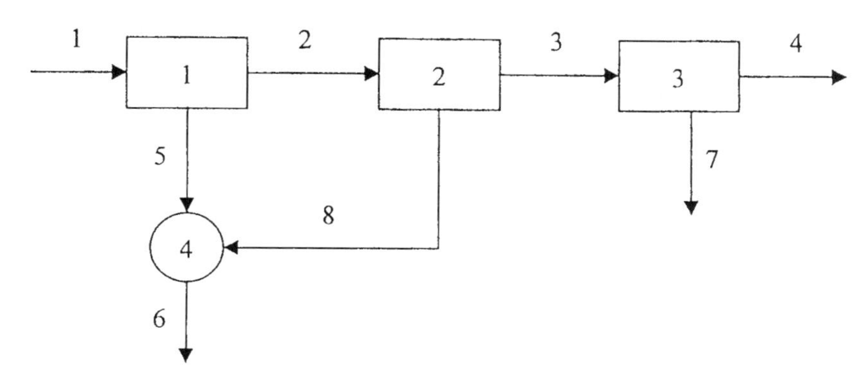

Figure 4-4. Mineral flotation process [5]. *Reproduced with permission of the Canadian Society for Chemical Engineering.*

Although, the flow rate of stream 8 being an unmeasured variable should be classified as a Category II variable, it can also be classified as a Category III variable *because all its compositions are unmeasured.*

The constraints imposed for this process are the flow balances and component flow balances around each unit. Normalization equations are not imposed because we have eliminated the composition variables corresponding to the unmeasured gangue in each stream. This leads to a reduced number of variables and constraints in the data reconciliation problem.

We start with an initial guess for the flows in streams 2 to 7 as given in Table 4-6, and apply the iterative procedure to obtain the reconciled values. Row 2 of Table 4-6 shows the reconciled estimates of the flows and mineral concentrations obtained using Crowe's method. It is observed that the estimate for zinc concentration in stream 8 is negative. For comparison, the reconciled estimates obtained using a nonlinear programming technique (NLP) (described in the next chapter) are also listed in the last row of Table 4-6. Again, the estimate for zinc concentration in stream 8 is infeasible. This points to the need for imposing bound constraints in the data reconciliation problem which we be discussed in the next chapter. The maximum difference between the (feasible) mineral concentrations between the two solutions is about 2–7%. Since Crowe's method uses an objective function different from the standard DR problem, its estimates will be less accurate than those obtained using the NLP approach.

Table 4-6
Measured and Reconciled Data of Mineral Flotation Process

Method	Var.	Stream							
		1	2	3	4	5	6	7	8
	F*	1	0.5	0.25	0.125	0.5	0.75	0.125	0.25
Measured	y_{Cu} %	1.928	0.450	0.128	0.090	19.88	21.43	0.513	35.36
	y_{Zn} %	3.81	4.72	5.36	0.41	7.09	4.95	52.10	–
	F	1	0.9229	0.9147	0.8324	0.0771	0.0853	0.0823	0.0081
Crowe	$\hat{x}_{Cu}$ %	1.9451	0.4498	0.1285	0.0906	19.834	21.431	0.512	0.2976
	$\hat{x}_{Zn}$ %	5.0356	4.8617	5.0461	0.4099	7.1167	4.9235	51.930	–15.91
	F	1	0.9253	0.9164	0.8287	0.0747	0.0836	0.0877	0.0089
NLP	$\hat{x}_{Cu}$ %	1.9122	0.4509	0.1301	0.0899	20.00	21.44	0.5098	35.554
	$\hat{x}_{Zn}$ %	4.2759	4.0584	5.3583	0.41	6.9694	4.95	52.116	–130.1

**Initial values of flows for streams 1 through 8 are listed in this row*

Simpson's Technique

The application of data reconciliation in the mineral processing industries, especially to mineral beneficiation circuits has been investigated about 30 years ago. Several methods for specifically solving DR problems arising in these industries have been developed. Among these, the method developed by Simpson et al. [6] is very efficient.

Before describing Simpson's technique, it is instructive to examine some of the process units encountered in the mineral processing industries and see in what respects they differ from the corresponding units in chemical process industries. In mineral beneficiation processes [7], the ore is first crushed to obtain particle sizes in the range 10–20 cm. The crushed particles are further reduced in size to between 10–300 microns in grinders. Generally, grinding is carried out with addition of water and/or recycled slurry. The particles containing the minerals are separated from the gangue particles in separation units that are either classifiers or flotation cells.

The slurry containing the mineral particles is referred to as the concentrate and that containing the gangue as tailings. Water may also be added to the separation units in order to maintain a desired pulp density. Figure 4-5a shows a schematic of such units, where the feed (F), tailings (O), and concentrate (U) contain both solids and liquid, but the stream denoted by W is a pure water stream. Although, these process units are similar to separators, all streams (except pure water streams) are two-phase streams and conservation equations have to be written for the overall slurry flow as well as for the solid or liquid flows through each of the units. We refer to these process units as two-phase separators. Similarly, there are two-phase mixers where mixing of these streams occurs.

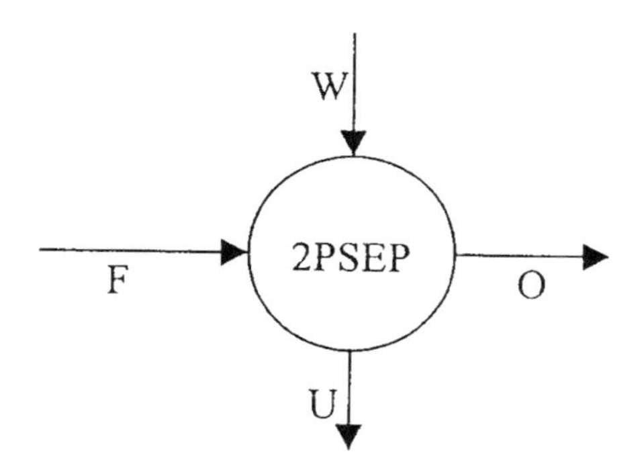

Figure 4-5a. Two-phase separator.

Two-Phase Separators

The type of equations written for these units differ from the simple separator in three respects. First, separate conservation relations have to be written for the overall slurry flow and the solid flow. Secondly, laboratory measurements of the pulp density of the slurry streams may also be available, which needs to be reconciled. Thirdly, the solids also contain gangue material which is not measured and only the mineral concentrations of the solids are measured. Thus normalization equations are not imposed since they are irrelevant.

In order to take into account the above aspects in the model equations, the following variables are associated with each stream *j*.

(1) The slurry flow rate (F_j)
(2) The flow rate of solid (S_j)
(3) The mineral concentrations (x_{jk}) expressed as a weight fraction of the slurry flow and
(4) The pulp density (ρ_j) which is the ratio of the solid to the total slurry flow in a stream.

Using the above variables and the notation in Figure 4-5a, the model equations for these units can be written as follows.

(i) Overall flow balance:

$$F + W - O - U = 0$$

(ii) Solids flow balance:

$$S_F - S_O - S_U = 0$$

(iii) Pulp density relations:

$$\rho_F F - S_F = 0$$

$$\rho_O O - S_O = 0$$

$$\rho_U U - S_U = 0$$

(iv) Mineral component balances:

$$F x_{Fk} - O x_{Ok} - U x_{Uk} = 0 \qquad k = 1 \ldots C$$

Two-Phase Mixers

These units are similar to mixers expect that two-phase streams are involved as shown in Figure 4-5b. As in the case of the two-phase separator, the stream W is a pure water stream. The equations are similar to those for a two-phase separator.

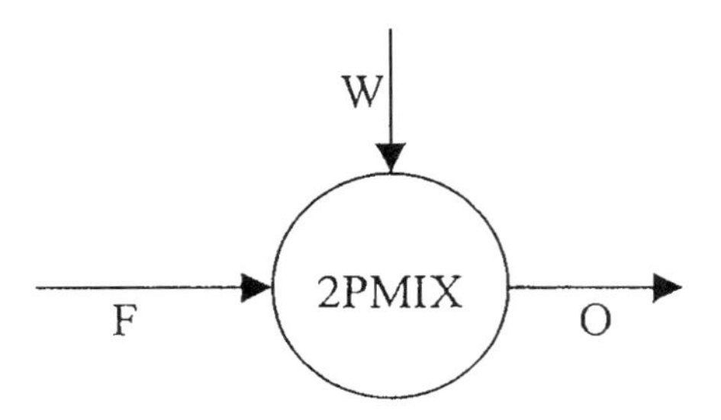

Figure 4-5b. Two-phase mixer unit.

Ball Mills

In crushers and grinders, the size distribution of the input and output streams are measured. Balance equations are written for the ore quantity within each size range. If a slurry stream is also recycled to the mill or water is added, then as in the case of two-phase mixers a slurry balance and pulp density relations have to be written. Figure 4-5c shows a schematic of a general mill.

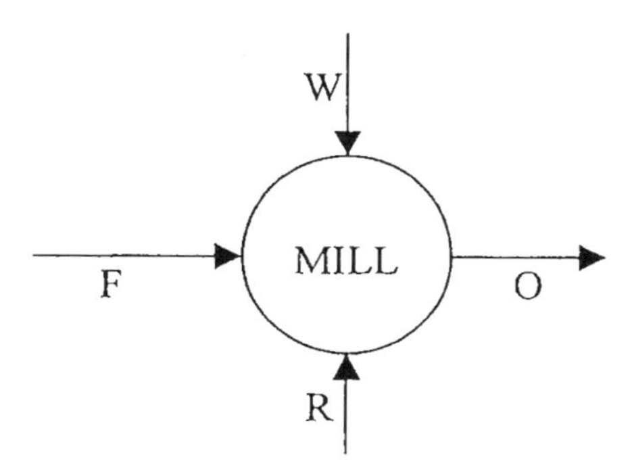

Figure 4-5c. Ball mill unit.

Let w_{ij} be the weight fraction of solids of size range *i* in stream *j*, and n_s be the number of size ranges. The following constraints are imposed for a mill in general.

(i) Overall flow balance:

$$F + W + R - O = 0$$

(ii) Overall solids flow balance:

$$S_F + S_R - S_O = 0$$

(iii) Solid flow balance for each size range:

$$S_F w_{iF} + S_R w_{iR} - S_O w_{iO} + \xi_i = 0 \qquad i = 1 \ldots n_s$$

where ξ_i is the increase (which could be negative or positive) in the weight of solids in size range *i* due to grinding. These are typically unknown quantities and have to be estimated as part of the reconciliation problem.

(iv) Pulp density relation:

$$\rho_F F - S_F = 0$$
$$\rho_O O - S_O = 0$$
$$\rho_U R - S_R = 0$$

(v) Normalization equations (for weight fractions):

$$\sum_{i=1}^{n_S} w_{iF} = 1$$

$$\sum_{i=1}^{n_S} w_{iR} = 1$$

$$\sum_{i=1}^{n_S} w_{iO} = 1$$

(vi) Mineral component balances:

$$Fx_{Fk} + Rx_{Rk} - Ox_{Ok} = 0 \qquad k = 1...C$$

It can be observed that the model equations for different units are bilinear. The method developed by Simpson et al. [6] exploits this feature for efficient solution of the DR problem. We describe this method for the simple case when the mineral beneficiation circuit consists only of two-phase mixers and separators. We assume that the measurements made are slurry flows (or liquid flows for pure liquid streams), pulp densities, mineral concentrations given in terms of mass fraction of mineral in wet solids. The objective function of the DR problem in this case is given by

$$\sum_{j=1}^{s_1} (\tilde{F}_j - F_j)^2 W_{F_j} + \sum_{j=1}^{s_2} \sum_{k=1}^{C} (\tilde{x}_{jk} - x_{jk})^2 W_{x_{jk}} + \sum_{j=1}^{s_2} (\tilde{\rho}_j - \rho_j)^2 W_{\rho_j} \quad (4\text{-}18)$$

where F_j is the overall flow rate of stream *j*, s_1 is the total number of streams, and s_2 is the number of slurry streams. It should be noted that all variables whether measured or not is included in the objective function.

Since there are no measured values associated with unmeasured variables, an initial estimate of these variables can be used in the objective function. Moreover, the weights for all unmeasured variables are chosen to be zero so that the objective function is identical to the standard DR objective function which minimizes the weighted sum square of adjustments made to *measured* variables.

In Simpson's method, the nonlinear data reconciliation problem is approximated by a linear data reconciliation problem by suitable choice of working variables and linearisation.

The pulp density relations are first used to substitute for the variables ρ_j in terms of the flow rates.

$$\rho_j = S_j/F_j \qquad j = 1 \ldots s_2 \tag{4-19}$$

With this substitution, the pulp density relations need not be considered in the DR problem. As seen before, the component balances are linear if we express them in terms of component flow rates. The variables x_{jk} in the objective function can also be expressed in terms of component flow variables as

$$x_{jk} = N_{jk}/F_j \qquad j = 1 \ldots s_2;\ k = 1 \ldots C \tag{4-20}$$

Using Equations 4-19 and 4-20, the objective function can be written as

$$\sum_{j=1}^{s_1} (\tilde{F}_j - F_j)^2 W_{F_j} + \sum_{j=1}^{s_2} \sum_{k=1}^{C} (\tilde{x}_{jk} - N_{jk} / F_j)^2 W_{x_{jk}} + \sum_{j=1}^{s_2} (\tilde{\rho}_j - S_j / F_j)^2 W_{\rho_j} \tag{4-21}$$

The second and third terms in the objective function are no longer quadratic in the flow variables. The objective function can be approximated by a quadratic function by using a first order approximation of the flow ratios around some estimates F_j^*, S_j^* and N_{jk}^*

$$N_{jk} / F_j \approx (N_{jk} - N_{jk}^*) / F_j^* - (F_j - F_j^*) N_{jk}^* / (F_j^*)^2 + N_{jk}^* / F_j^* = a_j N_{jk} + b_j F_j + x_{jk}^* \tag{4-22}$$

where a_j and b_{jk} are respectively equal to and $1/F_j^*$ and $-N_{jk}^*/(F_j^*)^2$ and

$$\begin{aligned} S_j / F_j &\approx (S_j - S_j^*) / F_j^* - (F_j - F_j^*) S_j^* / (F_j^*)^2 + S_j^* / F_j^* \\ &= a_j S_{jk} + c_j F_j + \rho_{jk}^* \end{aligned} \tag{4-23}$$

where c_j is equal to $-S_j^*/(F_j^*)2$.

The quadratic approximation of the objective function is therefore given by

$$\sum_{j=1}^{s_1} (\tilde{F}_j - F_j)^2 W_{F_j} + \sum_{j=1}^{s_2} \sum_{k=1}^{C} (\tilde{x}_{jk} - x_{jk}^* - a_j N_{jk} - b_{jk} F_j)^2 W_{x_{jk}} + \sum_{j=1}^{s_2} (\tilde{\rho}_j - \rho_j^* - a_j S_j - c_j F_j)^2 W_{\rho_j} \tag{4-24}$$

Since the constraints are linear in the flow and component flow variables, we now have an approximate linear DR problem corresponding to the objective 4-24 and flow balance constraints for overall slurry flow, solids flow and component flows. This linear DR problem can be solved using techniques described in the preceding chapter. This DR problem, however, can be solved more efficiently by reducing it to an unconstrained optimization problem by eliminating all the constraints together with a suitable choice of dependent variables. The dependent variables are so chosen that their relation to the independent variables is obtained easily. Graph theoretic concepts are exploited to achieve this.

Let us first consider the overall flows of streams. From the preceding chapter, it may be recalled that the concept of a spanning tree of a process graph is useful in determining the observability or estimability of unmeasured flow variables. A fundamental cutset of a stream which is a branch of the spanning tree provides a flow balance equation which relates the flow of that stream with the flows of streams which are chords of the fundamental cutset. These ideas can be used to conveniently choose the dependent variables. We construct a spanning tree of the process graph and choose the flows of the branch streams of the spanning tree as dependent variables and the chord stream flows as independent variables. It can be immediately deduced that the fundamental cutsets of the spanning tree can be used to relate the dependent and independent flow variables. These relationships can be expressed as

$$F_{bi} = \sum_{j \in \text{chords}} p_{ij} F_{cj} \quad i = 1 \ldots \text{branches} \tag{4-25}$$

where F_{bi} is the flow of branch stream *i*, F_{cj} is the flow of chord stream *j*, p_{ij} is a coefficient which is 0 if chord *j* is not part of the fundamental cutset of branch stream *i*, +1 or −1 if chord *j* is in the fundamental cutset of branch *i*, and the flow directions of chord *j* and branch *i*, are opposite or same with respect to each other, respectively. Thus, the dependent branch flow variables can be eliminated from the objective function, Equation 4-24 by using Equation 4-25.

If we consider the solid flows, the above ideas can again be applied since the solids flows are related in exactly the same manner as the overall flows. This is true also of component flows of streams for each component. Thus, the solids flows and component flows of branch streams can be chosen as dependent variables and related to the corresponding chord flows similar to Equation 4-25. The solution of the resulting unconstrained optimization problem can be obtained by setting the derivatives of the objective function with respect to the chord flows to zero and solving the resulting linear equations. Complete details of the linear equations to be solved at each iteration are given in Simpson et al. [6]. It should be noted that in this technique initial estimates of chord flows (total, slurry and component flows) only have to be guessed since the branch flow estimates can always be calculated using Equation 4-25.

Example 4-5

We illustrate Simpson's method for the mineral flotation process considered in Example 4-4. The process graph of Figure 4-4 is constructed by including the environment node and is shown in Figure 4-6. A spanning tree of this process graph is constructed and is shown in Figure 4-7.

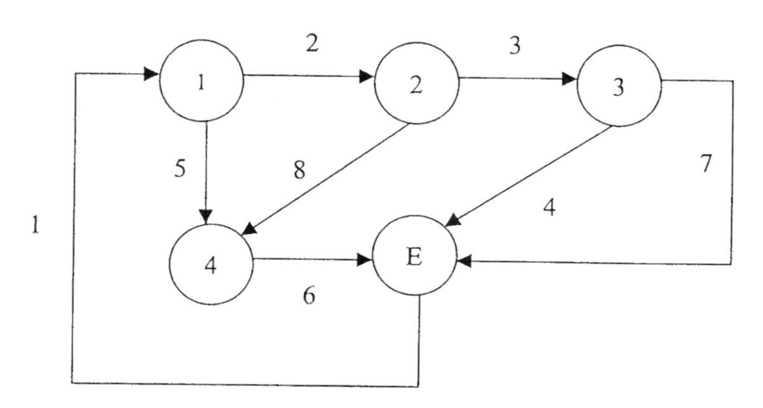

Figure 4-6. Graph of mineral flotation process.

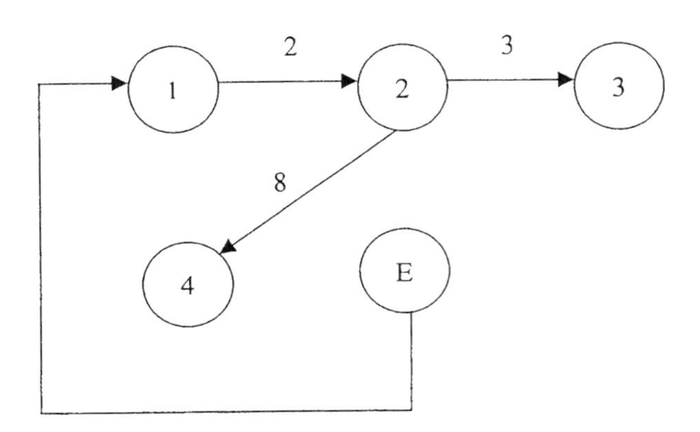

Figure 4-7. Spanning tree of mineral flotation process.

The choice of this spanning tree implies that the flows of branch streams 1, 2, 3, and 8 are the dependent variables. The fundamental cutsets with respect to this spanning can be easily obtained and are given by the sets [1, 4, 6, 7], [2, 4, 5, 6, 7], [3, 4, 7], and [8, 5, 6].

Based on the stream flow directions, the matrix of coefficients, p_{ij}, can now be constructed and is given by

$$\mathbf{P} = \begin{bmatrix} 1 & 0 & 1 & 1 \\ 1 & -1 & 1 & 1 \\ 1 & 0 & 0 & 1 \\ 0 & -1 & 1 & 0 \end{bmatrix}$$

where the rows of **P** correspond to the branch streams, 1, 2, 3, and 8 while the columns of **P** correspond to the chord streams 4, 5, 6, and 7 arranged in order. We start with initial estimates of chord flow variables and compute the branch flows using Equation 4-25, which are shown in Table 4-7. For these estimates, the coefficients a_j and b_{jk} are computed. The solution of the reduced unconstrained DR problem gives new estimates for the chord flows which are used for the next iteration. The iterations are carried out until convergence. The final reconciled estimates obtained are shown in Table 4-7.

Table 4-7
Reconciled Data of Mineral Flotation Process Using Simpson's Method

Method	Var.	Stream							
		1	2	3	4	5	6	7	8
Initial Estimates	F*	1.25	1.15	1.1	0.9	0.1	0.15	0.2	0.05
	N_{Cu}	3.399	1.413	0.183	0.0810	1.986	3.216	0.102	1.23
	N_{Zn}	11.532	10.823	10.789	0.369	0.709	0.7425	10.42	0.0335
Simpson	F	1	0.9246	0.9160	0.8424	0.0754	0.0840	0.0736	0.0087
	$\hat{x}_{Cu}$ %	1.9189	0.4505	0.1258	0.0918	19.932	21.462	0.515	34.773
	$\hat{x}_{Zn}$ %	4.5575	4.3564	4.5278	0.4097	7.0231	4.8803	51.657	−13.77

Generalization of Bilinear Data Reconciliation Techniques

In describing the above methods, we have reconciled all measured values. It is more common in industrial applications to hold some of the measured variables constant during the process of reconciliation. A simple way to accomplish this is to assign a very high weight (or very small standard deviation) in the objective function to measurements that have to be kept constant. This will force the adjustment made to these measurements to be negligibly small.

Both Crowe's method and Simpson's method have been described for processes involving primarily mixers and separators. If splitters, reactors, or grinding mills are present in the process, then these methods have to be suitably modified because the type of equations imposed for these units do not conform to those of other units such as mixers and separators. Crowe [3] has outlined the modifications necessary to take splitters into account. For this purpose, the splitter equations are formulated using split-fraction variables.

As pointed out earlier, the use of split-fraction variables leads to a trilinear structure for the component balances. In order to use Crowe's method for the bilinear problem, the split-fraction variables are estimated in an outer iterative loop. For each guess of the split-fraction variables, a bilinear problem results which can be solved using Crowe's method. In general, a constrained optimization technique is required to obtain updated estimates of the split-fraction variables at each iteration which robs Crowe's method of much of its efficiency. If only one splitter is present in the process, however, then a univariate optimization method such as the golden section search [8] can be used for this purpose.

Treatment of Enthalpy Flows

Although both Crowe's method and Simpson's method were developed to solve multicomponent data reconciliation problems, it is possible to extend these techniques to take into account enthalpy balances and to reconcile temperature variables. In general, the enthalpy of a stream is a nonlinear function of the stream temperature and composition. However, if the enthalpy of a stream can be assumed to be a linear function of temperature and independent of composition, then simultaneous material and energy balance reconciliation also give rise to a bilinear problem. Even if the enthalpy of a stream is a nonlinear function of temperature but is independent of composition, the methods discussed in this chapter can be used with minor modifications.

An important subsystem that satisfies this assumption is that of a crude preheat train of a refinery, where the enthalpy of a petroleum stream is related to the temperature and physical properties such as API gravity and normal boiling point of the stream. For the purposes of this chapter, we will make this assumption and describe the modifications necessary to apply Crowe's method or Simpson's method for simultaneous material and energy balance reconciliation. As before, we first describe the energy balances for different types of process units.

Mixer Enthalpy Balance

$$\sum_{j=1}^{S} F_j H(T_j) - F_o H(T_o) = 0$$

where H(T) is the specific enthalpy of the stream which is assumed to be only a function of temperature.

Splitter Enthalpy Balance

$$T_i - T_j = 0 \qquad j = 1 \ldots S$$

or in terms of specific enthalpies of streams

$$H(T_i) - H(T_j) = 0 \qquad j = 1 \ldots S$$

Heat Exchanger

By definition, we assume for a heat exchanger the data for both the cold and hot side fluids need to be reconciled or estimated. We also assume that the streams are single-phase fluids.

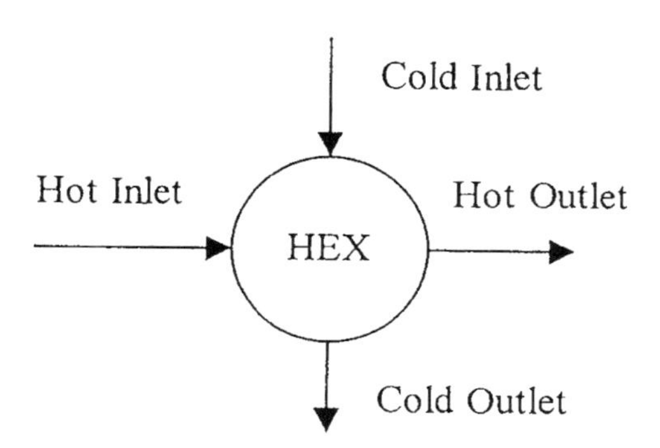

Figure 4-8. Heat exchanger.

The equations for this unit shown in Figure 4-8 are

(i) Flow balances for hot and cold side fluids:

$$F_{hi} - F_{ho} = 0$$

$$F_{ci} - F_{co} = 0$$

(ii) Enthalpy balance:

$$F_{hi}\,H_h\,(T_{hi}) - F_{ho}H_h\,(T_{ho}) + F_{ci}H_c\,(T_{ci}) - F_{co}H_c\,(T_{co}) = 0$$

(iii) Component flow balances:

$$F_{hi}x_{hi,k} - F_{ho}x_{ho,k} = 0 \qquad k = 1...C$$

$$F_{ci}x_{ci,k} - F_{co}x_{co,k} = 0 \qquad k = 1...C$$

(iv) Normalization equations for outlet streams:

$$\sum_{k=1}^{C} x_{ho,k} = 1$$

$$\sum_{k=1}^{C} x_{co,k} = 1$$

Heaters or Coolers

A heater or a cooler is a heat exchanger for which data of only the process stream are reconciled, while data of the utility stream are assumed to be unavailable or unimportant. The constraints for this unit are a subset of the constraints of a heat exchanger. The material, energy, and normalization equations for the process stream only are written.

Crowe's method can be easily extended to include enthalpy balances and temperature variables in the reconciliation problem as suggested by Sanchez and Romagnoli [9]. The specific enthalpy variables can be treated in a similar manner as composition variables. The enthalpy flow of different streams can be classified into the three categories in a similar manner as component flow variables. The objective function will now contain terms for the adjustments made to enthalpy flows of Category I streams and specific enthalpies of Category II streams. The two-step Crowe's projection technique described earlier can be applied to also obtain the reconciled values of specific enthalpies of all streams. If the specific enthalpy is a nonlinear function of temperature, then the temperature estimate of each stream can be recovered from the specific enthalpy. In general, this may require the solution of a one-dimensional nonlinear equation for each stream.

Simpson's method has also been extended to include splitters, as well as to treat enthalpy balances along with flow and component balances [10]. It should be cautioned, however, that generalizing these methods to include other types of process units such as a flash drum (which can also be described by bilinear equations for ideal thermodynamics) may not be a trivial exercise. A significant disadvantage of these methods is that at present they cannot take into account simple bounds on process variables. This can seriously limit the use of these methods in industrial applications where it is required to obtain feasible estimates of process variables.

SUMMARY

- A complete set of independent constraints has to be imposed for each process unit in formulating data reconciliation problems. Different independent sets of equations can be imposed for a process unit, but some are more convenient than others.
- It is important to include normalization constraints on compositions to ensure that the reconciled estimates satisfy them.
- The constraints of bilinear data reconciliation problems contain products of two variables (flow and composition or flow and temperature).
- Special methods have been developed for solving bilinear data reconciliation problems. These are efficient but cannot handle all types of process units and they also cannot take into account feasibility constraints such as bounds on variables.
- Nonlinear data reconciliation techniques can be used to solve bilinear problems. These are less efficient, but do not have any other limitations.

REFERENCES

1. Reklaitis, G. V. *Introduction to Material and Energy Balances.* New York: John Wiley & Sons, 1983.
2. Meyer, M., M. Enjalbert, and B. Koehret. "Data Reconciliation on Multicomponent Networks Using Observability and Redundancy Classification," in *Computer Applications in Chemical Engineering* (edited by H. Th. Bussemaker and P. D. Iedema). Amsterdam: Elsevier: 1990.
3. Crowe, C. M. ""Reconciliation of Process Flow Rates by Matrix Projection. II: The Nonlinear Case." *AIChE Journal* 32 (1989): 616–623.
4. Rao, R. R., and S. Narasimhan. "Comparison of Techniques for Data Reconciliation of Multicomponent Processes." *Ind. & Eng. Chem. Res.* 35 (1996): 1362–1368.
5. Smith, H. W., and N. Ichiyen. "Computer Adjustment of Metallurgical Balances." *CIM Bull.* (1973): 97–100.

6. Simpson, D. E., V. R. Voller, and M. G. Everett. "An Efficient Algorithm for Mineral Processing Data Adjustment." *Int. J. Miner. Proc.* 31 (1991): 73–96.

7. Wills, B. A. *Mineral Processing Technology,* 4th ed. Oxford: Pergamon, 1989.

8. Edgar, T. F., and D. M. Himmelblau. *Optimization of Chemical Processes.* New York: McGraw-Hill, 1988.

9. Sanchez, M., and J. Romagnoli. "Use of Orthogonal Transformations in Data Classification-Reconciliation." *Computers Chem. Engng.* 20 (1996): 483–493.

10. Siraj, S. M. *An Efficient Decomposition Strategy for General Data Reconciliation.* M.Tech thesis, IIT Kanpur, India, 1995.

5

Nonlinear Steady-State Data Reconciliation

The steady-state conservation constraints that are used to describe most chemical processes are nonlinear in nature. If we are interested in only overall flow balance reconciliation of such processes, then linear data reconciliation techniques described in Chapter 3 are sufficient. Moreover, under some restrictions some of these processes can be solved using bilinear data reconciliation techniques as described in Chapter 4.

If we wish, however, to take into consideration thermodynamic equilibrium relationships and complex correlations for thermodynamic and physical properties, then nonlinear data reconciliation techniques must be used. Moreover, in Chapters 3 and 4, we have considered only *equality constraints* corresponding to material and energy conservation and have not imposed even simple bounds on the variables. The reconciled estimates of variables can therefore become infeasible. For example, negative reconciled estimates for flows or compositions can be obtained. If we impose bounds on the estimates of variables or other feasibility constraints, then these give rise to inequality constraints in the data reconciliation problem which can be solved only using a nonlinear data reconciliation solution technique.

FORMULATION OF NONLINEAR DATA RECONCILIATION PROBLEMS

Equilibrium Flash Data Reconciliation Example

We will use a simplified version of the example of a single stage flash unit drawn from MacDonald and Howat [1] to illustrate the formulation and solution of nonlinear data reconciliation problems. Figure 5-1 shows an isothermal flash unit with a feed stream containing propane, n-butane and n-pentane. The steady-state constraint equations for this unit are as given below:

Component Balances: $$Fz_i - Lx_i - Vy_i = 0 \quad i = 1, 2, 3 \qquad (5\text{-}1)$$

Normalization Equations: $$\sum_{i=1}^{3} x_i - 1 = 0 \qquad (5\text{-}2)$$

$$\sum_{i=1}^{3} y_i - 1 = 0 \qquad (5\text{-}3)$$

$$\sum_{i=1}^{3} z_i - 1 = 0 \qquad (5\text{-}4)$$

Equilibrium Relations: $$y_i = P_i^{sat}(T)x_i / P \quad i = 1, 2, 3 \qquad (5\text{-}5)$$

For simplicity, Raoult's law has been used to describe the equilibrium relations. The saturation pressure is obtained using the Antoine equation which is given by:

Antoine Equation: $$\ln P_i^{sat} = A_i + B_i / (T + C_i) \quad i = 1,2,3 \qquad (5\text{-}6)$$

The nonlinear data reconciliation problem is to reconcile the measurements of the flow rate, temperature, pressure, and compositions of the feed, liquid and vapor product streams so as to satisfy the constraints 5-1 through 5-6.

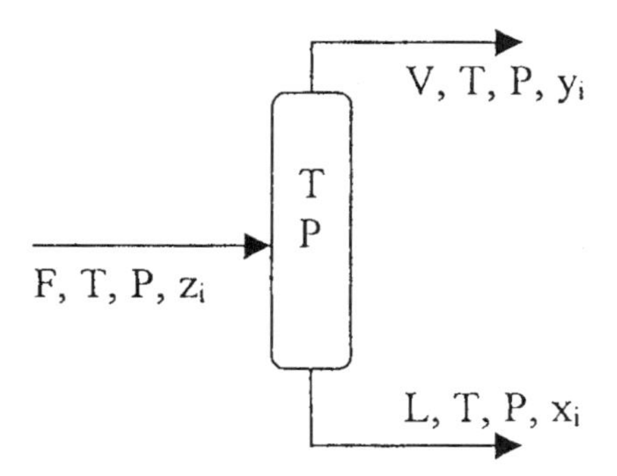

Figure 5-1. Equilibrium flash unit.

General Problem Formulation

As in the linear case, it is assumed that the random measurement errors follow a normal distribution with zero mean and a variance-covariance matrix Σ. The general nonlinear data reconciliation problem can be formulated as a least-squares minimization problem as follows :

$$\min_{\mathbf{x},\mathbf{u}} (\mathbf{y}-\mathbf{x})^T \Sigma^{-1} (\mathbf{y}-\mathbf{x}) \tag{5-7}$$

subject to

$$\mathbf{f}(\mathbf{x},\mathbf{u}) = \mathbf{0} \tag{5-8}$$

$$\mathbf{g}(\mathbf{x},\mathbf{u}) \le \mathbf{0} \tag{5-9}$$

where

f : m × 1 vector of equality constraints;
g : q × 1 vector of inequality constraints;
Σ : n × n variance-covariance matrix;
u : p × 1 vector of unmeasured variables;
x : n × 1 vector of measured variables;
y : n × 1 vector of measured values of measurements of variables **x**.

The equality constraints defined by Equation 5-8 typically include all material and energy conservation relations, thermodynamic equilibrium constraints, constitutive equations for material behavior similar to Equations 5-1 through 5-6 of the equilibrium flash example. The inequality constraints given by Equation 5-9 may be as elementary as upper and lower bounds on variables or complex feasibility constraints related to equipment operation.

In the above formulation, we have tacitly assumed that the variables $\mathbf{x}$ are directly measured. However, this does not impose any limitation. If measurements are functions (linear or nonlinear) of the variables (for example pH is a function of H^+ concentration), then we can always define a new state variable for pH which is directly measured and the relationship between pH and H^+ concentration can be included as part of the equality constraint set. We will first consider the solution techniques for nonlinear data reconciliation problems in which only equality constraints are present in the process model. Two solution techniques and their variants are discussed in the following section.

SOLUTION TECHNIQUES FOR EQUALITY CONSTRAINED PROBLEMS

The minimization of Equation 5-7 subject to the equality constraints of Equation 5-8 can be achieved by using a general purpose nonlinear optimization technique. However, since the objective function is quadratic in nature, efficient techniques have been developed to solve the problem. The estimates obtained by solving this optimization problem can be shown to be *maximum likelihood estimates* (MLE). It should be noted, however, that these estimates may be biased, whereas, in the linear case the estimates are unbiased.

Methods Using Lagrange Multipliers

The equality constrained nonlinear data reconciliation problem can be solved by using the classical method of Lagrange multipliers [2]. The Lagrangian for the problem is given by

$$L(\mathbf{x},\mathbf{u},\lambda) = (\mathbf{y}-\mathbf{x})^T \Sigma^{-1}(\mathbf{y}-\mathbf{x}) + 2\lambda^T\mathbf{f}(\mathbf{x},\mathbf{u}) \tag{5-10}$$

The solution of the data reconciliation problem can be obtained by setting the partial derivatives of Equation 5-10 with respect to the variables **x**, **u**, and λ to zero—the necessary conditions for an optimal solution point of Problem defined by Equations 5-7 and 5-8—and solving the resulting equations. The following equations are obtained:

$$\frac{\partial L}{\partial \mathbf{x}} = -\Sigma^{-1}(\mathbf{y}-\mathbf{x}) + \mathbf{J}_x^T \lambda = 0 \tag{5-11}$$

$$\frac{\partial L}{\partial \mathbf{u}} = \mathbf{J}_u^T \lambda = \mathbf{0} \tag{5-12}$$

$$\frac{\partial L}{\partial \lambda} = \mathbf{f}(\mathbf{x},\mathbf{u}) = \mathbf{0} \tag{5-13}$$

where

$$\mathbf{J}_x = \frac{\partial \mathbf{f}}{\partial \mathbf{x}} \tag{5-14}$$

$$\mathbf{J}_u = \frac{\partial \mathbf{f}}{\partial \mathbf{u}} \tag{5-15}$$

are the Jacobian matrices containing the partial derivatives of the nonlinear functions **f** with respect to **x** and **u**, respectively.

Since the constraints are nonlinear, solving for **x**, **u** and λ involves an iterative numerical procedure. The system of normal equations 5-11 through 5-13 can be solved by any simultaneous equation solver [3]. Stephenson and Shewchuck [4] used a Newton-Raphson iterative method which is based on a quasi-Newton linearization of the nonlinear model. Their algorithm takes advantage of the sparsity of the Jacobian matrix and the invariance of the partial derivatives of the linear terms in the model equations which makes the computations more efficient for large systems. Serth et al. [5] reported a similar approach but with a different nonlinear equation solver.

Madron [6] suggested an iterative approach for solving the normal equations 5-11 through 5-13 based on successive linearization. Let $\hat{\mathbf{x}}_k$ and $\hat{\mathbf{u}}_k$ represent the estimates of the variables obtained at the start of iteration *k*. A linear approximation can be obtained for the nonlinear constraint from the Taylor's expansion of the function **f(x,u)** in Equation 5-8 and retaining only the constant term and the first-order derivative term.

$$\mathbf{f}(\mathbf{x},\mathbf{u}) = \mathbf{f}(\hat{\mathbf{x}}_k,\hat{\mathbf{u}}_k) + \mathbf{J}_x^k(\mathbf{x}-\hat{\mathbf{x}}_k) + \mathbf{J}_u^k(\mathbf{u}-\hat{\mathbf{u}}_k) = \mathbf{0} \tag{5-16}$$

where the Jacobian matrices $\mathbf{J}_x^k$ and $\mathbf{J}_u^k$ are as defined by Equations 5-14 and 5-15, with the superscript k indicating that they are evaluated at the estimates $\hat{\mathbf{x}}_k$, $\hat{\mathbf{u}}_k$. The Jacobian matrices that appear in Equations 5-11 and 5-12 are also replaced by their estimated values at iteration k. The resulting set of equations are now linear. In Madron's procedure, these linear equations are de-coupled by eliminating vector **x** from Equations 5-12 and 5-13 using Equation 5-11.

From Equation 5-11,

$$\mathbf{x} = \mathbf{y} - \Sigma\left(\mathbf{J}_x^k\right)^T \boldsymbol{\lambda} \tag{5-17}$$

Using Equations 5-16 and 5-17 in Equations 5-12 and 5-13, and rearranging we obtain the following linear equations involving **u** and λ

$$\begin{bmatrix} \mathbf{J}_x^k\Sigma(\mathbf{J}_x^k)^T & \mathbf{J}_u^k \\ (\mathbf{J}_u^k)^T & \mathbf{0} \end{bmatrix}\begin{bmatrix} \lambda \\ \mathbf{u} \end{bmatrix} = \begin{bmatrix} -\mathbf{f}(\hat{\mathbf{x}}_k,\hat{\mathbf{u}}_k) + \mathbf{J}_x^k\hat{\mathbf{x}}_k + \mathbf{J}_u^k\hat{\mathbf{u}}_k - \mathbf{J}_x^k\mathbf{y} \\ \mathbf{0} \end{bmatrix} \tag{5-18}$$

Equation 5-18 can be solved for obtaining the new estimates for **u** and λ. The estimates for λ are used in Equation 5-17 to obtain the new estimates for **x**. This procedure is repeated using these new estimates as initial guesses for the next iteration. A disadvantage with all these methods is that the inclusion of Lagrange multipliers λ in the solution increases the size of the problem which require more computational time.

To reduce the size of the problem, Madron [6] proposed a Gauss-Jordan elimination process of the original linear/linearized constraint matrices ($\mathbf{J}_x \mid \mathbf{J}_u$ for the nonlinear case). The structure of the resulting matrix provides useful information for variable classification.

Method of Successive Linear Data Reconciliation

A simpler way to handle the nonlinear data reconciliation is to successively solve a series of linear data reconciliation problems by linearization of the nonlinear constraints. A linear approximation to the nonlinear constraints are obtained as in Equation 5-16. We thus obtain a linear data reconciliation problem for minimizing 5-7 subject to linear equality constraints of Equation 5-16 which can be solved using the technique described in Chapter 3. Britt and Luecke [5] proposed an alternative solution procedure for the linearized problem. Their solution for the estimates which have to be used at the next iteration are given by

$$\hat{\mathbf{u}}_{k+1} = \hat{\mathbf{u}}_k - \left[\left(\mathbf{J}_u^k\right)^T \mathbf{R}^{-1} \mathbf{J}_u^k\right]^{-1} \left(\mathbf{J}_u^k\right)^T \mathbf{R}^{-1} \left\{\mathbf{f}(\hat{\mathbf{x}}_k, \hat{\mathbf{u}}_k) + \mathbf{J}_x^k (\mathbf{y} - \hat{\mathbf{x}}_k)\right\} \tag{5-19}$$

$$\hat{\mathbf{x}}_{k+1} = \mathbf{y} - \Sigma\left(\mathbf{J}_x^k\right)^T \mathbf{R}^{-1} \left\{\mathbf{f}(\hat{\mathbf{x}}_k, \hat{\mathbf{u}}_k) + \mathbf{J}_x^k (\mathrm{y} - \hat{\mathrm{x}}_k) + \mathbf{J}_u^k (\hat{\mathbf{u}}_{k+1} - \hat{\mathbf{u}}_k)\right\} \tag{5-20}$$

where

$$\mathbf{R} = \mathbf{J}_x^k \Sigma \left(\mathbf{J}_x^k\right)^T \tag{5-21}$$

Equations 5-19 through 5-20 were derived by Britt and Luecke [2] for parameter estimation in nonlinear regression. It was adapted for nonlinear data reconciliation by Knepper and Gorman [7] and also used later by MacDonald and Howat [1]. The algorithm requires initial estimates for all variables contained in the vectors **x** and **u.** The measured values **y** can be used to initialize the variables **x.** Britt and Luecke also designed a simplified algorithm that can be used to initialize the unmeasured parameters **u.** At each iteration, the function **f(x,u)** and the Jacobian matrices $\mathbf{J}_x$ and $\mathbf{J}_u$ are re-evaluated with the new estimates. The iterations are continued until $\| \mathbf{u}_{k+1} - \mathbf{u}_k \|$ and $\| \mathbf{x}_{k+1} - \mathbf{x}_k \|$ satisfy a small tolerance criterion. If convergence is achieved, the solution might not be a global minimum solution. This difficulty is common to most nonlinear least-squares estimation problems.

A variant of the above algorithm suggested by Knepper and Gorman [6] in order to reduce the computational time, is to hold the Jacobian matrices constant at the initial estimates and re-compute them only after the constraints are satisfied (*constant direction approach*). This approach, however, is characterized by slow convergence.

Another variant suggested by MacDonald and Howat [1] is a de-coupled procedure, in which the estimates for **u** are held constant and Equation 5-20 is repeatedly used until the estimates for **x** converge. Equation 5-19 is now used to obtain new estimates for **u,** and the procedure is repeated until all the estimates converge. MacDonald and Howat [1] demonstrated through application to a non-equilibrium flash unit that the coupled algorithm provides marginally more accurate estimates at the expense of a greater computational time. The de-coupled procedure can be a useful computational scheme when the nonlinear equations are implicit in the parameters.

The method of Britt and Leucke [2] and their variants described above have some limitations. Equations 5-19 and 5-20 involve the inverse of two matrix products, **R** and $(\mathbf{J}_u^k)^T\mathbf{R}^{-1}(\mathbf{J}_u^k)$. In order for the inverse of the matrix products to exist, the following conditions should be satisfied:

(i) The matrix $\mathbf{J}_x^k$ should have full row rank
(ii) The matrix $\mathbf{J}_u^k$ should be of full column rank.

The second of the above conditions imply that all the unmeasured variables should be observable. This is identical to the condition seen in the case of linear systems in Chapter 3, where it was shown that in order for all unmeasured variables to be observable, the columns of the constraint matrix corresponding to these variables must be linearly independent. Even if this condition is met, the first condition may not be satisfied by some processes depending on which of the variables are measured (see Exercise 5-1). Thus, the above methods cannot be applied in general for all processes.

Exercise 5-1. *Consider the flotation process described in Example 4-4. Generate the submatrix of the Jacobian corresponding to measured variables (this can be done analytically). Show that the row corresponding to the flow balance for node 4 in this submatrix consists only of zeros and hence prove that this matrix does not have full row rank.*

An approach that can be used in general is based on the use of Crowe's projection matrix technique [8] to solve the linear data reconciliation problem in each iteration. The basic steps involved in this approach are as follows:

Step 1. Start with the measured values as initial estimates for the variables **x** and initial estimates for **u** which are provided by the user.

Step 2. Evaluate the Jacobian matrices of the nonlinear constraints with respect to variables **x** and **u** at the current estimates.

Step 3. Compute the projection matrix $\mathbf{P}^k$ such that it satisfies

$$\mathbf{P}^k \mathbf{J}_u^k = \mathbf{0} \tag{5-22}$$

The projection matrix can be obtained using QR factorization of the matrix $\mathbf{J}_u^k$ as described in Chapter 3.

Step 4. Compute new estimates for **x** using

$$\hat{\mathbf{x}} = \mathbf{y} - \Sigma(\mathbf{P}^k \mathbf{J}_x^k)^T \left[\mathbf{P}^k \mathbf{J}_x^k \Sigma (\mathbf{P}^k \mathbf{J}_x^k)^T\right]^{-1} \mathbf{P}^k \left[\mathbf{J}_x^k \mathbf{y} + \mathbf{f}(\hat{\mathbf{x}}_k, \hat{\mathbf{u}}_k) - \mathbf{J}_x^k \hat{\mathbf{x}}_k\right] \tag{5-23}$$

Step 5. Compute the new estimates for **u** through Equation 3-24 utilizing the QR factorization of matrix $\mathbf{J}_u^k$.

Step 6. Stop if the new estimates are not significantly different from those obtained in the preceding iteration. Otherwise using these new estimates repeat the procedure starting with Step 2.

Pai and Fisher [9] were the first to use a procedure similar to the algorithm described above. The additional modifications that their algorithm incorporates are:

(a) A Broyden's update procedure [10] for updating the Jacobian matrices rather than recalculating them at each iteration, in order to reduce the computational effort involved.

(b) A line search procedure after Step 5 based on a penalty function method to compute the estimates to be used for starting the next iteration. The penalty function $\|\mathbf{f}(\mathbf{x},\mathbf{u})\| + \alpha(\mathbf{y}-\mathbf{x})^T\Sigma^{-1}(\mathbf{y}-\mathbf{x})$ was used, where α is an arbitrary number $0 \leq \alpha \leq 1$.

The modification described in (a) can improve computational efficiency of the algorithm and has been demonstrated for small problems. However, the modification in (b) is of questionable utility. This is due to the fact that the objective function of data reconciliation is quadratic and the solution for the estimates obtained in Steps 4 and 5 are optimal, although they do not satisfy the nonlinear constraints. A line search procedure for modifying these estimates can improve feasibility with respect to the nonlinear constraints but at the cost of sacrificing optimality. This may not lead to an overall reduction in the computational effort. The right choice for the parameter α which plays the role of a subjective weight for the least-squares objective function is also difficult to come up with. Pai and Fisher used $\alpha = 0.1$.

Swartz [11] first recommended the use of QR factorization for separating the estimation of the measured and unmeasured variables at each iteration. If the problem is highly nonlinear and the size of the problem is large, this can become computationally inefficient. Ramamurthi and Bequette [12] reported an increase in the computational time with the noise level (magnitude of gross errors) in measurements. Because of the successive linearization process, more iterations are usually required to converge a problem with large errors in data. This procedure is very popular, however, for data reconciliation problems involving overall mass and energy balance equations.

In the preceding chapter, the reconciled values for the binary distillation column reported in Table 4-3 and the reconciled values reported in the last row of Table 4-6 were obtained using the successive linear data reconciliation algorithm described above. It can be observed from the results of Table 4-6, that the reconciled estimate of zinc concentration in stream 8 has a high negative value, which is absurd. Thus, this method cannot be guaranteed to give feasible estimates in all cases.

NONLINEAR PROGRAMMING (NLP) METHODS FOR INEQUALITY CONSTRAINED RECONCILIATION PROBLEMS

The major limitation of the methods described in the preceding section is their inability to handle inequality constraints. In many situations, especially when significant gross errors exist, standard data reconciliation can cause the effect of the gross errors to spread over all estimates. If sufficient redundancy does not exist, the estimates for variables which have small values contain significant errors. In some cases, infeasible estimates such as negative values for flow rates or compositions are obtained. In order to tackle this problem, it is necessary to impose limits or bounds on the variables. These inequality constraints on measured and unmeasured variables take the form

$$\mathbf{x}_{min} \leq \mathbf{x} \leq \mathbf{x}_{max} \tag{5-24}$$

$$\mathbf{u}_{min} \leq \mathbf{u} \leq \mathbf{u}_{max} \tag{5-25}$$

In extreme cases, other types of feasibility constraints have to be imposed. For example, when data reconciliation is applied to heat exchanger networks for reconciling flows and temperatures, it is possible that the estimates violate thermodynamic feasibility, such as the temperature estimate of the hot stream being lower than the corresponding cold stream temperature estimate. In order to combat this, an inequality constraint should be imposed that forces the temperature of the hot stream to be greater than the corresponding cold stream temperature at both ends of every exchanger. These type of feasibility constraints can be cast in the form of Equation 5-9.

The solution to the nonlinear data reconciliation problem when constraints such as Equation 5-2 or 5-24 through 5-25 are imposed, can be obtained by using general purpose nonlinear programming techniques. A detailed description of such techniques is beyond the scope of this text and the reader is referred to excellent books on this subject such as Gill et al. [13] and Edgar and Himmelblau [14]. We discuss below the broad features of two popular nonlinear programming techniques especially with reference to their application for solving nonlinear data reconciliation problems.

Sequential Quadratic Programming (SQP)

The *sequential quadratic programming technique* [15, 16, 17] solves a nonlinear optimization problem by successively solving a series of quadratic programming problems. At each iteration, a quadratic program approximation of the general optimization problem is obtained by a quadratic function approximation of the objective function, and a linear approximation of the constraints both using Taylor's series expansion around the current estimates. In the case of the data reconciliation problem defined by Equations 5-7 through 5-9, the objective function is already quadratic and only the constraints have to be linearized. The resulting quadratic program at iteration $k+1$ is formulated as

$$\underset{\mathbf{s}}{\text{Min}}\,(\nabla\phi)^{T}\mathbf{s}+\mathbf{s}^{T}\mathbf{B}\mathbf{s} \qquad (5\text{-}26)$$

subject to

$$f_i(\hat{\mathbf{z}}_k)+\left[\nabla f_i(\hat{\mathbf{z}}_k)\right]^{T}\mathbf{s}=\mathbf{0} \qquad i=1\ldots m \qquad (5\text{-}27)$$

$$g_j(\hat{\mathbf{z}}_k)+\left[\nabla g_j(\hat{\mathbf{z}}_k)\right]^{T}\mathbf{s}\le\mathbf{0} \qquad j=1\ldots q \qquad (5\text{-}28)$$

where

$\mathbf{z}$ is the vector of original variables ($\mathbf{x}$, $\mathbf{u}$);

$\mathbf{s} = \mathbf{z} - \hat{\mathbf{z}}_{\mathbf{k}}$ is the direction of search for iteration $k+1$;

$\nabla\phi$, ∇f_i, ∇g_j are, respectively, the gradients (derivatives with respect to variables $\mathbf{z}$) of the objective function, equality constraint i, and inequality constraint j, all evaluated at the current estimate $\hat{\mathbf{z}}_k$.

$\mathbf{B}$ is the Hessian (matrix of second order derivatives of the objective function with respect to the variables $\mathbf{z}$) evaluated at the current estimate $\hat{\mathbf{z}}_{\mathbf{k}}$.

In the quadratic program formulation, all variables are included in the objective function. Comparing with the data reconciliation objective function Equation 5-7, it can be deduced that

$$\nabla\phi=\begin{bmatrix}-2\,\Sigma^{-1}(\mathbf{y}-\hat{\mathbf{x}}_k)\\ \mathbf{0}\end{bmatrix} \qquad (5\text{-}29)$$

$$\mathbf{B} = 2\begin{bmatrix}\Sigma^{-1} & \mathbf{0}\\ \mathbf{0} & \mathbf{0}\end{bmatrix} \qquad (5\text{-}30)$$

Note that the Hessian matrix is constant and is also singular if there are unmeasured variables in the process.

The solution of the quadratic program gives the search direction for obtaining the estimates. A one-dimensional search is performed in direction $\mathbf{s}_k$ at each iteration k, so that the new value for $\mathbf{z}$ at the next iteration is

$$\hat{\mathbf{z}}_{k+1} = \hat{\mathbf{z}}_k + \alpha_k\mathbf{s}_k \qquad (5\text{-}31)$$

where α_k is a step-length parameter between 0 and 1. The step length is obtained by minimizing a penalty function (similar to the Lagrangian). The procedure is repeated using the new estimates until convergence.

There are several issues of particular interest in solving data reconciliation problems using SQP. Generally, in SQP the exact Hessian matrix (or its inverse) is not computed at each iteration because of the computational burden it entails. Instead, an approximate inverse of the Hessian matrix (or its square root) is obtained by a symmetric Broyden's update technique. In the case of data reconciliation, Equation 5-29 shows that the Hessian matrix is constant and, therefore, there is no need of updating it. Secondly, Equation 5-29 also shows that the Hessian matrix is positive semi-definite, if unmeasured variables are present. Therefore, the QP solver that is used should be capable of handling positive semi-definite Hessian matrices. Thirdly, the solution obtained using the QP is used as a search direction and the optimum step length in this direction is obtained by minimizing a penalty function. If the objective function contains nonlinear terms of higher order than quadratic, then this line minimization gives estimates which are more optimal and less infeasible. In the case of data reconciliation, however, since the objective function is quadratic, a step length of unity gives the optimal estimates that satisfy the linearized approximation of the constraints.

In this case, line minimization will *improve the feasibility with respect to the nonlinear constraints by sacrificing optimality.* It is debatable whether this can lead to an overall reduction in the number of iterations required for convergence. Thus, even though a general purpose SQP technique can be used, by exploiting the special features as discussed above it is possible to develop a more efficient tailor-made SQP technique for solving nonlinear data reconciliation problems.

Methods for solving quadratic programs are available in HARWELL library, SQPHP [17] or QPSOL [18]. An efficient SQP solver, denoted as RND-SQP, has been recently developed by Vasantharajan and Biegler [19]. In this technique, a *reduced* QP is solved at each iteration by partitioning the variables into a dependent and independent set, the number of dependent variables being equal to the number of equality constraints. Using the linearized equality constraints, the dependent variables are expressed in terms of independent variables and are thus eliminated from the QP subproblem. Furthermore, the QP subproblem contains the inequality constraints only. The solution for the independent variables obtained by solving the reduced QP is used to obtain the solution for the dependent variables. This technique can also be adapted for data reconciliation to obtain a very efficient method.

Generalized Reduced Gradient (GRG)

The GRG optimization technique solves a nonlinear optimization problem essentially by solving a series of successive linear programming problems. At each iteration, a linear program (LP) approximation is obtained by linearizing the objective function and constraints. The LP subproblem is formulated as in Equations 5-26 through 5-28 with the difference that the second (quadratic) term in Equation 5-26 is not present.

The GRG technique differs from SQP in one fundamental aspect. At each iteration, the GRG method requires estimates which satisfy the nonlinear constraints, whereas in SQP the estimates at each iteration need not be feasible with respect to the nonlinear constraints. Instead of a line minimization as in SQP, the solution of the LP subproblem is adjusted using an iterative procedure such as Newton-Raphson in order to obtain estimates that satisfy the nonlinear constraints.

The LP subproblem is itself solved using a standard algorithm by partitioning the variables into *dependent (basic)* variables and *independent (nonbasic)* variables; the dependent variables are implicitly determined by the independent variables, making the objective function a function of only the nonbasic variables. The nonbasic variables are split further into superbasic variables, which lie between their bounds and nonbasic variables which are at their bounds. A one-dimensional search is performed in the direction of the gradient of the superbasic variables (hence the term "reduced gradient"). Various commercial GRG algorithms differ in the methods they use to carry out the search and to regain a feasible point with respect to the nonlinear constraints [20, 21].

An interesting issue concerns the manner in which unmeasured variables are handled in both SQP and GRG. In both these approaches, no distinction is made between measured and unmeasured variables. A question worth considering is whether Crowe's projection matrix method for decoupling the estimation of measured and unmeasured variables can be gainfully exploited in each iteration of the nonlinear programming techniques. The answer is that Crowe's projection technique cannot be utilized for eliminating unmeasured variables if there are bounds on these variables, because the estimates for unmeasured variables obtained using this technique may violate the bounds. Furthermore, both successive quadratic programming (RND-SQP) and GRG employ a projection technique to eliminate not just unmeasured variables but a set of dependent variables (equal to number of equality constraints which are usually more than the number of unmeasured variables). An analogy between this and Simpson's technique discussed in Chapter 4 may also be drawn.

Example 5-1

We illustrate the necessity for including bounds in data reconciliation for obtaining feasible estimates using the mineral flotation process considered in Example 4-4 for which the measured data are listed in the first row of Table 4-6. The last row in this table also gives the reconciled estimates obtained using a successive linear data reconciliation solution procedure along with Crowe's projection (method of Pai and Fisher [9] discussed in the preceding section). Since this method cannot handle bounds, they were not imposed. These reconciled estimates show that an absurdly large negative value for the estimate of the zinc concentration in stream 8 is obtained. The same problem was solved using SQP by imposing lower bound of 0.1% and upper bound of 100% on all concentrations and lower bound of 0 and upper bound of 1 on all flows. The reconciled estimates obtained are shown in Table 5-1.

Table 5-1
Reconciled Data of Mineral Flotation Process using SQP

Method	Var.	Stream							
		1	2	3	4	5	6	7	8
	F	1	0.9267	0.9157	0.8448	0.0733	0.0843	0.0709	0.0110
SQP	$\hat{x}_{Cu}$%	1.9042	0.4526	0.1316	0.1	20.267	21.164	0.5080	27.126
	$\hat{x}_{Zn}$%	4.5580	4.3728	4.4242	0.4101	6.9002	4.95	52.266	0.1

In this problem, we are able to obtain feasible estimates by including bounds in the nonlinear DR problem. The concentrations of Cu in stream 4 and Zn in stream 8 are at the lower bounds in the reconciled solution. By comparing these results with Table 4-6, we note that the reconciled estimates for all other variables are not significantly different.

VARIABLE CLASSIFICATION FOR NONLINEAR DATA RECONCILIATION

In Chapter 3, we have reviewed the methods of observability and redundancy variable classification for linear data reconciliation. Many of those methods [11, 22–29] are also applicable to nonlinear data reconciliation (particularly for problems with bilinear constraints).

For problems with higher levels of nonlinearity, a usual procedure is to perform a model linearization first, and apply the variable classification methods for the linear models. Albuquerque and Biegler [30] recently described such a procedure. Although designed for variable classification in connection with dynamic data reconciliation problems, their method can be used for steady state nonlinear data reconciliation problems as well. In this approach, an LU decomposition was used to build a projection matrix, in order to separate the unmeasured variables from the measured ones. The variable classification rules are very similar with those described by Swartz [11] with a QR decomposition algorithm. Since the LU decomposition is part of some NLP solving methods for data reconciliation, we briefly describe the Albuquerque and Biegler algorithm for variable classification here.

Equation 5-16 describing a linearized model, can also be written in an abbreviated form such as:

$$\mathbf{J}_x\mathbf{x} + \mathbf{J}_u\mathbf{u} = \mathbf{c} \tag{5-32}$$

where we grouped all constant terms from the model linearization into a global constant vector **c.** We also dropped the subscript *k,* by assuming that the linearization is performed about the final solution point. In order to eliminate the unmeasured variables **u,** a projection matrix **P** such that $\mathbf{P}\mathbf{J}_u = \mathbf{0}$ should be constructed. Let us assume that an LU decomposition of matrix is performed as follows:

$$\mathbf{E}\mathbf{J}_u\Pi = \mathbf{L}\begin{bmatrix} \mathbf{U}_1 & \mathbf{U}_2 \\ \mathbf{0} & \mathbf{0} \end{bmatrix} \tag{5-33}$$

where **E** and Π are some permutation matrices, **L** is a lower triangular matrix, $\mathbf{U}_1$ is an upper triangular matrix of rank *r* (the column rank of matrix $\mathbf{J}_u$) and $\mathbf{U}_2$ is some rectangular matrix. If $\mathbf{J}_u$ is of full row-rank, the zero rows in the upper triangular matrix would not exist. Furthermore, if $\mathbf{J}_u$

is of full column-rank, $\mathbf{U}_2$ would not exist and no unobservable variables would exist. A projection matrix $\mathbf{P}$ for matrix $\mathbf{J}_u$ can be created as follows:

$$\mathbf{P} = [\mathbf{0} \mid \mathbf{I}]\mathbf{L}^{-1} \tag{5-34}$$

Similarly to the rules derived by Swartz [11] or Crowe [22], *observability* requires a zero row of $\mathbf{U}_I^{-1}\mathbf{U}_2$. Furthermore, the nonredundant measured variables will have zero columns in the matrix $\mathbf{PJ}_x$. All other measured variables are declared redundant. It should be noted that these methods depend on the actual values of the measurements and may give rise to incorrect classification due to numerical problems.

Alternatively, graph-theoretic theorems and algorithms have been developed by Kretsovalis and Mah [23, 24] for observability and redundancy classification in general processes, which are based on *algebraic solvability of the constraint equations rather than an actual solution of the DR problem.* However, this method has the limitation that it does not take into account all types of process units (such as flash units) in its analysis. Specialized graph-theoretic observability and redundancy classification algorithms for bilinear (multicomponent) processes have also been developed by other researchers, as indicated in Chapter 3.

***Exercise 5-2.** Prove the observability and redundancy rules for the LU decomposition approach described above.*

Several issues regarding the variable classification issues in connection with SQP solving algorithm are important and will be briefly mentioned here. First, SQP requires initial estimates of all variables (measured and unmeasured). If there are unobservable variables, SQP will still be able to provide estimates for all variables. It uses the initial estimates of unmeasured variables (just sufficient number to make all other unmeasured variables observable) as "specifications" and performs the data reconciliation.

Which of the unmeasured variables will be chosen as the specified ones is implicit in the numerical method of solution (basically when the choice of independent and dependent variables are made based on the columns of the linearized constraint matrix at each iteration). The only way we will get to know which of the unobservable unmeasured variables have been chosen as specified is by examining the final results. If the reconciled estimate of an unmeasured variable is equal to the initial estimate provided by the user, then the unmeasured variable is unobservable (Note that there may be other unobservable variables that may have been back-calculated based on the choice of specified variables which we cannot figure out from the results).

Similarly, a redundant measurement can be identified by examining the reconciled values. If the reconciled value of a measured variable is equal to the measured value, then the measurement is nonredundant. In some rare cases, the initial estimate and final reconciled estimate may be the same due to numerical values (small variance, etc.). Again, we may not be able to pinpoint the cause for zero adjustment precisely. Thus, the only way to perform observability/redundancy variables is through comprehensive algorithms cited above.

The RND-SQP algorithm [19] automatically generates a reduced quadratic program by eliminating all equality constraints (mass, energy, component balances) and an equal number of variables from the original problem. Therefore, this gives the smallest reconciliation problem. There is no need to identify redundant variables because the RND-SQP uses an LP type technique to separate the variables into dependent/independent variables and eliminates all dependent variables (a mixture of measured and unmeasured variables) from the problem to construct reduced QP at each iteration. But if a redundancy analysis is required for sensor placement or other reasons, a separate redundancy analysis by one of the methods mentioned above can be performed.

COMPARISON OF NONLINEAR OPTIMIZATION STRATEGIES FOR DATA RECONCILIATION

Nonlinear programming codes are already commercially available and they have proved to be numerically robust and reliable for large-scale industrial problems. They perform best when rigorous models are used [31, 32]. Nonlinear programming allows a complete formulation for data reconciliation, as described by Equations 5-7 through 5-9.

Tjoa and Biegler [33] have developed an efficient hybrid SQP method specifically tailored to solve nonlinear data reconciliation problems. The data reconciliation software package RAGE developed by Ravikumar et al. [34] also uses an SQP solver which has been specially adapted to solve data reconciliation problems. Liebman and Edgar [35] compared the generalized reduced gradient (the GRG2 version of Lasdon and Waren [21]) with the successive linear (SL) data reconciliation solution method, and found that the NLP method was more robust at the expense of the computational time.

While GRG2, a feasible path method, requires the convergence of the constraints at each iteration, SQP—the infeasible path method, satisfies the constraints only at the end when convergence is achieved. SQP and other infeasible path methods (such as MINOS, another generalized reduced gradient method developed by Murtagh and Saunders [36]) usually require less computational time than feasible path methods. Ramamurthi and Bequette [12] compared SQP, GRG and SL methods for data reconciliation purposes. Their findings are summarized as follows:

1. Successive linearization yields significant biases, particularly in the unmeasured variables, while the NLP approaches yield little bias in both measured and unmeasured estimates.
2. Computational time increases with the magnitude of the measurement error for SL, but not for SQP or GRG.
3. Computational time is a strong function of desired accuracy for SL, but not for SQP or GRG.
4. NLP algorithms are more efficient and more robust for highly nonlinear problems. SQP is more efficient, while GRG is more reliable.

SUMMARY

- The constraints of a nonlinear data reconciliation problem can contain equality constraints (material balances, energy balances, equilibrium constraints, and property correlations) and inequality constraints (bounds, thermodynamic feasibility constraints).
- Nonlinear data reconciliation problems which contain only equality constraints can be solved using iterative techniques based on successive linearization and analytical solution of the linear data reconciliation problem.
- Nonlinear data reconciliation problems containing inequality constraints can be solved only using nonlinear constrained optimization techniques.
- The Generalized Reduced Gradient (GRG) and Successive Quadratic Programming (SQP) methods are two competitive nonlinear optimization techniques used for solving nonlinear data reconciliation problems.
- If bounds on unmeasured variables are imposed, then unmeasured variables should not be eliminated using Crowe's projection technique to obtain a reduced problem.
- It is necessary to impose bounds on variables in certain problems to obtain feasible estimates.

REFERENCES

1. MacDonald, R. J, and C. S. Howat. "Data Reconciliation and Parameter Estimation in Plant Performance Analysis." *AIChE Journal* 34(no. 1, Jan. 1980): 1–8.

2. Britt, H. I., and R. H. Luecke. "The Estimation of Parameters in Nonlinear Implicit Models." *Technometrics* 15 (no. 2, 1973): 233–247.

3. Dennis, J. E. Jr., and R. B. Schnabel. *Numerical Methods for Unconstrained Optimization and Nonlinear Equations.* Englewood Cliffs, N.J.: Prentice-Hall, 1983.

4. Stephenson, G. R., and C. F. Shewchuck. "Reconciliation of Process Data with Process Simulation." *AIChE Journal* 32 (no. 2, Feb. 1986): 247–254.

5. Serth, R. W., C. M. Valero, and W. A. Heenan. "Detection of Gross Errors in Nonlinearly Constrained Data: A Case Study." *Chem. Eng. Comm.* 51 (1987): 89–104.

6. Madron, F. *Process Plant Performance: Measurement and Data Processing for Optimization and Retrofits.* Chichester, West Sussex, England: Ellis Horwood Limited Co., 1992
7. Knepper, J. C., and J. W. Gorman. "Statistical Analysis of Constrained Data Sets." *AIChE Journal* 26 (no. 2, Mar. 1980): 260–264.
8. Crowe, C. M., Y.A.G. Campos, and A. Hrymak. "Reconciliation of Process Flow Rates by Matrix Projection. I: Linear Case." *AIChE Journal* 29 (no. 6, 1983): 881–888.
9. Pai, C.C.D., and G. Fisher. "Application of Broyden's Method to Reconciliation of Nonlinearly Constrained Data." *AIChE Journal* 34 (no. 5, 1988): 873–876.
10. Broyden, C. G. "A Class of Methods for Solving Nonlinear Simultaneous Equations." *Math. Comp.* 19 (1965): 577.
11. Swartz, C.L.E. "Data Reconciliation for Generalized Flowsheet Applications," presented at the Amer. Chem. Society National Meeting, Dallas, Tex., 1989.
12. Ramamurthi, Y., and B. W. Bequette. "Data Reconciliation of Systems with Unmeasured Variables Using Nonlinear Programming Techniques," presented at the AIChE Spring National Meeting, Orlando, Fla., 1990.
13. Gill, P. E., W. Murray, and M. H. Wright. *Practical Optimization.* London and New York: Academic Press, 1981.
14. Edgar, T. F., and D. M. Himmelblau. *Optimization of Chemical Processes.* New York: McGraw-Hill, 1988.
15. Han, S. P. "A Globally Convergent Method for Nonlinear Programming." *J. Optimization Theory Control* 22 (1977): 297.
16. Powell, M.J.D. "A Fast Algorithm for Nonlinearly Constrained Optimization Calculations." *Dundee Conf. Numer. Analysis,* 1977.
17. Chen, H-S., and M. A. Stadtherrr. "Enhancements of Han-Powell Method for Succesive Quadratic Programming," *Computers Chem. Engng.* 8 (no. 3/4, 1984): 229–234.
18. Gill, P. E., W. Murray, M. A. Saunders, and M. H. Wright. *User's Guide for SOL/QPSOL; A Fortran Package for Quadratic Programming.* Technical Report SOL 83-7, 1983.
19. Vasantharajan, S., and L. T. Biegler. "Large-Scale Decomposition for Successive Quadratic Programming." *Computers Chem. Engng.* 12 (no. 11, 1988): 1087–1101.

20. Abadie, J. "The GRG Method for Nonlinear Programming," in *Design and Implementation of Optimization Software,* Sijthoff and Noordhoff. Holland: H. Greenberg, Ed., 1978.

21. Lasdon, L. S., and A. D. Waren, "Generalized Reduced Gradient Software for Linearly and Nonlinearly Constrained Problems," in *Design and Implementation of Optimization Software,* Sijthoff and Noordhoff, Holland: H. Greenberg, Ed., 1978.

22. Crowe, C. M. "Observability and Redundancy of Process Data for Steady State Reconciliation." *Chem. Eng. Science* 44 (no. 12, 1989): 2909–2917.

23. Kretsovalis, A., and R.S.H. Mah. "Observability and Redundancy Classification in Generalized Process Networks. I: Theorems." *Computers Chem. Engng.* 12 (1988): 671–688.

24. Kretsovalis, A., and R.S.H. Mah. "Observability and Redundancy Classification in Generalized Process Networks. II: Algorithms." *Computers Chem. Engng.* 12 (1988): 689–703.

25. Meyer, M., B. Koehret, and M. Enjalbert. "Data Reconciliation on Multicomponent Network Process." *Computers Chem. Engng.* 17 (no. 8, 1993): 807–817.

26. Romagnoli, J., and G. Stephanopoulos. "On Rectification of Measurement Errors for Complex Chemical Plants." *Chem. Eng. Science* 35 (1980): 1067–1081.

27. Romagnoli, J., and G. Stephanopoulos. "A General Approach to Classify Operational Parameters and Rectify Measurement Errors for Complex Chemical Processes." *Comp. Appl. to Chem. Engng.* (1980): 153–174

28. Sanchez, M., A. Bandoni, and J. Romagnoli. "PLADAT: A Package for Process Variable Classification and Plant Data Reconciliation." *Computers Chem. Engng.* 616 (Suppl., 1992): S499–S506.

29. Sanchez, M., and J. Romagnoli. "Use of Orthogonal Transformations in Data Classification-Reconciliation." *Computers Chem. Engng.* 20 (no. 5, 1996): 483–493.

30. Albuquerque, J. S., and L. T. Biegler. "Data Reconciliation and Gross Error Detection for Dynamic Systems." *AIChE Journal* 42 (no. 10, 1996): 2841–2856.

31. Nair, P., and C. Jordache. "On-line Reconciliation of Steady-State Process Plants Applying Rigorous Model-Based Reconciliation," presented at the AIChE Spring National Meeting, Orlando, Fla., 1990.

32. Nair, P., and C. Jordache. "Rigorous Data Reconciliation is Key to Optimal Operations." *Control for the Process Industries,* Vol. IV, no. 10, pp. 118–123. Chicago: Putnam Publ., 1991.

33. Tjoa, I. B., and L. T. Biegler. "Simultaneous Solution and Optimization Strategies for Parameter Estimation of Differential-Algebraic Equation Systems." *Ind. & Eng. Chem. Research* 30 (no. 2, 1991): 376–385.

34. Ravikumar, V., S. R. Singh, M. O. Garg, and S. Narasimhan. "RAGE—A Software Tool for Data Reconciliation and Gross Error Detection," in *Foundations of Computer-Aided Process Operations* (edited by D.W.T. Ripping, J. C. Hale, and J. F. Davis). Amsterdam: CACHE/Elsevier, 1994, 429–436.

35. Liebman, M. J., and T. F. Edgar. "Data Reconciliation for Nonlinear Processes," presented at the AIChE Annual Meeting, Washington, D.C., 1988.

36. Murtagh, B. A., and M. A. Saunders. *MINOS 5.0 User's Guide.* Report SOL 83-20, Dept. of Operation Research, Stanford University, Calif., 1983.

6

Data Reconciliation in Dynamic Systems

THE NEED FOR DYNAMIC DATA RECONCILIATION

In the preceding chapters, data reconciliation has been applied to a single vector of measurements of process variables. This vector could be the measurements made at any time instant corresponding to a single snapshot of the process. It is more likely, however, that steady-state data reconciliation is applied to a vector containing the average values of the measurements made over a period of time of, say, a few hours. This approach is satisfactory if the reconciled data is required for applications such as steady-state simulation, or on-line optimization where the optimal set points are calculated once every few hours.

If we consider applications such as regulatory control which require accurate estimates of process variables frequently, then data reconciliation may have to be applied to measurements made at every sampling instant. In this case, it can no longer be assumed that the variables obey steady-state material and energy balance relationships. Storage capacities and transportation lags should be taken into account, and dynamic material and energy balances that relate the variables must be used.

Estimation of process variables which uses measurements and dynamic relationships between the variables have been developed long before the subject of data reconciliation was born. We discuss some of these important estimation techniques along with recent advances in dynamic data reconciliation under the broad umbrella of dynamic data reconciliation in this chapter. Since the area of dynamic data reconciliation is fairly nascent and will continue to evolve, the intent of this chapter is to only introduce the reader to this topic.

Before we proceed further, it is useful to explicitly describe what we mean by a dynamic state of a process. Two features characterize a dynamic state of a process:

1. The true values of process variables change with time and thus the measurements of these variables are also functions of time even if we entertain the extreme possibility that measurement errors are absent.
2. Due to continuously changing inputs, the accumulation within a process unit also changes continuously and has to be taken into account.

The above features characterize both operations around a nominal steady state as well as process transients that take the process from one nominal steady state to another.

Different techniques are available for developing a dynamic model of a process. These techniques are described under *model identification* in several textbooks [1, 2]. We will consider only discrete time models as opposed to continuous models because we will be dealing with measurements made at discrete time instants which are conveniently treated using digital computers. Furthermore, we will consider state space models as opposed to input-output models due to their inherent advantages. We will begin our description with linear discrete system models before moving on to nonlinear systems.

LINEAR DISCRETE DYNAMIC SYSTEM MODEL

A linear, discrete, state-space dynamic model of a process is usually described by the following equations:

$$\mathbf{x}_k = \mathbf{A}_k \mathbf{x}_{k-1} + \mathbf{B}_k \mathbf{u}_{k-1} + \mathbf{w}_{k-1} \quad (6\text{-}1)$$

$$\mathbf{y}_k = \mathbf{H}_k \mathbf{x}_k + \mathbf{v}_k \quad (6\text{-}2)$$

where

$\mathbf{x}_k$: $n \times 1$ vector of state variables
$\mathbf{u}_k$: $p \times 1$ vector of manipulated inputs
$\mathbf{w}_k$: $s \times 1$ vector of random disturbances
$\mathbf{y}_k$: $m \times 1$ vector of measurements
$\mathbf{v}_k$: $m \times 1$ vector of random errors in measurements

The subscript k represents time instant $t = kT$ when the variables are sampled or measured, T being the sampling period. The matrices $\mathbf{A}_k$, $\mathbf{B}_k$, and $\mathbf{H}_k$ are matrices of appropriate dimensions whose coefficients are known at all times. If the coefficients of these matrices do not change with time, then the resulting model is known as a linear time-invariant (LTI) system model. It is also customary to use deviation variables rather than actual variables in the model equations. Thus, the state variables, $\mathbf{x}_k$, represent the differences between the true values of the variables and their nominal steady-state values. Similarly, the variables $\mathbf{u}_k$ and $\mathbf{y}_k$ also represent deviation variables. In this chapter, we implicitly assume that all the variables are deviation variables.

Equation 6-1 describes the dynamic evolution of the state variables while Equation 6-2 is the measurement model which describes the relationship between the measurements and the state variables. The standard assumptions made about the random disturbances $\mathbf{w}_k$ and the random errors $\mathbf{v}_k$ are that they are normally distributed variables with statistical properties given by

$$E[\mathbf{w}_k] = E[\mathbf{v}_k] = 0 \tag{6-3}$$

$$\mathrm{Cov}[\mathbf{w}_k] = \mathbf{R}_k \tag{6-4}$$

$$\mathrm{Cov}[\mathbf{v}_k] = \mathbf{Q}_k \tag{6-5}$$

$$\mathrm{Cov}[\mathbf{w}_k, \mathbf{w}_j] = \mathrm{Cov}[\mathbf{v}_k, \mathbf{v}_j] = 0 \qquad j \neq k \tag{6-6}$$

$$\mathrm{Cov}[\mathbf{w}_k, \mathbf{v}_j] = \mathbf{0} \tag{6-7}$$

Equations 6-3 and 6-4 imply that the random variables $\mathbf{w}_k$ and $\mathbf{v}_k$ have zero mean and covariance matrices given by $\mathbf{R}_k$ and $\mathbf{Q}_k$, respectively. Equation 6-6 implies that the disturbances at different times are not correlated, and similarly the measurement errors at different time instants are not correlated. Furthermore, Equation 6-7 stipulates that the disturbances and measurement errors are not cross-correlated.

The random errors in measurements, $\mathbf{v}_k$ arises due to several reasons as explained in Chapters 1 and 2. On the other hand, the causes of random disturbances, $\mathbf{w}_k$, in the state evolution equation can be best explained only if we consider a first principles model derived from the differential mass and energy balances of the process. In this case, random fluctuations in the process feed characteristics such as its flow, temperature, pressure and composition can be modeled as disturbances.

Any random errors in the control inputs arising due to electrical noise in the transmission lines of the controller or due to imprecise actuator positioning can also be modeled as random disturbances. On the other hand, if an input-output model of the process is identified from the process data, then it may not be possible to separate the effects of random measurement errors and random disturbances. In this case, the differences between the model predictions and actual measurements can be attributed to the combined effect of measurement errors, process feed disturbances, and errors between the actual and computed manipulated inputs.

A linear system model of the form given by Equations 6-1 and 6-2 can be derived for any process from the differential equations that describe the mass and energy conservation relations of a process (also known as a first principles model). Alternatively, model identification techniques may be used for obtaining a dynamic model from the outputs or response of a process to given inputs. We illustrate the development of a first principles dynamic model for a simple level control process taken from Bellingham and Lees [3].

Example 6-1

A simple level control process is shown in Figure 6-1, which has a feed (F_1) and two outputs (F_2 and F_3). The valve V_1 is kept open at a fixed position, while valve V_2 is manipulated to control the level of the tank (instead of directly computing the new valve position, it is assumed that the adjustment, *a,* to the valve position x is computed at each time). The tank level and position of the valve V_2 are measured, denoted by measurements Z_1 and Z_2, respectively.

The differential equation describing the mass balance for this process is given by

$$A\frac{dh}{dt} = F_1 - F_2 - F_3$$

The outlet flow rates are related to the tank level and valve positions by

$$F_3 = K_{01}h$$

$$F_2 = K_{02}h + K_{03}x$$

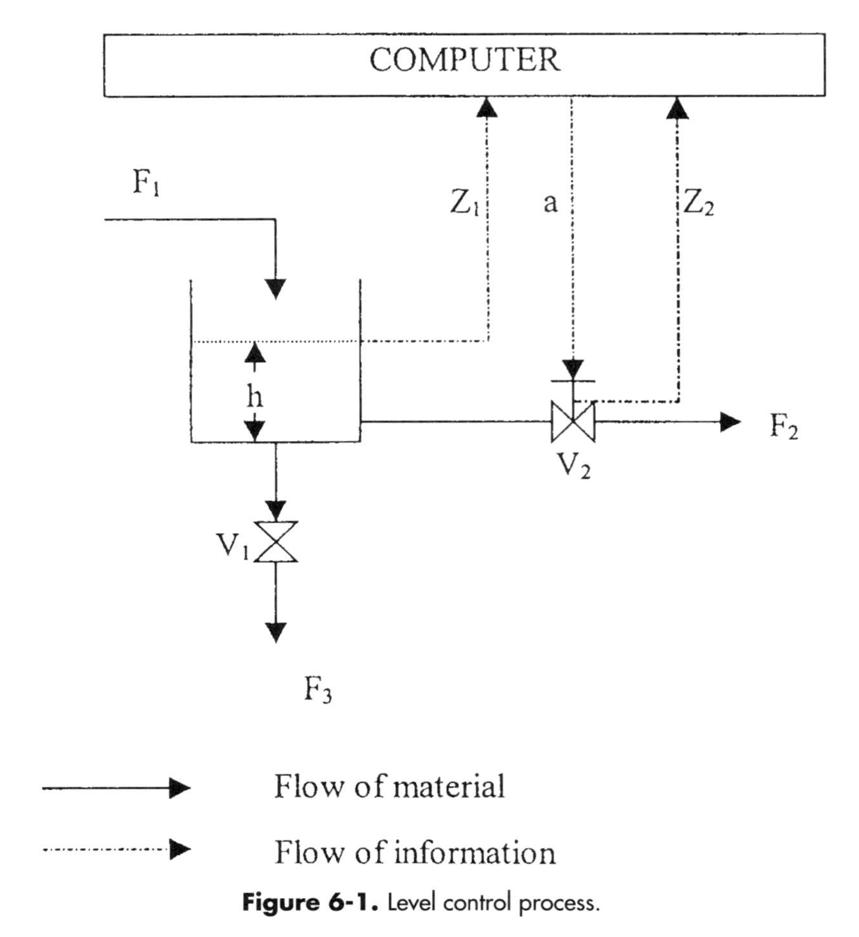

Figure 6-1. Level control process.

Substituting the above relations in the mass balance equation we get

$$A\frac{dh}{dt} = F_1 - (K_{01} + K_{02})h - K_{03}x$$

If we assume a uniform sampling interval *T*, between measurements and use the subscript *k* to represent the variables at sampling instant *kT*, then the discrete equivalent of the above differential equation can be obtained using the method described in [4].

$$h_{k+1} = \alpha h_k - [(1-\alpha)K_{03}/(K_{01}+K_{02})]x_k + [(1-\alpha)/(K_{01}+K_{02})]F_1$$

where

$$\alpha = \exp\left[-(K_{02} + K_{03})T / A\right]$$

In deriving the above discrete representation, it is implicitly assumed that the valve position x is constant at the value x_k during the time interval *kT* to *(k+1)T.* It is also assumed that the random disturbance in the feed F_1 is piecewise constant (of constant magnitude within each sampling interval, but the magnitude being random from interval to interval). If the adjustment to valve position a_k (computed by the controller after measurements are made at sampling instant *k*) is implemented at the beginning of the next sampling interval, then the valve position at each sampling instant is given by

$$x_{k+1} = x_k + a_k + e_{k+1}$$

where e_{k+1} is the random error in positioning the valve.

Table 6-1
Parameter Values for Level Control Process

Parameter	Value	Units
K_{01}	7.2	cm^2/min
K_{02}	34.78	cm^2/min
K_{03}	1156.0	cm^2/min
A	280.0	cm^2
T	2.0	sec

Table 6-1 gives the values for different constants used for this process. Using these values, the state-space model of the level control process is obtained as

$$\begin{bmatrix} h_{k+1} \\ x_{k+1} \end{bmatrix} = \begin{bmatrix} 0.995 & -0.1373 \\ 0 & 1 \end{bmatrix} \begin{bmatrix} h_k \\ x_k \end{bmatrix} + \begin{bmatrix} 0 \\ 1 \end{bmatrix} a_k + \begin{bmatrix} 0.00012 & 0 \\ 0 & 1 \end{bmatrix} \begin{bmatrix} F_{1,k+1} \\ e_{k+1} \end{bmatrix}$$

$$\begin{bmatrix} z_{1,k+1} \\ z_{2,k+1} \end{bmatrix} = \begin{bmatrix} 0.631 & 0 \\ 0 & 1.57 \end{bmatrix} \begin{bmatrix} h_{k+1} \\ x_{k+1} \end{bmatrix} + \begin{bmatrix} v_{1,k+1} \\ v_{2,k+1} \end{bmatrix}$$

where $v_{1,k+1}$ and $v_{2,k+1}$ are the random errors in the measurements of the level and valve position (in terms of volts), respectively.

The manipulated inputs $\mathbf{u}_k$ are obtained using a control law which is generally a function of the measurements when the variables that need to be controlled are also measured. For a simple proportional linear control law we can write the control law as

$$\mathbf{u}_k = \overline{\mathbf{C}}_k(\mathbf{y}_k - \mathbf{y}_{sp,k}) \tag{6-8}$$

where $\mathbf{y}_{sp,k}$ represents the deviation or change from the current operating set points, and is equal to $\mathbf{0}$ if there is no change from current set points. In some cases when it is difficult or expensive to measure the controlled variables an inferential control strategy is used. The manipulated inputs in this case are a function of the state estimates. Even in the case when controlled variables are measured, it may be better to base the control law on estimates of these variables since these are likely to be more accurate, if the estimator is designed properly. As mentioned in the preceding section, a primary reason for dynamic data reconciliation is to derive estimates which can be used for better control. We therefore assume a control law of the form

$$\mathbf{u}_k = \overline{\mathbf{C}}_k(\hat{\mathbf{x}}_k - \mathbf{x}_{sp,k}) \tag{6-9}$$

where $\hat{\mathbf{x}}_k$ are estimates of the true values of state variables and $\mathbf{x}_{sp,k}$ are changes in the set points of state variables from current set points. In order to achieve good control, it is therefore required to estimate the state variables as accurately as possible.

OPTIMAL STATE ESTIMATION USING KALMAN FILTER

We first deal with the problem of optimal estimation of state variables for a process that can be described by a linear model of the form given by Equations 6-1 and 6-2, and which satisfies the assumptions of Equations 6-3 through 6-7. We will also assume that the manipulated inputs at each time are known constant values, and ignore for the time being the fact that these are functions of the state estimates. (We will address this issue in a subsequent section.) The optimal linear state estimator called the Kalman filter which we describe in this section can be derived using different theoretical formulations and an excellent treatment may be found in Sage and Melsa [5]. We use a least squares formulation approach because it helps us to readily compare this with data reconciliation.

Initial estimates of state variables are assumed to be available which possess the following statistical properties:

$$\hat{\mathbf{x}}_0 = E[\mathbf{x}_0] \tag{6-10}$$

$$\text{Cov}[\hat{\mathbf{x}}_0] = \mathbf{P}_0 \tag{6-11}$$

Given a set of measurements, $\mathbf{Y}_k = (\mathbf{y}_1, \mathbf{y}_2, \ldots, \mathbf{y}_k)$, it is desired to obtain estimates of the state variables $\mathbf{x}_k$, which are *best* in some sense. We will denote these estimates using the notation $\hat{\mathbf{x}}_{k|k}$, which are interpreted as the state estimates at time k obtained using all measurements from time $t = 1$ to time $t = k$. It should be noted that by using all the measurements from initial time to derive the estimates, we are automatically exploiting *temporal redundancy* in the measured data.

The estimation problem that we are considering here is a special case of a more general problem in which it is desired to obtain estimates of state variables $\mathbf{x}_j$ for time j, using all measurements made from initial time to time k. The estimates so derived are denoted as $\hat{\mathbf{x}}_{j|k}$. The estimation problem is referred to as a prediction problem if $j > k$, a filtering problem if $j = k$, and a smoothing problem if $j < k$. Here we are concerned with the filtering problem.

Due to the presence of random disturbances in Equation 6-1, the true values of the state variables at every time instant are themselves random variables. Therefore, a probabilistic measure has to be used for determining the best estimates of the state variables. The best estimates of the state variables at time k are obtained by minimizing the following function:

$$\mathbf{J}_k = E[(\hat{\mathbf{x}}_{k|k} - \mathbf{x}_k)^T (\hat{\mathbf{x}}_{k|k} - \mathbf{x}_k)] \tag{6-12}$$

Equation 6-12 is the expected sum of squares of the differences between the estimates and true values of state variables, and is thus an extension of the well-known deterministic least squares objective function. The solution to the problem was first obtained by Kalman [6, 7] in a convenient recursive form and is generally now referred to as the Kalman filter. The Kalman filter equations are given by

$$\hat{\mathbf{x}}_{k|k} = \hat{\mathbf{x}}_{k|k-1} + \mathbf{K}_k(\mathbf{y}_k - \mathbf{H}_k\hat{\mathbf{x}}_{k|k-1}) \tag{6-13}$$

where

$$\hat{\mathbf{x}}_{k|k-1} = \mathbf{A}_k\hat{\mathbf{x}}_{k-1|k-1} + \mathbf{B}_k\mathbf{u}_{k-1} \tag{6-14}$$

$$\mathbf{K}_k = \mathbf{P}_{k|k-1}\mathbf{H}_k^T(\mathbf{H}_k\mathbf{P}_{k|k-1}\mathbf{H}_k^T + \mathbf{Q}_k)^{-1} \tag{6-15}$$

$$\mathbf{P}_{k|k-1} = \mathbf{A}_k\mathbf{P}_{k-1|k-1}\mathbf{A}_k^T + \mathbf{R}_k \tag{6-16}$$

$$\mathbf{P}_{k|k} = (\mathbf{I} - \mathbf{K}_k\mathbf{H}_k)\mathbf{P}_{k|k-1} \tag{6-17}$$

The matrices $\mathbf{P}_{k|k}$ and $\mathbf{P}_{k|k-1}$ are the covariance matrices of the estimates $\hat{\mathbf{x}}_{k|k}$ and $\hat{\mathbf{x}}_{k|k-1}$, respectively. Starting with initial estimates $\hat{\mathbf{x}}_0$ and $\mathbf{P}_0$, the above equations can be applied in the reverse order of Equations 6-16 to 6-13 to obtain the state estimates at each time *k*. The derivation of the Kalman filter equations is described clearly in Gelb [8] and Sage and Melsa [5]. The book edited by Sorenson [9] contains several papers on Kalman filtering and its applications including the original papers by Kalman.

Exercise 6-1. *Derive the Kalman filter equations by minimizing 6-12 for the process model Equations 6-1 and 6-2 and utilizing the statistical properties Equations 6-3 through 6-7, 6-10, and 6-11.*

The recursive form of the estimator equations considerably reduces the computational effort involved in obtaining the estimates. It can be observed from these equations that the effort spent in obtaining the state estimates at a time is effectively utilized to obtain the estimates at the next time instant. Equation 6-13 can also be interpreted as a predictor-corrector method for obtaining the estimates. The estimates $\hat{\mathbf{x}}_{k|k-1}$ are the predicted estimates of the state variables at time *k* based on all the measurements until time *k-1*.

The second term in this equation is the correction to this estimate based on the measurement at time *k*. The matrix $\mathbf{K}_k$ is known as the *Kalman filter gain,* and the difference $(\mathbf{y}_k - \mathbf{H}_k\hat{\mathbf{x}}_{k|k-1})$ is known as the *innovations.* The innovations are equivalent to measurement residuals of steady-state processes which were defined in Chapter 3, and play a crucial role in gross error detection, as will be seen in Chapter 9. The Kalman filter estimates possess the desirable statistical properties of being unbiased, that is,

$$E[\hat{\mathbf{x}}_{k|k}] = E[\mathbf{x}_k] \tag{6-18}$$

and also have the minimum variance among all unbiased estimators. Furthermore, it can also be shown that the Kalman filter estimates are the maximum-likelihood estimates (Sage and Melsa [5]). For a linear time invariant process, the Kalman gain matrix becomes constant after some time, which is also known as the *steady-state Kalman gain.*

Exercise 6-2. *Prove that the Kalman filter estimates are unbiased.*

Applications of Kalman filter in chemical engineering have been discussed by several authors. Fisher and Seborg [10] have applied Kalman filtering to a pilot scale multiple effect evaporator in the context of investigating various types of control strategies. Stanley and Mah [11] applied it to a subsection of a refinery for estimating flows and temperatures. The dynamic model used in this application is a heuristic random walk model for the state variables which is appropriate for describing processes that operate for long periods around a nominal steady state with occasional slow transitions to a new nominal steady state. The state variables were also forced to satisfy the steady-state material and energy balances. Through this approach, they attempted to exploit both spatial and temporal redundancy in the data for reconciliation purposes.

Makni et al. [12] recently used a similar technique for estimating flows and concentrations of a mineral beneficiation circuit. A first-order, identified-transfer-function model was used to describe the dynamics and steady-state material balances were also imposed, although the estimates were expected to satisfy them in a minimum least-square sense.

Example 6-2

We illustrate the application of the Kalman filter for obtaining optimal estimates for the level control process described in Example 6-1. In this process, the fluctuations in the feed flow rate and random error in positioning of the control valve are taken as state disturbances. The standard deviations of these random disturbances are assumed to be 250 cm^3/min and 0.05, respectively. The standard deviations of the errors in measurements of level and valve position are taken as 0.01 volts each. Based on the state space model derived in Example 6-1, the measurements corresponding to the closed loop behavior of the process are simulated. The control law used for this simulation is based on *observations* and is given by

$$a_k = 0.02057\tilde{z}_{1k} - 0.6369\tilde{z}_{2k}$$

where $\tilde{z}_{1k}$ and $\tilde{z}_{1k}$ are the measured values of level and valve position (in cm) obtained by dividing the actual measurements in volts by 0.631 and 1.57, respectively (see Example 6-1). A Kalman filter is used to estimate the states using the steady-state Kalman gain, since we have used a time invariant model. The steady-state Kalman gain obtained by solving a matrix Ricatti equation [9] is obtained as

$$K_{ss} = \begin{bmatrix} 0.2758 & -0.0291 \\ -0.0177 & 0.3406 \end{bmatrix}$$

Figure 6-2 shows the true, measured and estimated values of the level. It can be observed from Figure 6-2, that the estimates are closer to the true values as compared to the measurements. The measurement error variance, calculated from the sample data over the time period of 200 seconds, is 0.0294 cm^2, whereas the variance of the error in the estimate is 0.0023 cm^2. The variance of the difference between the true values and the set point (which is 0 in this case) is an indicator of the control perfor-

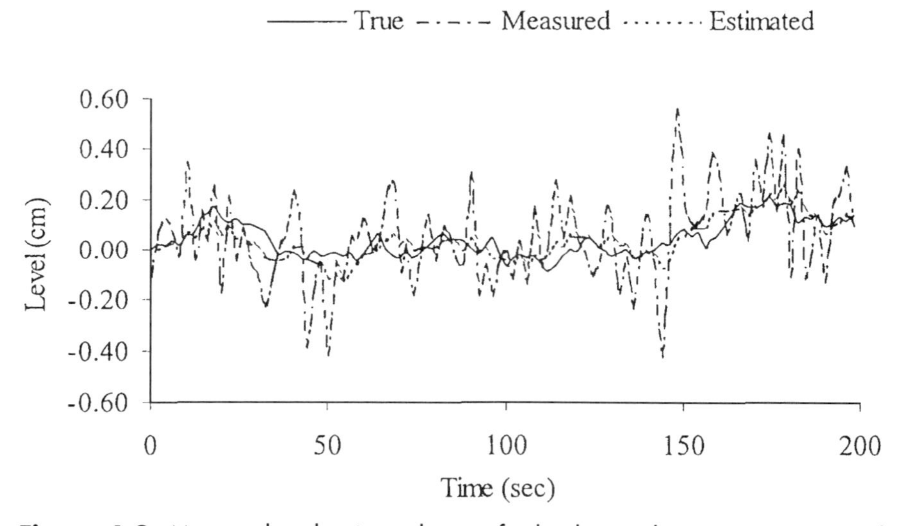

Figure 6-2. Measured and estimated states for level control process using control law based on measurements.

mance. For this case it was computed to be 0.0068 cm^2. The variance of the valve position to achieve this control is 0.1057 cm^2.

Analogy Between Kalman Filtering and Steady-State Data Reconciliation

Data reconciliation techniques were developed primarily for steady-state processes whereas the Kalman filter was developed independently for a linear dynamic process. Both these techniques can be derived using a weighted least-squares estimation procedure. In order to bring out the link between these two approaches, we prove that steady-state data reconciliation can be regarded as a special case of a Kalman filter. It can be recalled from Chapter 3 that for steady-state processes, the material and energy conservation relations are written as algebraic constraints. The differential dynamic form of these conservation relations can also be used to derive a discrete linear-state space model form of Equation 6-1 as shown in Example 6-1. As a special case, we can consider a disturbance free form of this equation by setting $\mathbf{w}_k$ to be identically equal to zero for all time to get:

$$\mathbf{x}_k = \mathbf{A}_k \mathbf{x}_{k-1} + \mathbf{B}_k \mathbf{u}_{k-1} \tag{6-19}$$

Let us define a new state vector $\mathbf{x}$ comprised of both $\mathbf{x}_k$ and $\mathbf{x}_{k-1}$ (where we have deliberately chosen to omit the time index k for ready comparison with steady-state reconciliation). Equation 6-19 can be rewritten as

$$\mathbf{Ax} = \mathbf{c} \tag{6-20}$$

where

$$\mathbf{x} = \begin{bmatrix} \mathbf{x}_k \\ \mathbf{x}_{k-1} \end{bmatrix} \tag{6-21}$$

$$\mathbf{A} = \begin{bmatrix} \mathbf{I} & -\mathbf{A}_k \end{bmatrix} \tag{6-22}$$

$$\mathbf{c} = \mathbf{B}_k \mathbf{u}_{k-1} \tag{6-23}$$

If all the state variables are assumed to be measured, then Equation 6-2 can be written as

$$\mathbf{y}_k = \mathbf{x}_k + \mathbf{v}_k \tag{6-24}$$

Let us also assume that we have unbiased estimates $\hat{\mathbf{x}}_{k-1|k-1}$ of the state variables $\mathbf{x}_{k-1}$ obtained from the preceding time instant and that its covariance matrix is $\mathbf{P}_{k-1|k-1}$. These estimates can be treated as additional *measurements* and can be written as

$$\hat{\mathbf{x}}_{k-1|k-1} = \mathbf{x}_{k-1} + \varepsilon_{k-1} \tag{6-25}$$

where ε_{k-1} are the random errors in the estimate of $\mathbf{x}_{k-1}$ with zero mean and covariance matrix $\mathbf{P}_{k-1|k-1}$. From the assumed properties for $\mathbf{v}_k$, we can easily prove that they are not correlated with ε_{k-1}. Combining Equations 6-24 and 6-25 we can write the modified measurement model as

$$\mathbf{y} = \mathbf{x} + \mathbf{e} \tag{6-26}$$

where

$$\mathbf{y} = \begin{bmatrix} \mathbf{y}_k \\ \hat{\mathbf{x}}_{k-1|k-1} \end{bmatrix} \tag{6-27}$$

$$\mathbf{e} = \begin{bmatrix} \mathbf{v}_k \\ \varepsilon_{k-1} \end{bmatrix} \tag{6-28}$$

where $\mathbf{v}$ is the vector of random errors with zero mean and covariance matrix Σ defined by

$$\Sigma = \begin{bmatrix} \mathbf{Q}_k & \mathbf{0} \\ \mathbf{0} & \mathbf{P}_{k-1|k-1} \end{bmatrix} \tag{6-29}$$

Equations 6-26 and 6-20 are similar to the steady-state measurement and constraint models, Equations 3-1 and 3-2. Applying the steady-state reconciliation solution to this model, we can obtain the estimates of $\mathbf{x}$ using Equation 3-8. Substituting for the different variables as defined by Equations 6-21 through 6-23, 6-27 and 6-29 in this solution we get

$$\begin{bmatrix} \hat{\mathbf{x}}_k \\ \hat{\mathbf{x}}_{k-1} \end{bmatrix} = \begin{bmatrix} \mathbf{y}_k \\ \hat{\mathbf{x}}_{k-1|k-1} \end{bmatrix} -$$

$$\begin{bmatrix} \mathbf{Q}_k \\ -\mathbf{P}_{k-1|k-1}\mathbf{A}_k^T \end{bmatrix} \left(\mathbf{Q}_k + \mathbf{A}_k \mathbf{P}_{k-1|k-1} \mathbf{A}_k^T\right)^{-1} \left(\mathbf{y}_k - \mathbf{A}_k \hat{\mathbf{x}}_{k-1|k-1} - \mathbf{B}_k \mathbf{u}_{k-1}\right) \tag{6-30}$$

Considering only the estimates of $\mathbf{x}_k$ in Equation 6-30 and using the predicted estimates defined by Equation 6-14 we obtain

$$\hat{\mathbf{x}}_k = \mathbf{y}_k - \mathbf{Q}_k\left(\mathbf{Q}_k + \mathbf{A}_k\mathbf{P}_{k-1|k-1}\mathbf{A}_k^T\right)^{-1}\left(\mathbf{y}_k - \hat{\mathbf{x}}_{k|k-1}\right) \tag{6-31}$$

The estimates given by Equation 6-31 can be shown to be identical to the Kalman filter estimates given by Equation 6-13 with $\mathbf{R}_k = \mathbf{0}$ and $\mathbf{H}_k = \mathbf{I}$ as in the case of the simplified model considered in this section. Since the Kalman filter also accounts for random disturbances in the process model, it may be regarded as an extension of linear steady-state reconciliation technique to dynamic processes. An interesting by-product of the analysis carried out in this section is that it can also be proved that the estimates $\hat{\mathbf{x}}_{k-1}$ given in Equation 6-30 are identical to the optimal smoothed estimates $\hat{\mathbf{x}}_{k-1|k}$ of the simplified model considered in this section.

> ***Exercise 6-3.*** *Prove that the Kalman filter estimates for the model given by Equations 6-19 and 6-24 are identical to the reconciled estimates given by Equation 6-31.*

Linear dynamic data reconciliation techniques have been applied for estimating the flows of a process by Darouach and Zasadzinski [13] and Rollins and Devanathan [14]. These authors have converted the linear differential equations to algebraic equations by replacing the derivative by a forward difference. The problem can now be solved using linear data reconciliation solution techniques similar to the procedure discussed in this section.

Since the problem dimension increases with time, efficient techniques for obtaining the estimates have been proposed by these authors. Bagajewicz and Jiang [15] also considered the problem of dynamic data reconciliation of process flows and tank holdups. Assuming the flows and tank holdups to be polynomial functions of time, these authors convert the differential equations into algebraic equations. Using a window of measurements, the coefficients of the polynomials are estimated.

Optimal Control and Kalman Filtering

Let us first consider the optimal control problem for a deterministic linear process which evolves according to Equation 6-1, but without any state disturbances. Let us also assume that the state variables are directly measured without any errors (that is, the true values of state variables are available). We wish to determine the optimal values of the manipulated inputs which minimize the performance index

$$\underset{\mathbf{u}_i}{\text{Min}}\, J_k^c = \sum_{i=1}^{n}\left(\mathbf{x}_i^T \mathbf{E}_i \mathbf{x}_i + \mathbf{u}_i^T \mathbf{F}_i \mathbf{u}_i\right) \tag{6-32}$$

where $\mathbf{E}_i$ and $\mathbf{F}_i$ are specified weight matrices ($\mathbf{E}_i$ are assumed to be non-negative definite symmetric matrices and $\mathbf{F}_i$ are assumed to be positive definite symmetric matrices). The first term in Equation 6-32 attempts to keep the state variables (deviations of the state variables from their current set-points) on their target value of zero while the second term attempts to minimize large values of the manipulated inputs. The weight matrices can be chosen based on the relative importance of the state variables and manipulated inputs.

The solution to the above problem [16] leads to a linear control law of the form

$$\mathbf{u}_i = \overline{\mathbf{K}}_i \mathbf{x}_i \tag{6-33}$$

where $\overline{\mathbf{K}}_i$ is dependent on the weight matrices used in Equation 6-32 and the system matrices.

Let us now consider the optimal control problem for a linear stochastic system which evolves according to Equation 6-1. In this case, the performance index for the optimal control problem can be written as

$$\underset{\mathbf{u}_i}{\text{Min}}\quad J_k^c = E\left[\sum_{i=1}^{n}\left(\mathbf{x}_i^T \mathbf{E}_i \mathbf{x}_i + \mathbf{u}_i^T \mathbf{F}_i \mathbf{u}_i\right)\right] \tag{6-34}$$

The optimal values of the manipulated inputs, which minimize 6-34, can be obtained as follows:

(1) Compute $\hat{\mathbf{x}}_{k|k-1}$, the predicted estimates of the state variables at time k, using the Kalman filter equations by treating the manipulated inputs prior to time k as known deterministic inputs.
(2) Compute the manipulated inputs at time k by using the estimates $\hat{\mathbf{x}}_{k|k-1}$ instead of $\mathbf{x}_k$ in Equation 6-33.

Despite the fact that the manipulated inputs are themselves functions of the state estimates, it is assumed that they are known deterministic inputs when deriving the state estimates using the Kalman filter. On the other hand, the optimal control law has been obtained for a deterministic system for which the true values of the state variables are assumed to be available, but is used for the stochastic system also. Essentially, this implies that the optimal estimation problem and optimal control problem has been separated. The proof that this procedure gives the optimal manipulated inputs for minimizing 6-34 follows from the Separation Theorem or Certainty Equivalence Principle [16].

Example 6-3

In order to investigate the effect of using a control law based on estimated states, the level control process described in the preceding two examples was simulated using a control law similar to that used in Example 6-2. The control law, however, was based on the estimates of the level and valve position obtained using a Kalman filter. The control law for this case is given by

$$a_k = 0.02057\hat{h}_k - 0.6369\hat{x}_k$$

The true, measured and simulated values of the level for this case are shown in Figure 6-3. We can compare the true values obtained in this case with those obtained in Example 6-2 and find that there is a marginal improvement in control performance. The variance of the error between the true values and set point is 0.0065 cm^2 which is marginally lower than that obtained when the control law is based on measured values. However, the variance in the valve position in this case is only 0.0523 cm^2 which is about $\frac{1}{20}$ of that obtained in the preceding example. This implies that we are able to achieve as good control as before with less change in the manipulated variable. This is due to the fact that the estimated states are more accurate than the measured values.

Kalman Filter Implementation

The matrices $\mathbf{P}_{k|k-1}$ and $\mathbf{P}_{k|k}$ occurring in the Kalman filter equations, by virtue of being covariance matrices, should normally be nonnegative definite, or in other words their eigenvalues should be nonnegative. If this is

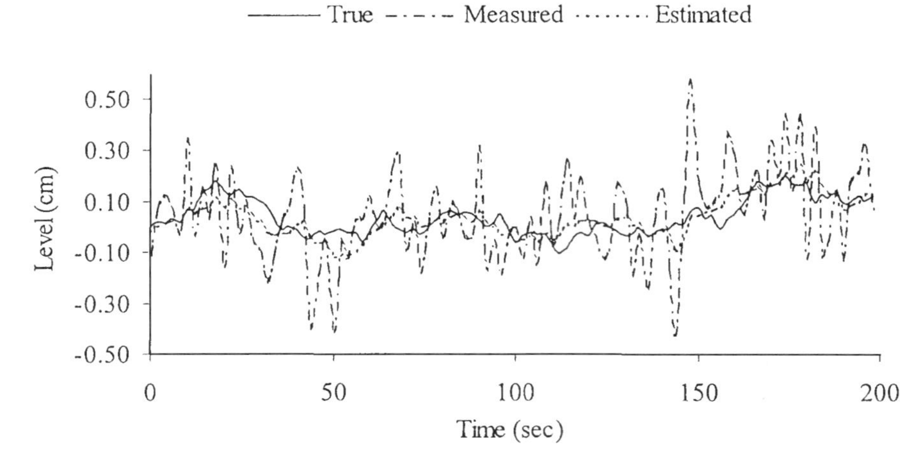

Figure 6-3. Measured and estimated states for level control process for control law based on estimates.

ensured, then the Kalman filter will be stable. However, if the Kalman filter is implemented as given by Equations 6-13 through 6-17, these matrices tend to lose their nonnegative definiteness character and the estimates tend to diverge, due to numerical inaccuracies in the computation.

A form known as the square-root covariance filter can be used to implement the Kalman filter. Equation 6-16, which is used to obtain $\mathbf{P}_{k|k-1}$, preserves the symmetry and positive definiteness character of this matrix, but Equation 6-17, used for obtaining the updated covariance matrix, can cause numerical problems since it involves the inversion of a matrix in the computation of the filter gain matrix. It is this equation which is recast in terms of square roots of the covariance matrices.

Furthermore, the computational efficiency is also increased by processing the measurements in a sequential manner rather than simultaneously, thus avoiding the need to compute the inverse of a matrix in Equation 6-15. We describe the steps involved in the implementation and refer the reader to Bagchi [17] and Borrie [18] for a detailed derivation of the algorithm.

Step 1. Starting with estimates $\hat{\mathbf{x}}_{k-1|k-1}$ and $\mathbf{P}_{k-1|k-1}$ apply Equations 6-14 and 6-16 to compute the one-step ahead predictions, $\hat{\mathbf{x}}_{k|k-1}$ and $\mathbf{P}_{k|k-1}$.

Step 2. Obtain the square roots $\mathbf{S}_{k|k-1}$ and Σ_k of the covariance matrices $\mathbf{P}_{k|k-1}$ and $\mathbf{Q}_k$, respectively defined by

$$\mathbf{P}_{k|k-1} = \mathbf{S}_{k|k-1}\mathbf{S}_{k|k-1}^{T} \tag{6-35}$$

$$\mathbf{Q}_{k} = \Sigma_{k}\Sigma_{k}^{T} \tag{6-36}$$

The square roots of the matrices can be obtained using Cholesky factorization [19].

Step 3. Compute the transformed measurements and transformed measurement matrix defined by

$$\Sigma_{k}\mathbf{y}_{k}^{*} = \mathbf{y}_{k} \tag{6-37}$$

$$\Sigma_{k}\mathbf{H}_{k}^{*} = \mathbf{H}_{k} \tag{6-38}$$

Since Σ_k is upper triangular, $\mathbf{y}_k^*$ and $\mathbf{H}_k^*$ can be computed without the explicit need to invert Σ_k.

> ***Exercise 6-4.*** *Prove* $\boldsymbol{y}_k^* = \boldsymbol{H}_k^* \boldsymbol{x}_k + \boldsymbol{v}_k^*$ *where* $\boldsymbol{v}_k^*$ *is normally distributed with zero mean and covariance matrix equal to identity matrix.*

Step 4. Since the transformed measurements are not correlated (see Exercise 6-4), they can be processed sequentially using the following equations:

$$\mathbf{t}_{ki} = \mathbf{S}_{k|k,i-1}\mathbf{h}_{k,i}^{*} \tag{6-39}$$

$$\beta_{ki} = 1/\left(\mathbf{t}_{ki}^{T}\mathbf{t}_{ki} + 1\right) \tag{6-40}$$

$$\mathbf{S}_{k|k,i} = \mathbf{S}_{k|k,i-1} - \beta_{ki}\mathbf{S}_{k|k,i-1}\mathbf{t}_{ki}\mathbf{t}_{ki}^{T} / \left(1+\sqrt{\beta_{ki}}\right) \tag{6-41}$$

where $\mathbf{S}_{k|k-1}$ is the square root of the updated covariance matrix $\mathbf{P}_{k|k}$ after processing the first i measurements, and $\mathbf{h}_{k,i}^*$ is row i of the transformed measurement matrix $\mathbf{H}_k^*$.

We initialize the computations of this step using

$$\mathbf{S}_{k|k,0} = \mathbf{S}_{k|k-1} \tag{6-42}$$

Thus, after all n measurements are processed, the updated covariance matrix is obtained as

$$\mathbf{P}_{k|k} = \mathbf{S}_{k|k,n}\mathbf{S}_{k|k,n}^{T} \tag{6-43}$$

Although, the Kalman gain matrix is not explicitly computed in the above sequential procedure, it can be computed if necessary using

$$\mathbf{K}_{k,i} = \beta_{ki}\mathbf{S}_{k|k,i}\mathbf{t}_{ki} \tag{6-44}$$

where $\mathbf{K}_{k,i}$ is column i of the gain matrix.

DYNAMIC DATA RECONCILIATION OF NONLINEAR SYSTEMS

Nonlinear State Estimations

The treatment of nonlinear processes presents several difficulties which are not encountered in linear systems. First, it is generally not possible to analytically obtain a discrete form representation of the process analogous to Equation 6-1 starting with a set of nonlinear differential equations describing the process. Secondly, it is mathematically difficult to treat random noise, if the state transition equations or measurement equations are nonlinear functions of the noise (see Borrie [18] for a detailed explanation).

Thus, the effect of noise in a nonlinear process is modeled as a linear additive term. Thirdly, even if the random noises are assumed to be normally distributed, neither the state variables nor the measurements follow a Gaussian distribution due to the nonlinearity of the equations. Thus, a probabilistic framework can be used only under some approximations (see Jazwinski [20] for a more complete treatment). A least-squares formulation, however, can always be used to derive the estimates.

Under the above limitations, the evolution of the state variables for a general nonlinear process is modeled by the following differential equation:

$$\dot{\mathbf{x}} = \mathbf{f}(\mathbf{t}, \mathbf{x}, \mathbf{u}) + \dot{\boldsymbol{\beta}} \tag{6-45}$$

$$\dot{\beta} = \mathbf{w}(t) \tag{6-46}$$

where $\mathbf{w}(t)$ is a white noise process with mean function zero and covariance matrix function $\mathbf{R}(t)\delta(t\text{–}\tau)$, where $\delta(t\text{–}\tau)$ is the Dirac delta function.

The variables are assumed to be sampled at discrete times $t = kT$ and the relation between the measurements and state variables are represented as

$$\mathbf{y}_k = \mathbf{h}(\mathbf{x}_k) + \mathbf{v}_k \tag{6-47}$$

where $\mathbf{v}_k$ are the random measurement errors which are assumed to be Gaussian with mean zero and covariance matrix $\mathbf{Q}_k$. As in the linear case, we assume that $\mathbf{w}(t)$ and $\mathbf{v}_k$ are not correlated with each other. Equations 6-45 and 6-47 describe a nonlinear continuous stochastic process with discrete measurements.

Based on a linear approximation of Equation 6-45 and Equation 6-47, at each time around the current state estimates, an *extended Kalman filter* can be used to obtain the state estimates recursively using the following equations which are analogous to Equations 6-45 and 6-47:

$$\hat{\mathbf{x}}_{k|k} = \hat{\mathbf{x}}_{k|k-1} + \mathbf{K}_k\left[\mathbf{y}_k - \mathbf{h}(\hat{\mathbf{x}}_{k|k-1})\right] \tag{6-48}$$

$$\dot{\hat{\mathbf{x}}}_{\tau|k-1} = \mathbf{f}(\tau, \hat{\mathbf{x}}_{\tau|k-1}, \mathbf{u}_{k-1}), \quad \hat{\mathbf{x}}_{k|k-1} = \hat{\mathbf{x}}_{\tau=kT|k} \tag{6-49}$$

$$\mathbf{K}_k = \mathbf{P}_{k|k-1}\mathbf{H}_k^T \; (\mathbf{H}_k\mathbf{P}_{k|k-1}\mathbf{H}_k^T + \mathbf{Q}_k)^{-1} \tag{6-50}$$

$$\dot{\mathbf{P}}_{\tau|k-1} = \mathbf{F}_{\tau|k-1}\mathbf{P}_{\tau|k-1} + \mathbf{P}_{\tau|k-1}\mathbf{F}_{\tau|k-1}^T,$$

$$\mathbf{P}_{k|k-1} = \mathbf{P}_{\tau=kT|k-1} + \mathbf{R}(t = kT) \tag{6-51}$$

$$\mathbf{P}_{k|k} = (\mathbf{I} - \mathbf{K}_k\mathbf{H}_k)\mathbf{P}_{k|k-1} \tag{6-52}$$

where

$$\mathbf{F}_{\tau|k-1} = \left.\frac{\partial \mathbf{f}(t, \mathbf{x}, \mathbf{u})}{\partial \mathbf{x}}\right|_{\tau, \hat{x}_{k-1|k-1}, u_{k-1}} \tag{6-53}$$

$$\mathbf{H}_k = \left.\frac{\partial \mathbf{h}(\mathbf{x})}{\partial \mathbf{x}}\right|_{\hat{x}_{k|k-1}} \quad (6\text{-}54)$$

Equations 6-49 and 6-51 are nonlinear differential equations that have to be numerically integrated to obtain the predicted estimates of the state variables and the predicted covariance matrix of estimates. Equation 6-51, which involves the solution of n^2 coupled differential equations, can be computationally demanding. These can be avoided by computing a state transition matrix $\mathbf{A}_k$ at each time based on a linear approximation of the nonlinear functions and assuming it to be constant during each sampling period (Wishner et al. [21]). With this additional approximation, Equation 6-16 can be used to obtain the predicted covariance matrix. The method described here represents one of many different approaches for developing recursive estimation techniques and these are described in Muske and Edgar [22].

Example 6-4

A *continuous stirred tank reactor* (CSTR) with external heat exchange [23], and in which a first order exothermic reaction (decomposition of a reactant A) occurs is used to illustrate the application of state estimation for a nonlinear process. The differential equations describing the change in concentration (of reactant A) and temperature in the reactor are given by

$$\frac{dA}{dt} = \frac{q}{V}(A_0 - A) - kA$$

$$\frac{dT}{dt} = \frac{q}{V}(T_0 - T) - \frac{\Delta H_R A_r}{\rho C_p T_r} kA - \frac{UA_R}{\rho C_p V}(T - T_c)$$

where A_0 and T_0 are the feed concentration and temperature, respectively and A, T are the reactor concentration and temperature, respectively. The concentration and temperature variables are scaled using factors A_r and T_r, respectively. The reaction rate constant is given by

$$k = k_0 \exp(-E_A / TT_r)$$

Table 6-2
Parameter Values for CSTR

Parameter	Value	Units
q	10.0	cm^3/s
V	1000.0	cm^3
ΔH_R	-27,000	cal/gmol
ρ	0.001	g/cm^3
C_p	1.0	cal/(gm K)
U	5.0×10^{-4}	cal/(cm^2 s K)
A_R	10.0	cm^3
T_c	340.0	K
k_0	7.86×10^{12}	s^{-1}
E_a/R	14,090	K
A_0	6.5	
T_0	3.5	
A_r	1.0×10^{-6}	$gmol/cm^3$
T_r	100	K

The values of all parameters are listed in Table 6-2. Corresponding to these values, it can be verified that the steady-state reactor concentration is 0.1531 and steady-state reactor temperature 4.6091.

It is assumed that for this process, the reactor concentration and temperature are measured using a sampling period of 2.5 s, and that the standard deviations of random errors in these measurements are 0.0077 and 0.2305 (5% of the steady-state values), respectively. The open-loop response of this process is simulated for a step change in the feed concentration from 6.5 to 7.5, and an extended Kalman filter is used to estimate the reactor concentration and temperature. In the implementation of the extended Kalman filter, the predicted state estimates and predicted covariance matrix of estimation errors at each sampling instant are obtained by integrating the differential equations (Equations 6-49 and 6-51) using a 4th order Runge-Kutta method.

The true, measured and estimated values of reactor concentration and temperature, respectively, are shown in Figures 6-4 and 6-5. It can be observed that the estimated values are very close to the true values (in the figures, the estimated values almost coincide with the true values). The variances of errors in the measurements of reactor concentration and temperature, calculated from the sample data over the time period of 250 s, are 5.9×10^{-5} and 0.0534, respectively. In comparison, the variances of errors in the estimated concentration and temperature are only 1.36×10^{-7} and 2.52×10^{-8}, respectively.

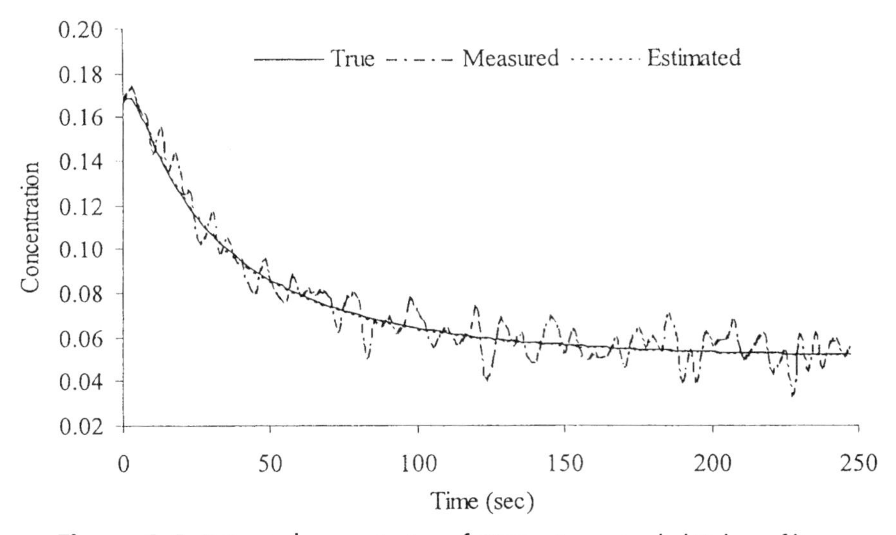

Figure 6-4. Estimated concentration of CSTR using extended Kalman filter.

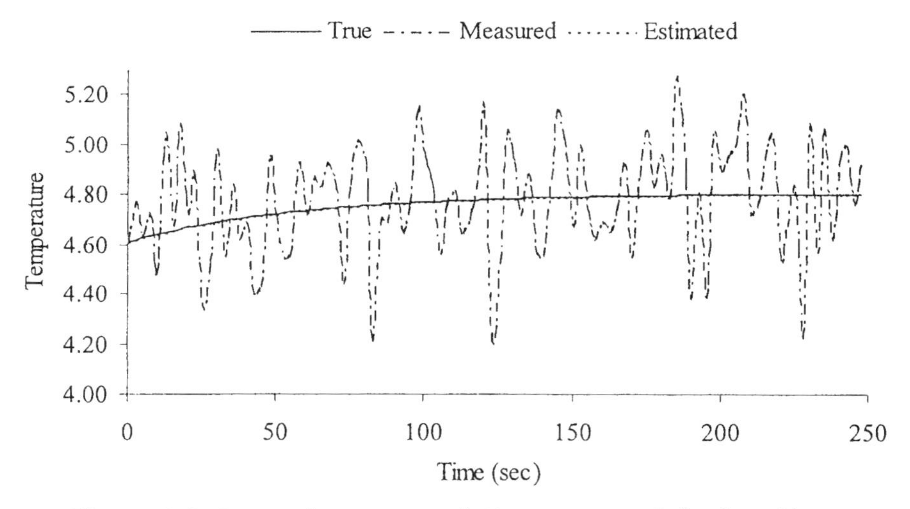

Figure 6-5. Estimated temperature of CSTR using extended Kalman filter.

Nonlinear Data Reconciliation Methods

It was demonstrated earlier that Kalman filter is equivalent to data reconciliation, if we assume that the state transition equations are not corrupted by noises or random disturbances. A similar progression from nonlinear filtering to data reconciliation can be made by neglecting the random noise term in Equation 6-45. There are other key differences, however, in the formulation and solution of *nonlinear dynamic data reconciliation* problems as compared to nonlinear filtering problems. Liebman et al. [23] and later Ramamurthi et al. [24] formulated the nonlinear dynamic data reconciliation (NDDR) problem and also proposed solution strategies. We first start with a general statement of the problem as posed by Liebman et al. [23] before discussing these solution techniques.

The NDDR problem may be formulated as

$$\underset{\mathbf{x}}{\text{Min}}\ J = \sum_{j=t_0}^{t_N} \left(\mathbf{y}_j - \mathbf{x}_j\right)^T \mathbf{Q}_j^{-1} \left(\mathbf{y}_j - \mathbf{x}_j\right) +$$

$$\sum_{j=t_0}^{t_N} \left(\mathbf{u}_{cj} - \mathbf{u}_j\right)^T \mathbf{Q}_{uj}^{-1} \left(\mathbf{u}_{cj} - \mathbf{u}_j\right) \qquad (6\text{-}55)$$

subject to

$$\dot{\mathbf{x}} = \mathbf{f}(\mathbf{x}); \qquad \mathbf{x}(\mathbf{t}_0) = \hat{\mathbf{x}}_0 \qquad (6\text{-}56)$$

$$\mathbf{h}(\mathbf{x}) = 0 \qquad (6\text{-}57)$$

$$\mathbf{g}(\mathbf{x}) \leq \mathbf{0} \qquad (6\text{-}58)$$

There are several features to the above formulation that need elaboration. Firstly, the manipulated input variables **u** are included as part of the objective function and are estimated at each time step, although they are assumed to be constant within each sampling period. The computed values of the manipulated inputs, $\mathbf{u}_{cj}$ at each time *j,* using Equation 6-9 or any other control law, are different from the actual manipulated inputs to the process due to inherent errors in the actuators. Thus, the computed values of the manipulated inputs serve as measurements, and the true values of these variables have to be estimated.

This formulation is more general as compared to the model used in filtering, where the manipulated inputs are assumed to be known exactly. Secondly, the state variables are assumed to be directly measured (or equivalently, the matrix $\mathbf{H}_k$ is assumed to be an identity matrix). This does not impose any limitation because by using a simple transformation, the problem can still be formulated as above.

If a measurement is a nonlinear function of the state variables, then we can introduce a new artificial state variable corresponding to this measurement and the nonlinear relation between the artificial state variable and the actual state variables can be included as part of the equality constraints of Equation 6-57. This transformation is similar to the treatment of indirectly measured variables in steady-state data reconciliation (see Chapter 7). Lastly, the inequality constraints, Equation 6-58 allows bounds on state variables and other feasibility constraints to be also included. It should be noted that filtering methods cannot handle inequality constraints and can therefore give rise to infeasible estimates much like linear steady-state reconciliation methods. Thus, the formulation given by Equations 6-55 through 6-58 is extremely general and practically useful.

The general formulation of the NDDR problem comes with a price. It is no longer possible to develop a recursive solution technique as in filtering. Furthermore, a close look at the objective function reveals that all the state variables from initial time up to current time are being simultaneously estimated at each sampling instant. This leads to an ever growing increase in the number of variables with time that have to be estimated, which is not practically acceptable.

In order to reduce the computational burden, a *moving window* approach was adopted [23, 24]. In this approach, at each time t only a *window* of measurements from time t-N to time t are used to estimate all the state variables within this time window of size N. The objective function to be minimized is the weighted sum squared differences between the measurements and state estimates within this time window. The estimate obtained for the state variables at time t from this optimization are used to compute the manipulated inputs. The procedure is repeated at the next sampling instant, giving rise to the term "moving window."

The solution strategy used to solve the estimation problem at each time of the moving window approach requires some explanation due to the presence of nonlinear differential equations, Equations 6-56 along with algebraic equations Equations 6-57 and 6-58. Liebman et al. [23] converted the differential equations into algebraic equality constraints by discretizing them using *orthogonal collocation.* In this technique, the state variable functions (of time) within each sampling period are expressed as a weighted sum of the state variable values at different time instants, within this sampling period, representing the collocation nodal points. The weights used in this representation are the orthogonal polynomials. Although, a sampling period can be subdivided into several elements, for convenience one element is used per sampling interval. With this choice, the state variable functions within each sampling interval *j* can be written as

$$\mathbf{x}(t) = \sum_{i=1}^{n_c} l_i(t)\mathbf{x}_i^j \qquad (6\text{-}59)$$

where $l_i(t)$ are orthogonal basis polynomials, n_c is the order and are $\mathbf{x}_i^j$ the state variable values at the *i*th collocation point in sampling interval *j*. The end points of this interval 1 and n_c correspond to the sampling instants. Using Equation 6-59, the derivatives can also be expressed in terms of the state variable values at different time instants. Equations 6-56 can now be forced to be satisfied at all the collocation points resulting in the following algebraic equations for each sampling interval *j*.

$$\mathbf{D}\mathbf{x}^j - Ts(\mathbf{x}^j, \mathbf{u}_{j-1}) \qquad (6\text{-}60)$$

where $\mathbf{x}^j$ is the vector of all state variables at all collocation points in sampling interval *j*. Equation 6-60 can be written for each of the *N* sampling intervals in the window chosen with the additional stipulation that the variable values at the end of a sampling interval are equal to those at the beginning of the next interval. A nonlinear optimization technique such as GRG or SQP discussed in Chapter 5 can be used to minimize 6-55 subject to 6-57, 6-58, and 6-60. It should be noted that the number of variables in this optimization problem is more than the number of state and input variables at the *N* sampling instants within the window, since we are also simultaneously estimating the state variables at the collocation points within each interval. More details on the type of orthogonal polynomial used, the size of the problem, and the structure of the derivative matrix **D** are available in Liebman et al. [23].

In order to reduce the computational effort required by the nonlinear programming strategy described above, Ramamurthi et al. [24] proposed a *successively linearized horizon estimation* (SLHE) method in which Equations 6-56 and 6-57 are both linearized around a given reference trajectory for the state variables. The reference values at each sampling instant *j* are used to obtain the linearized form of these equations for the sampling period *j*. If inequality constraints are not included, then an analytical solution for the estimates of the state variables at the beginning of the time window can be obtained which is then used to numerically integrate the differential equations to obtain the state estimates at other sampling instants within the window. Although this method is efficient, it can give rise to infeasible estimates because it cannot handle inequality constraints.

In the above discussion, we have not explicitly included unmeasured variables or parameters as part of the model equations. The nonlinear programming methods can also be used to simultaneously estimate both the measured states and unmeasured parameters. Simultaneous state and parameter estimation in dynamic processes have been considered by Kim et al. [25, 26], who refer to it as *error-in-variables* method (EVM) estimation.

In summary, nonlinear dynamic data reconciliation strategies have several advantages over classical filtering techniques as discussed in this section, but they do not address the problem of random noise in the state equations which can be caused by unmeasured disturbances to the process. They are also computationally more demanding because a recursive form of the estimator has not been developed. Currently, these techniques have not been applied to industrial processes and further developments are required before they can be applied in practice.

Example 6-5

The nonisothermal CSTR described in Example 6-4 is used to illustrate the application of nonlinear dynamic data reconciliation technique. Measurements corresponding to the open-loop response of this process for a step change in the feed concentration from 6.5 to 7.5 at initial time were simulated as in Example 6-4. Using a window length of 10 sampling periods, a nonlinear dynamic data reconciliation technique is applied to estimate the concentration and temperature in the reactor.

Lower and upper bounds on concentration were imposed as 0.01 and 0.2, respectively, and on temperature as 4.0 and 5.0, respectively. Since an open-loop simulation is performed in this example, the objective function of data reconciliation is the weighted sum square of differences between measured and estimated values over the past 10 sampling periods—that is, the second term in Equation 6-55 is absent. The optimization at every sampling instant t is carried out by making initial guesses of the temperature and concentration at time $t-10$.

The differential equations describing the CSTR are integrated from time *t–10* to time *t* by using a 4th-order Runge-Kutta method to obtain the estimates of the state variables at all sampling instants within this time period. The objective function value is computed corresponding to these estimates and the state estimates at time *t–10* are iterated upon until a minimum value of the objective function is obtained subject to the constraints of upper and lower bounds on the initial estimates. This approach differs from the method of Ramamurthi et al. [24] in that the nonlinear differential equations are not linearized, but are explicitly integrated.

It should be noted that in this approach bounds are imposed only on the state estimates at the start of the time window and it is possible that the state estimates at other sampling instants obtained by explicit integration may violate the bounds. It has the advantage, however, that it is more efficient than the method of Liebman et al. [23]. The estimated concentration and temperatures obtained using this approach are shown in Figures 6-6 and 6-7, respectively.

It can be observed that the estimated states are very close to the true values. In order to ensure convergence of the optimization problem at

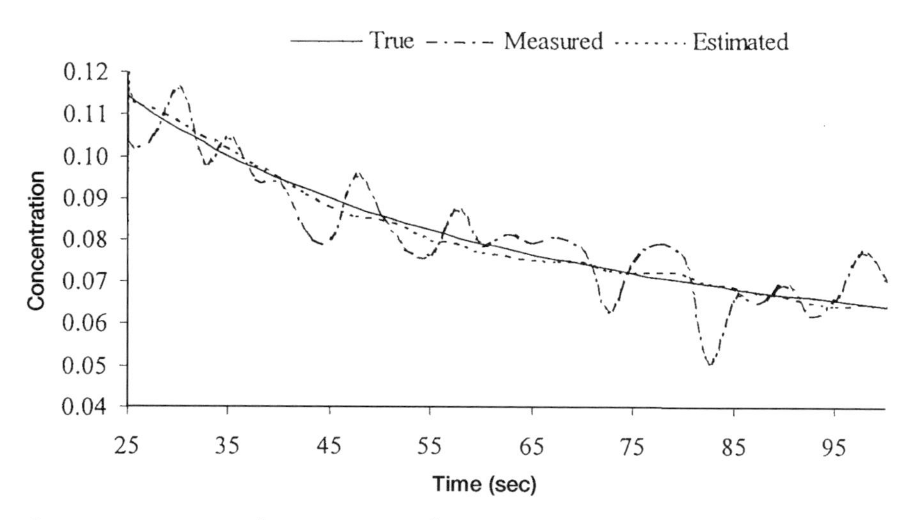

Figure 6-6. Estimated concentration of CSTR using dynamic data reconciliation for window length of 10.

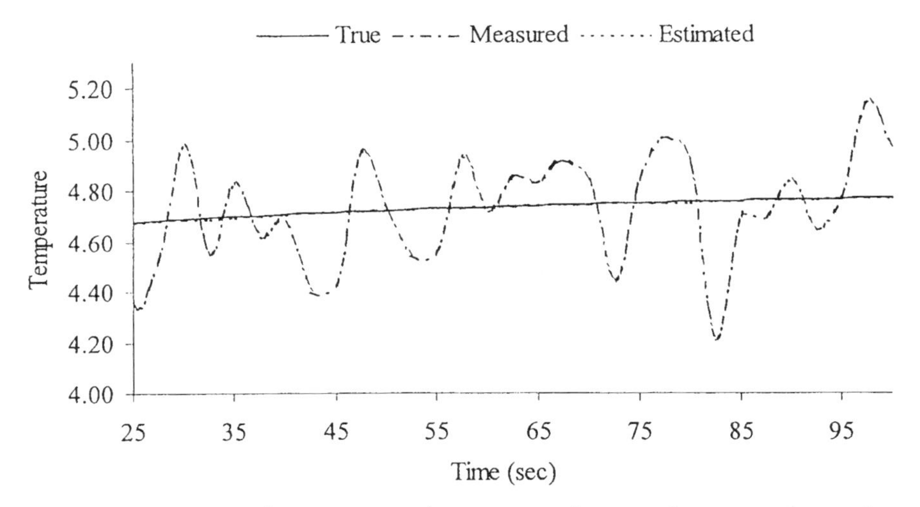

Figure 6-7. Estimated temperature of CSTR using dynamic data reconciliation for window length of 10.

each time, it was found that bounds on variables had to be imposed. Comparing these estimates with those obtained using the extended Kalman filter in Example 6-4, it is observed that the extended Kalman filter gives better results and is also computationally more efficient. Thus, it is better to use NDDR technique for estimating the states only when extended Kalman filtering techniques do not give estimates that satisfy bounds on variables.

SUMMARY

- Dynamic data reconciliation is important for process control applications.
- In order to exploit temporal redundancy in data, dynamic models for the evolution of the state variables have to be used in conjunction with measurements.
- The Kalman filter can be used to estimate state variables in linear dynamic systems.
- If disturbances in state variables are ignored, then the Kalman filter is equivalent to data reconciliation.
- Use of estimated states instead of measurements can lead to better control.
- State estimation in nonlinear dynamic systems can be performed using extended Kalman filters or its variants. Feasibility restrictions on variables cannot be handled by these methods.
- Nonlinear optimization methods can be used for dynamic data reconciliation in nonlinear processes. They can account for inequality constraints but are less efficient than extended Kalman filters.

REFERENCES

1. Ljung, L. *System Identification—Theory for the User.* Englewood Cliffs, N.J.: Prentice-Hall, 1987.
2. Soderstrom, T., and P. Stoica. *System Identification.* Englewood Cliffs, N.J.: Prentice-Hall, 1989.
3. Bellingham, B., and F. P. Lees. "The Detection of Malfunction Using a Process Control Computer: A Kalman Filtering Technique for General Control Loops." *Trans. Inst. Chem. Eng.* 55 (1977): 253–265.
4. Franklin, G. F, J. D. Powell, and M. L. Workman. *Digital Control of Dynamic Systems.* Reading, Mass.: Addison-Wesley, 1980.
5. Sage, A. P., and J. L. Melsa. *Estimation Theory with Applications to Communications and Control.* New York: McGraw-Hill, 1971.
6. Kalman, R. E. "A New Approach to Linear Filtering and Prediction Problems." *Trans. ASME J. Basic Eng.* 82D (1960): 35–45.

7. Kalman, R. E. "New Results in Linear Filtering and Prediction Problems," *Trans. ASME J. Basic Eng.* 83D (1961): 95–108.

8. Gelb, A. *Applied Optimal Estimation.* Cambridge, Mass.: MIT Press, 1974.

9. Sorenson, H. W. *Kalman Filtering: Theory and Applications.* New York: IEEE Press, 1985.

10. Fisher, D. G., and Seborg, D. E. *Multivariable Computer Control—A Case Study.* Amsterdam: North Holland, 1976.

11. Stanley, G. M., and R.S.H. Mah. "Estimation of Flows and Temperatures in Process Networks." *AIChE Journal* 23 (1977): 642–650.

12. Makni, S., D. Hodouin, and C. Bazin. "A Recursive Node Imbalance Method Incorporating a Model of Flowrate Dynamics for On-line Material Balance of Complex Flowsheets." *Minerals Eng.* 8 (1995): 753–766.

13. Darouach, M., and M. Zasadzinski. "Data Reconciliation in Generalized Linear Dynamic Systems." *AIChE Journal* 37 (1991): 193–201.

14. Rollins, D. K., and S. Devanathan "Unbiased Estimation in Dynamic Data Reconciliation." *AIChE Journal* 39 (1993): 1330–1334.

15. Bagajewicz, M., and Q. Jiang. "An Integral Approach to Dynamic Data Reconciliation." *AIChE Journal* 43 (1997): 2546–2558.

16. Anderson, B.D.O., and J. B. Moore. *Optimal Control: Linear Quadratic Methods.* Englewood Cliffs, N.J.: Prentice-Hall, 1989.

17. Bagchi, A. *Optimal Control of Stochastic Systems.* Hertfordshire, UK: Prentice-Hall, 1993.

18. Borrie, J. A. *Stochastic Systems for Engineers—Modeling Estimation and Control.* Hertfordshire, UK: Prentice-Hall, 1992.

19. Press, W. H., B. P. Flannery, S. A. Teukolsky, and W. T. Vetterling. *Numerical Recipes.* New York: Cambridge University Press, 1986.

20. Jazwinski, A. H. *Stochastic Processes and Filtering Theory.* New York: Academic Press, 1970.

21. Wishner, R. P., J. A. Tabaczynski, and M. Athans. "Comparative Study of Three Nonlinear Filters." *Automatica* 5 (1969): 478–495.

22. Muske, K. R., and T. F. Edgar. "Nonlinear State Estimation." In *Nonlinear Process Control* (edited by M. A. Henson and D. E. Seborg), New Jersey: Prentice-Hall, 1997, 311–370.

23. Liebman, M. J., T. F. Edgar, and L. S. Lasdon. "Efficient Data Reconciliation and Estimation for Dynamic Processes Using Nonlinear Programming Technqiues." *Computers Chem. Engng.* 16 (no. 10/11, 1992): 963–986.

24. Ramamurthi, Y., P. B. Sistu, and B. W. Bequette. "Control Relevant Dynamic Data Reconciliation and Parameter Estimation." *Computers Chem. Engng.* 17 (no. 1, 1993): 41–59.

25. Kim, I. W., M. J. Liebman, and T. F. Edgar. "Robust Error in Variables Estimation using Nonlinear Programming Techniques." *AIChE Journal* 36 (1990): 985–993.

26. Kim, I. W., M. J. Liebman, and T. F. Edgar. "A Sequential Error in Variables Estimation Method for Nonlinear Dynamic Systems." *Computers Chem. Engng.* 15 (1991): 663–670.

7

Introduction to Gross Error Detection

PROBLEM STATEMENTS

The technique of data reconciliation crucially depends on the assumption that only random errors are present in the data and systematic errors either in the measurements or the model equations are not present. If this assumption is invalid, reconciliation can lead to large adjustments being made to the measured values, and the resulting estimates can be very inaccurate and even infeasible. Thus it is important to identify such systematic or gross errors before the final reconciled estimates are obtained.

In the first chapter, it was pointed out that reconciliation can be performed only if constraints are present. The same statement can be made with regard to the detection of gross errors. Without the availability of constraints as a counter-check of the measurements, gross error detection cannot be carried out. Therefore, both data reconciliation and gross error detection techniques exploit the same information available from measurements and constraints. These techniques, therefore, go hand-in-hand in the processing of data.

There are two major types of gross errors, as indicated in Chapter 2. One is related to the instrument performance and includes measurement bias, drifting, miscalibration, and total instrument failure. The other is constraint model-related and includes unaccounted loss of material and energy resulting from leaks from process equipment or model inaccuracies due to inaccurate parameters. Various techniques have been designed for the detection and elimination of these two types of gross errors. Before describing these techniques, at the outset it is better to clearly state the requirements of a gross error detection strategy. This also leads to a better understanding of the variety of techniques that have been proposed, their interrelationships and the achievable results from their usage.

Any comprehensive gross error detection strategy should preferably possess the following capabilities:

- Ability to detect the presence of one or more gross errors in the data (**the detection problem**)
- Ability to identify the type and location of the gross error (**the identification problem**)
- Ability to locate and identify multiple gross errors which may be present simultaneously in the data (**the multiple gross error identification problem**)
- Ability to estimate the magnitude of the gross errors (**the estimation problem**)

Not all gross error detection strategies may fulfill all of the above requirements. The last of the above requirements, although useful, is not absolutely necessary. A gross error detection strategy can be analyzed in terms of the component methods it uses to tackle the three main problems of detection, identification, and multiple gross error identification, and the performance of the strategy is a strong function of these component methods. In this chapter, we focus on the first two components of a gross error detection strategy, that of detection and identification of a single gross error. Methods for multiple gross error detection are discussed in the following chapter.

BASIC STATISTICAL TESTS FOR GROSS ERROR DETECTION

This component of a gross error detection strategy simply attempts to answer the question of whether gross errors are present in the data or not. It does not provide any clues on either the number of gross errors, their types, or their locations. We reiterate the fact that all detection methods either directly or indirectly utilize the fact that gross errors in measurements cause them to violate the model constraints. If measurements do not contain any random errors, then a violation of any of the model constraints by the measured values can be immediately interpreted as due to the presence of gross errors. This is a purely deterministic method.

We have assumed, however, and rightly so, that all measurements do contain random errors due to which we cannot expect the measurements to strictly satisfy any of the model constraints even if gross errors are absent. Thus, an allowance has to be made for the violation of the constraints due to random errors. Under an assumed probability distribution for the random errors, a probabilistic approach is used for resolving this problem. Some basics of probability distributions and statistical hypothesis testing are explained in Appendix C.

The basic principle in gross error detection is derived from the detection of outliers in statistical applications. The random error inherently present in any measurement is assumed to follow a normal distribution with zero mean and known variance. The *normalized error* (the difference between the measured value and the expected mean value divided by its standard deviation) follows a standard normal distribution. Most normalized errors fall inside a $(1 - \alpha)$ confidence interval at a chosen level of significance α. Any value (normalized error) which falls outside that confidence region is declared an outlier or a gross error.

A number of statistical tests are derived from this basic statistical principle and are able to detect *gross* errors. But not all statistical tests are able to identify different types and location of gross errors. Some basic statistical tests are able to detect only measurement errors (biases). Other statistical tests can only detect process model errors or leaks. On the other hand, the generalized likelihood ratio test, which is derived from maximum likelihood estimation principle in statistics, can be used to detect both instrument problems and process leaks.

The next two sections describe the two basic classes of statistical tests used for gross error detection. Next, a derived class of statistical tests known as the *principal component tests* is also presented and compared with the basic statistical tests. These tests are based on a special type of linear transformation of the residual vectors used in the basic tests. For the sake of clarity in this chapter, the principal component tests are presented in a separate section.

The most commonly used statistical techniques for detecting gross errors are based on hypothesis testing. In a gross error detection case, the *null hypothesis*, H_0, is that no gross error is present, and the *alternative hypothesis*, H_1, is that one or more gross errors are present in the system. All statistical techniques for choosing between these two hypotheses make use of a *test statistic* which is a function of the measurements and constraint model. The test statistic is compared with a prespecified threshold value and the null hypothesis is rejected or accepted, respectively, depending on whether the statistic exceeds the threshold or not. The threshold value is also known as the test criterion or the *critical value* of the test.

The outcome of hypothesis testing is not perfect. A statistical test may declare the presence of gross errors, when in fact there is no gross error (H_0 is true). In this case, the test commits a *Type I error* or gives rise to a *false alarm.* On the other hand, the test may declare the measurements to be free of error, when in fact one or more gross errors exists (*Type II error*). The *power* of a statistical test, which is the probability of correct detection, is equal to *1–Type II* error probability. The power and Type I error probability of any statistical test are intimately related.

By allowing a larger Type I error probability, the power of a statistical test can be increased. Therefore, in designing a statistical test, the power of the test must be balanced against the probability of false detection. If the probability distribution of the test statistic can be obtained under the assumption of the null hypothesis, then the test criterion can be selected so that the probability of Type I error is less than or equal to a specified value α. The parameter α is also referred to as the *level of significance* for the statistical test.

The different statistical tests for gross error detection and the choice of the test criterion are described in the following section. For the sake of simplicity, we will analyze the basic statistical tests assuming steady-state conditions and linear models. The applicability of such gross error tests to nonlinear models will be further discussed in Chapter 8.

We will assume that the linear constraint model is given by

$$\mathbf{Ax} = \mathbf{c} \tag{7-1}$$

where $\mathbf{A}$ is the linear constraint matrix and the vector $\mathbf{c}$ contains known coefficients. Typically, for linear flow processes, $\mathbf{c}$ is a zero vector unless some of the variables are known exactly. We have deliberately included this vector in Equation 7-1 for ease of comparison with the linearized form of nonlinear constraints which will be treated later. As in the previous chapters, the measurement errors are assumed to be distributed normally with known covariance matrix Σ.

Four basic statistical tests have been developed and widely applied for gross error detection. To simplify the description of these tests, a linear model with all variables measured will be first assumed. This does not exclude the application of such statistical tests to linear models with unmeasured variables, since, as shown in Chapter 3, linear models with unmeasured variables can be reduced to linear models with all measured variables by using a projection matrix.

The first two tests are based on the *vector of balance residuals,* **r**, which is given by

$$\mathbf{r} = \mathbf{A}\mathbf{y} - \mathbf{c} \tag{7-2}$$

In the absence of gross errors, the vector **r** follows a multivariate normal distribution with zero mean value and variance-covariance matrix **V** given by

$$\mathbf{V} = \text{cov}(r) = \mathbf{A}\Sigma\mathbf{A}^T \tag{7-3}$$

Therefore, under H_0, $\mathbf{r} \sim N(\mathbf{0}, \mathbf{V})$, etc. In the presence of gross errors, the elements of residual vector **r** reflects the degree of violation of process constraints (material and energy conservation laws). On the other hand, matrix **V** contains information of the process structure (matrix **A**) and the measurement variance-covariance matrix, Σ. The two quantities, **r** and **V**, can be used to construct statistical tests which can detect the existence of gross errors.

The Global Test (GT)

The *global test,* which was the first test proposed [1, 2, 3], uses the test statistic given by

$$\gamma = \mathbf{r}^T \mathbf{V}^{-1}\mathbf{r} \tag{7-4}$$

Under H_0, the above statistic follows a χ^2-distribution with ν degrees of freedom, where ν is the rank of matrix **A**. If the test criterion is chosen as $\chi^2_{1-\alpha,\nu}$, where $\chi^2_{1-\alpha,\nu}$ is the critical value of χ^2 distribution at the chosen α level of significance, then H_0 is rejected and a gross error is detected, if $\gamma \geq \chi^2_{1-\alpha,\nu}$. This choice of the test criterion ensures that the probability of Type I error for this test is less than or equal to α. The global test combines all the constraint residuals in obtaining the test statistic, and therefore gives rise to a multivariate or collective test.

A point worth mentioning here is that the global test statistic given by Equation 7-4 is also equal to the minimum objective function value of the data reconciliation problem. This can be verified easily by substituting the solution for the reconciled estimates given by Equation 3-8 in the objective function given by Equation 3-6. This result is used later in analyzing the techniques used for gross error identification.

> ***Exercise 7-1.*** *Prove that the global test statistic and the optimal data reconciliation objective function values are equal.*

Example 7-1

Consider the flow reconciliation of the heat exchanger with bypass process shown in Figure 1-2. Let us assume that all flows are measured and the true, measured and reconciled values (assuming no gross errors) are as given in Table 7-1, where the flow measurement of stream 2 contains a positive bias of 4 units. The standard deviations of all measurement errors are assumed as unity.

Table 7-1
Reconciliation of Data Containing a Gross Error for Process of Figure 1-2

Stream Number	True Flow Values	Measured Flow Values	Reconciled Flow Values
1	100	101.91	100.89
2	64	68.45	65.83
3	36	34.65	35.05
4	64	64.20	65.83
5	36	36.44	35.05
6	100	98.88	100.89

The constraint matrix for this process is given by

$$A = \begin{bmatrix} 1 & -1 & -1 & 0 & 0 & 0 \\ 0 & 1 & 0 & -1 & 0 & 0 \\ 0 & 0 & 1 & 0 & -1 & 0 \\ 0 & 0 & 0 & 1 & 1 & -1 \end{bmatrix}$$

where the rows correspond to flow balances for the splitter, heat exchanger, bypass valve, and mixer in order and the columns correspond to the six streams in order. The constraint residuals for the given measurements can be computed as [–1.19, 4.25, –1.79, 1.76]. The covariance matrix of constraint residuals is given by

$$\mathbf{V} = \begin{bmatrix} 3 & -1 & -1 & 0 \\ -1 & 2 & 0 & -1 \\ -1 & 0 & 2 & -1 \\ 0 & -1 & -1 & 3 \end{bmatrix}$$

Using Equation 7-4 the global test statistic is computed to be 16.674. This can be verified to be equal to the sum square of the differences between the reconciled and measured values (optimum DR objective function value). The test criterion at 5% level of significance drawn from the chi-square distribution with 4 degrees of freedom is equal to 9.488. Thus the global test rejects the null hypothesis and a gross error is detected.

The Constraint or Nodal Test (NT)

The vector **r** can also be used to derive test statistics, one for each constraint i, given by

$$z_{r,i} = \frac{|r_i|}{\sqrt{V_{ii}}} \qquad i = 1, 2, \ldots m \tag{7-5}$$

or, written in vector form

$$\mathbf{z}_r = [\mathrm{diag}(\mathbf{V})]^{-1/2}\mathbf{r} \tag{7-6}$$

where diag(**V**) is a diagonal matrix whose diagonal elements are V_{ii}. The *nodal* or *constraint test* [4, 5] uses the test statistics $z_{r,i}$ for gross error detection. It can be proved that $z_{r,i}$ follows a standard normal distribution, N (0,1) under H_0. If any of the test statistics $z_{r,i}$ (or equivalently, the maximum test statistic) exceeds the test criterion $Z_{1-\alpha/2}$ where $Z_{1-\alpha/2}$ is the critical value of the standard normal distribution for α level of significance (for the two-sided test), a gross error is detected.

Unlike the global test, the constraint test processes each constraint residual separately and gives rise to m univariate tests. Since multiple tests are performed using the same critical value, it increases the probability that one of the tests may be rejected even if no gross errors are present. In other words, the probability of Type I error will be more than the specified value of α. If we wish to control the Type I error probability, the following modified level of significance β, proposed by Mah and Tamhane [6] (derived from Sidak inequality [7]) can be used.

$$\beta = 1 - (1 - \alpha)^{1/m} \tag{7-7}$$

For any specified value of α, the modified value β can be computed using Equation 7-7 and the test criterion for all the constraint tests can be chosen as $Z_{1-\beta/2}$. This will ensure that the probability that any one of the constraints tests will be rejected under H_0 is less than or equal to α. It should be noted that α is only an upper bound on the Type I error probability and in order to ensure that the Type I error probability is exactly equal to α, the test criterion has to be chosen by trial and error using simulation. Alternatively, Rollins and Davis [8] proposed the use of a critical value based on the Bonferroni confidence interval which is given by

$$\beta = \alpha / m \tag{7-8}$$

For large values of *m*, Equation 7-7 reduces to Equation 7-8.

> ***Exercise 7-2.*** *Prove that by using a test criterion based on the modified level of significance given by Equation 7-7, the Type I error probability of the constraint test will be less than or equal to* α.

It is possible to obtain other forms of the constraint test by using a linear transformation of the constraint residuals. However, not all of these forms possess the same power to detect gross errors. Crowe [9] obtained a particular form of the constraint test which has the *maximum power.* The test statistics of the *maximum power constraint test* are given by

$$z_{r,i}^{*} = \frac{\left|\left[\mathbf{V}^{-1}\mathbf{r}\right]_i\right|}{\sqrt{\left[\mathbf{V}^{-1}\right]_{ii}}} \tag{7-9}$$

or, written in vector form,

$$\mathbf{z}_r^* = [\text{diag}(\mathbf{V}^{-1})]^{-1/2}\mathbf{V}^{-1}\mathbf{r} \qquad (7\text{-}10)$$

The test criterion is chosen to be the same as in the case of the standard constraint test. If there is a gross error in the process, then it can be shown that the expected value of the maximum among the test statistics given by Equation 7-9 is greater than the expected value of the maximum among the test statistics given by Equation 7-5. This implies that if there is a gross error, then the constraint test based on the test statistics of Equation 7-9 has a greater probability of detecting it than the test based on the statistics of Equation 7-5. If the constraint test statistics are derived using any other linear transformation of the residuals, we can show that they do not possess this property. Thus, the constraint test based on the statistics of Equation 7-9 has *maximal power* property (MP).

> ***Exercise 7-3.*** *Prove that the expected value of $z^*_{r,i}$ is greater than or equal to the expected value of $z_{r,i}$, if a gross error of any magnitude b is present in constraint i. Also prove for this case that the expected value of $z^*_{r,i}$ is greater than or equal to $z^*_{r,i}$ for all j. Extend this result to show that the statistics given by Equation 7-9 have more power for detecting a gross error in the constraints than constraint test statistics derived using any other linear transformation of the constraint residuals.* ***Hint:*** *The expected value of* $\mathbf{r}$ *in this case is* $b\mathbf{e}_i$*, where* $\mathbf{e}_i$ *is a unity vector with value 1 in position i and zero elsewhere. Use this result and Cauchy-Swartz inequality:* $|\mathbf{v}^T\mathbf{w}| \leq (\mathbf{v}^T\mathbf{v})^{1/2}\,(\mathbf{w}^T\mathbf{w})^{1/2}$.

Example 7-2

For the flow process considered in Example 7-1, the constraint residuals and its covariance matrix were computed. From these, the constraint test statistics can be obtained as [0.687, 3.0052, 1.2657, 1.0161]. The standard normal test criterion at 5% level of significance is 1.96. Thus, only the test for constraint residual 2 is rejected.

The Measurement Test (MT)

The third test is based on the *vector of measurement adjustments,*

$$\mathbf{a} = \mathbf{y} - \hat{\mathbf{x}} \tag{7-11}$$

where $\hat{\mathbf{x}}$ are the reconciled estimates obtained using Equation 3-8. Using this solution, the measurement adjustments can also be written as

$$\mathbf{a} = \Sigma \mathbf{A}^T \mathbf{V}^{-1} \mathbf{r} \tag{7-12}$$

which, under H_0, follows a multivariate normal distribution: N ($\mathbf{0}$, $\overline{\mathbf{W}}$), where:

$$\overline{\mathbf{W}} = \text{cov}(\mathbf{a}) = \Sigma \mathbf{A}^T \mathbf{V}^{-1} \Sigma \tag{7-13}$$

The following test statistics,

$$z_{a,j} = \frac{|a_j|}{\sqrt{\overline{W}_{jj}}} \qquad j = 1, 2, \ldots n \tag{7-14}$$

known as the *measurement test statistics,* follow a standard normal distribution, N (0,1) under H_0. Tamhane [10] has shown that for a nondiagonal covariance matrix Σ, a vector of test statistics with maximal power for detecting a single gross error is obtained by premultiplying $\mathbf{a}$ by Σ^{-1} which gives

$$\mathbf{d} = \Sigma^{-1} \mathbf{a} \tag{7-15}$$

Under H_0, $\mathbf{d}$ is also normally distributed with zero mean and a covariance matrix

$$\mathbf{W} = \text{cov}(\mathbf{d}) = \mathbf{A}^T \left(\mathbf{A} \Sigma \mathbf{A}^T \right)^{-1} \mathbf{A}^T \tag{7-16}$$

Mah and Tamhane [6] proposed the following test statistics,

$$z_{d,j} = \frac{|d_j|}{\sqrt{W_{jj}}} \qquad j = 1, 2, \ldots n \tag{7-17}$$

known as the *maximum power* (MP) *measurement test,* which follows a standard normal distribution, $N(0,1)$ under H_0. Similar to the constraint test, the measurement test also involves multiple univariate tests. Using similar arguments as before, we can show that the probability of Type I error will be less than or equal to α, if the test criterion is chosen as $Z_{1-\beta/2}$, where β is given by Equation 7-7 or 7-8 with m being replaced by n, the number of univariate measurement tests.

> ***Exercise 7-4.*** *Prove that $z_{d,j}$ have the maximum power for detecting a gross error in one of the measurements.* ***Hint:*** *Follow a similar proof as used for solving Exercise 7-3.*

> ***Exercise 7-5.*** *Show that for diagonal Σ, $z_{a,j} = z_{d,j}$.*

> ***Exercise 7-6.*** *Let $\mathbf{a}_i$ and $\mathbf{a}_j$ be two columns of matrix* **A***. If there is a constant c such that $\mathbf{a}_i = c\,\mathbf{a}_j$, show that $z_{d,i} = z_{d,j}$.*

Example 7-3

From the measured and reconciled values listed in Table 7-1, the measurement adjustments can be computed as [1.0233, 2.6167, –0.4033, –1.6333, 1.3867, –2.0067]. The covariance matrix of measurement adjustments is given by

$$\overline{W} = \begin{bmatrix} 0.6667 & -0.1667 & -0.1667 & -0.1667 & -0.1667 & -0.3333 \\ -0.1667 & 0.6667 & 0.1667 & -0.3333 & 0.1667 & -0.1667 \\ -0.1667 & 0.1667 & 0.6667 & 0.1667 & -0.3333 & -0.1667 \\ -0.1667 & -0.3333 & 0.1667 & 0.6667 & 0.1667 & -0.1667 \\ -0.1667 & 0.1667 & -0.3333 & 0.1667 & 0.6667 & -0.1667 \\ -0.3333 & -0.1667 & -0.1667 & -0.1667 & -0.1667 & 0.6667 \end{bmatrix}$$

The measurement test statistics are therefore obtained as [1.2533, 3.2047, 0.494, 2.0004, 1.6983, 2.4577]. Because, in this example, the measurement error covariance matrix is diagonal, the MP measurement test statistics are also the same. For a 5% level of significance, the standard normal test criterion is 1.96. From these we observe that the measurement tests for measurements 2, 4, and 6 are rejected. The modified level of significance given by Sidak's inequality (Equation 7-7) is equal to 0.0085, while that based on Bonferroni confidence interval (Equation 7-8) is equal to 0.0083. Corresponding to these modified significance levels, the test criteria are 2.6315 and 2.6396, respectively. Thus, if we use the modified levels of significance, only the test for measurement 2 is rejected.

Additional examples for GT, NT and MT tests are found in Crowe et al. [11] and Tamhane and Mah [12].

The Generalized Likelihood Ratio (GLR) Test

A fourth test for detecting gross errors in steady-state processes is the generalized likelihood ratio (GLR) test based on the maximum likelihood ratio principle used in statistics. In contrast to other tests, the formulation of this test requires a model of the process in the presence of a gross error, also known as the *gross error model.* As shown in the next section, this test can identify different types of gross error for which a gross error model is provided. The procedure has been illustrated for gross errors caused by measurement biases and process leaks by Narasimhan and Mah [13].

The gross error model for a bias of unknown magnitude b in measurement j is given by

$$\mathbf{y} = \mathbf{x} + \varepsilon + b\mathbf{e}_j \tag{7-18}$$

where $\mathbf{e}_j$ is a unity vector with value 1 in position j and zero elsewhere.

On the other hand, leakage of material should be modeled as part of the constraints. A mass flow leak in a process node i of unknown magnitude b can be modeled by

$$\mathbf{Ay} - b\mathbf{m}_i = \mathbf{c} \tag{7-19}$$

The elements of vector $\mathbf{m}_i$ are relatively easy to define when only total flow balances are involved. If the leak is from a process unit i, then only the flow constraint for this unit vector is affected and thus $\mathbf{m}_i$ is identical to $\mathbf{e}_i$. However, if the constraints also include component balances and energy balances (with precisely known composition and temperature values), then the vector $\mathbf{m}_i$ can only be defined approximately using engineering judgment. A recommendation made by Narasimhan and Mah [13] is to choose the elements of $\mathbf{m}_i$ as follows:

(a) Corresponding to the total mass flow constraint of unit i, $\mathbf{m}_i$ has a value of unity in the ith position.
(b) Corresponding to the energy flow constraint associated with node i, the value of of the ith element in the vector $\mathbf{m}_i$ can be chosen as the average specific enthalpy of the streams incident to node i. The same can be applied to a component flow constraint for the node i, by replacing specific enthalpy with concentration.
(c) The elements in $\mathbf{m}_i$ not associated with constraints of node i are chosen to be zero.

Pure energy or component flow losses in node i can also be modeled by Equation 7-19 by choosing the corresponding element in $\mathbf{m}_i$ to be unity and all other elements to be zero.

Using the gross error models, it is possible to derive the statistical distribution of the constraint residuals under H_1, when a gross error either in the measurements or constraints is present. It has already been proved that under H_0, the constraint residuals follow a normal distribution with zero mean and covariance matrix given by Equation 7-3. Under H_1, the constraint residuals still follow a normal distribution with covariance matrix given by Equation 7-3, but the expected value depends on the type of gross error present. If a gross error due to a bias of magnitude b is present in measurement j, then we can show that

$$E[\mathbf{r}] = b\mathbf{A}\mathbf{e}_j \tag{7-20}$$

On the other hand, if a gross error due to a process leak is present in node i, then we can show that

$$E[\mathbf{r}] = b\mathbf{m}_i \tag{7-21}$$

Therefore, when a gross error due to a bias or a process leak is present, we can write

$$E[\mathbf{r}] = b\mathbf{f}_k \tag{7-22}$$

where

$$\mathbf{f}_k = \begin{cases} \mathbf{A}\mathbf{e}_j & \text{for a bias in measurement } j \\ \mathbf{m}_i & \text{for a process leak in node } i \end{cases} \tag{7-23}$$

The vectors $\mathbf{f}_k$ are also referred to as *gross error signature vectors.* If we define μ as the unknown expected value of **r**, we can formulate the hypotheses for gross error detection as

$$\begin{aligned} H_0 &: \ \mu = \mathbf{0} \\ H_1 &: \ \mu = b\mathbf{f}_k \end{aligned} \tag{7-24}$$

where H_0 is the null hypothesis that no gross errors exist and H_1 is the alternative hypothesis that either a process leak or a measurement bias is present. The alternative hypothesis has two unknowns, b and $\mathbf{f}_k$. The parameter b can be any real number and $\mathbf{f}_k$ can be any vector from the set F, which is given by

$$F = \left\{ \mathbf{Ae}_j, \mathbf{m}_i : i = 1, \ldots, m; j = 1, \ldots, n \right\} \tag{7-25}$$

where m is the number of nodes or process units, and n is the number of measured variables.

In order to test the two hypotheses given by Equation 7-24, one can use the likelihood ratio test. The likelihood ratio test statistic in our case is given by

$$\lambda = \sup \frac{\Pr\{\mathbf{r}|H_1\}}{\Pr\{\mathbf{r}|H_0\}} \tag{7-26}$$

where $\Pr(\mathbf{r}|H_0\}$, $\Pr\{\mathbf{r}|H_1\}$ are the probabilities of obtaining residual vector **r** under H_0 and H_1 hypothesis respectively; the supremum ("sup" in Equation 7-26) is computed over all possible values of the parameters present in the hypotheses. Using the normal probability density function for **r**, we can write Equation 7-26 as

$$\lambda = \sup_{b, \mathbf{f}_k} \frac{\exp\left\{-0.5(\mathbf{r} - b\mathbf{f}_k)^T \mathbf{V}^{-1}(\mathbf{r} - b\mathbf{f}_k)\right\}}{\exp\left\{-0.5\mathbf{r}^T\mathbf{V}^{-1}\mathbf{r}\right\}} \tag{7-27}$$

Since the expression on the right hand side of Equation 7-27 is always positive, we can simplify the calculation by choosing as the test statistic

$$T = 2\ln\lambda = \sup_{b, \mathbf{f}_k}\left[\mathbf{r}^T\mathbf{V}^{-1}\mathbf{r} - (\mathbf{r} - b\mathbf{f}_k)^T\mathbf{V}^{-1}(\mathbf{r} - b\mathbf{f}_k)\right] \tag{7-28}$$

The computation of T proceeds as follows. For any vector $\mathbf{f}_k$ we compute the estimate b* of b, which gives the supremum in Equation 7-28. Thus, we obtain the maximum likelihood estimate

$$b^* = \left(\mathbf{f}_k^T \mathbf{V}^{-1} \mathbf{f}_k\right)^{-1} \left(\mathbf{f}_k^T \mathbf{V}^{-1} \mathbf{r}\right) \tag{7-29}$$

Substituting b* in Equation 7-28 and denoting the corresponding value of T by T_k we get

$$T_k = d_k^2 / C_k \tag{7-30}$$

where

$$d_k = \mathbf{f}_k^T \mathbf{V}^{-1} \mathbf{r} \tag{7-31}$$

$$C_k = \mathbf{f}_k^T \mathbf{V}^{-1} \mathbf{f}_k \tag{7-32}$$

This calculation is performed for every vector $\mathbf{f}_k$ in set F and the test statistic T is therefore obtained as

$$T = \sup_k T_k \qquad k = 1 \ldots m + n \tag{7-33}$$

Let **f*** be the vector that leads to the supremum in Equation 7-33. The test statistic T is compared with a prespecified threshold T_{cr} and a gross error is detected if T exceeds T_{cr}. We can interpret T_k as a test statistic for the presence of gross error *k*. Since T is the maximum among T_k, the GLR test detects a gross error if any of the test statistics T_k exceeds the critical value. Thus the GLR test, like the measurement test and the constraint test, performs multiple univariate tests to detect a gross error. The distribution of T_k, under H_0, can be shown to be a central chi-square distribution with one degree of freedom. Therefore, in order to maintain the Type I error probability of the GLR test less than or equal to a given value α, we can choose the test criterion as $\chi^2_{1-\beta,1}$, the upper 1- β quantile of the chi-square distribution with one degree of freedom, where β is given by

$$\beta = 1 - (1 - \alpha)^{1/(m+n)} \tag{7-34}$$

It can be easily observed by comparing Equations 7-7 and 7-34 that β, and α, are related by similar expressions, with the exponent being equal to the reciprocal of the number of multiple univariate tests being performed as part of the test to detect a gross error.

> ***Exercise 7-7.*** *Prove that T_k follows a central chi-square distribution with one degree of freedom.*

> ***Exercise 7-8.*** *Prove that the square root of the GLR test statistics T_k can be obtained using the linear transformation of the constraint residuals $\mathbf{F}^T\mathbf{V}^{-1}\mathbf{r}$, where the columns of matrix $\mathbf{F}$ are the gross error vectors $\mathbf{f}_k$.*

Example 7-4

If we consider gross errors caused by measurement biases for the simple flow process used in the preceding examples, the gross error signature vector for a bias in measurement i is the ith column of the constraint matrix which is given in Example 7-1. The GLR test statistics computed from the constraint residuals and its covariance matrix given in Example 7-2 are [1.5708, 10.2704, 0.244, 4.0017, 2.8843, 6.0401]. It can be verified that the GLR test statistics are the square of the MP measurement test statistics computed in Example 7-3. The test criteria at 5% level of significance and at the two modified levels of significance—Sidak and Bonferroni—are simply the square of the standard normal test criteria, [3.8415, 6.925, 6.9676], respectively. Hence, the GLR tests for measurements 2, 4, and 6 are rejected at the 5% level of significance, while only the test for measurement 2 is rejected at the modified levels of significance.

If we also wish to test for leaks in all the four nodes, then the signature vectors for these four gross errors are simply the unit vectors. The GLR test statistics for these four gross errors are given by [1.5708, 13.2496, 0.3844, 6.0401]. The GLR tests for leaks in nodes 2 and 4 are rejected at the 5% level of significance while the test for a leak in node 2 alone is rejected at the modified levels of significance. It can be observed that the GLR test statistic for a leak in node 1 (splitter node) is the same as the test statistic for a bias in measurement 1. This is due to the fact that the gross error signature vectors for these two gross errors are identical. The same observation can be made concerning a leak in node 4 and a bias in measurement 6.

Comparison of the Power of Basic Gross Error Detection Tests

As described in the preceding sections, several statistical tests have been developed for detecting gross errors in measurements caused by biases in the measuring instruments or gross errors in steady-state conservation constraints due to unknown leaks. In order to obtain the best performance, it is important to apply the test which has the maximum power (the probability of detecting the presence of a gross error when one is actually present) without increasing the probability of Type I error (probability of wrongly detecting a gross error when none is present).

Thus, an important question that can be asked is which among the above four tests gives the maximum power for detecting a single gross error in the data. This question has not been adequately addressed so far. Most of the works which compare the performance of different gross error detection strategies only consider the overall performance which includes all the components of detection, identification and multiple error detection, but does not compare the detection component part of the strategy in isolation. We provide some results that partially answer this question.

In making this comparison, we have to consider only the MP test for the constraint and measurement tests, besides the global test and the GLR test. We can further simplify our task by making use of theoretical results that have been derived by Crowe [9] and Narasimhan [14] to show that among the constraint, measurement and GLR tests, the GLR test has the maximum power to detect a single gross error. The proof of this result follows.

Lemma 7-1: The GLR test is more powerful than an MP measurement test or an MP constraint test, based on any singular or nonsingular linear transformation of the constraint residuals for detecting a single gross error.

It can be observed from Equations 7-12, 7-15, and 7-17 that the MP measurement test statistics are obtained using a linear transformation of constraint residuals. If we consider the positive square root of the GLR test statistics (without loss of generality), then we can show that the GLR test statistics are also obtained using a linear transformation of the constraint residuals (see Exercise 7-8).

Therefore, the MP constraint test, MP measurement test and the GLR test all derive test statistics based on a linear transformation of the constraint residuals. The question can then be posed as to which linear transformation of the constraint residuals gives the most powerful tests. In order to answer this question, we can consider an arbitrary linear transformation of the constraint residuals given by

$$\mathbf{r}^* = \mathbf{Y}\mathbf{r} \tag{7-35}$$

Let a gross error of magnitude b either due to a measurement bias or a leak be present with corresponding gross error vector $\mathbf{f}_k$. Then using Equation 7-22 the expected value of the transformed constraint residuals is obtained as

$$E[\mathbf{r}^*] = b\mathbf{Y}\mathbf{f}_k \tag{7-36}$$

The covariance matrix of the transformed constrained residuals is given by

$$\text{cov}(\mathbf{r}^*) = V^* = \mathbf{Y}\mathbf{V}\mathbf{Y}^T \tag{7-37}$$

A test can then be devised based on the transformed constrained residuals with test statistics given by

$$z_i^* = \frac{|r_i^*|}{\sqrt{V_{ii}^*}} \tag{7-38}$$

Equation 7-38 can also be written as

$$z_i^* = \frac{|\mathbf{e}_i^T \mathbf{r}^*|}{\sqrt{\mathbf{e}_i^T \mathbf{V}^* \mathbf{e}_i}} \tag{7-39}$$

It can be easily verified that by choosing $\mathbf{Y}$ to be $\mathbf{V}^{-1}$, $\mathbf{A}^T\mathbf{V}^{-1}$, or $\mathbf{F}^T\mathbf{V}^{-1}$ (where $\mathbf{F}$ is a matrix of vectors $\mathbf{f}_k$ defined by Equation 7-23), respectively, the MP constraint test, the MP measurement test or the GLR test statistics are obtained. In order to prove that the GLR test has the maximum power, we have to prove that the maximum among the expected values of the GLR test statistics is greater than or equal to the expected value of any of the test statistics given by Equation 7-39. For this, we first prove that the maximum among the expected values of the GLR test statistics is attained by T_k, that is,

$$E[\sqrt{T_k}] \geq E[\sqrt{T_i}] \tag{7-40}$$

where, from Equations 7-30 and 7-35 through 7-37 with $\mathbf{Y} = \mathbf{F}^T\mathbf{V}^{-1}$, the expected values of $\sqrt{T_k}$ and $\sqrt{T_i}$ are respectively

$$E[\sqrt{T_k}] = b\sqrt{\mathbf{f}_k^T \mathbf{V}^{-1} \mathbf{f}_k} \tag{7-41}$$

$$E[\sqrt{T_i}] = b\sqrt{\mathbf{f}_i^T \mathbf{V}^{-1} \mathbf{f}_k} \tag{7-42}$$

The above results can be easily established using the Cauchy-Schwartz inequality.

$$|\mathbf{v}^T \mathbf{w}| \leq \left(\mathbf{v}^T \mathbf{v}\right)^{1/2} \left(\mathbf{w}^T \mathbf{w}\right)^{1/2} \tag{7-43}$$

and by defining vectors $\mathbf{v}$ and $\mathbf{w}$ given by

$$\mathbf{v} = \mathbf{R}\mathbf{f}_i, \text{ and } \mathbf{w} = \mathbf{R}\mathbf{f}_k \tag{7-44}$$

where $\mathbf{R}$ is a matrix such that $\mathbf{R}^T\mathbf{R} = \mathbf{V}^{-1}$.

In order to prove $E[\sqrt{T_k}] \geq E[z_i^*]$ over all *i,* first we need to define $E[z_i^*]$, which according to Equations 7-39, 7-36, and 7-37 is

$$E\left[z_i^*\right] = b\mathbf{e}_i^T \mathbf{Y} \mathbf{f}_k \,/\, \mathbf{e}_i^T \mathbf{Y} \mathbf{V} \mathbf{Y}^T \mathbf{e}_i \tag{7-45}$$

Then, we can again make use of the Cauchy-Schwartz inequality by defining matrices

$$\mathbf{R}^T\mathbf{R} = \mathbf{V}^{-1} \text{ and } \mathbf{P} = \mathbf{Y}\mathbf{R}^{-1} \tag{7-46}$$

and identifying the vectors **v** and **w** to be

$$\mathbf{v} = \mathbf{R}\mathbf{f}_k, \text{ and } \mathbf{w} = \mathbf{P}^T\mathbf{e}_i \tag{7-47}$$

Since the MP constraint and MP measurements tests are obtained using a particular linear transformation of the constraint residuals, based on the above results we can claim that on an average we can expect the GLR test to give higher power for detecting the presence of a single gross error than either the MP constraint test or MP measurement test. If we assume that only gross errors due to measurement biases can be present in the system, then the GLR test becomes identical to the MP measurement test.

On the other hand, if we assume that only gross errors which affect only one constraint (for example, leaks in overall flow balance constraints) can be present, then the GLR test becomes identical to the MP constraint test. However, if we allow for both types of gross errors to be present in the system, then the GLR test is more powerful than either of the other two tests. It should be cautioned, though, that an implicit assumption has been made that we precisely know the gross error vectors $\mathbf{f}_k$ for the different types of gross errors which can occur in the process. This assumption may not be valid if there are uncertainties in the distribution model or gross error model. Moreover, these results are valid only if we assume that, at most, one gross error is present.

Exercise 7-9. *Prove that MP measurement test statistic* $z_{d,i} = (T_j)^{1/2}$ *when only gross errors in measurements are allowed.*

Exercise 7-10. *Prove that MP constraint test statistic* $z^*_{r,i} = (T_i)^{1/2}$ *when only gross errors in constraints which affect a single constraint are allowed.* ***Hint:*** *The gross error vectors for these types of gross errors are* $\mathbf{e}_i$.

It is now only necessary to examine whether the global test or the GLR test gives higher power to detect the presence of a single gross error. We note from Equation 7-4 that the global test is also based on the constraint residuals and thus it uses the same information as the GLR test for detecting gross errors. But there is a fundamental difference in the manner in which this information is processed. The global test performs a single multivariate test for detecting a gross error, whereas the GLR test performs multiple univariate tests, one for each possible gross error hypothesized for detecting if any one of them is present. The question is which one of these processing schemes gives a higher power. This problem has also been studied in the statistical literature. Unfortunately, it is difficult to obtain a unique answer to this question theoretically.

We can perform simulation studies of selected processes to evaluate the power of the two tests. Before attempting such a comparison, however, it must be ensured that both these tests give the same Type I error probability. This implies that the criterion for each test has to be chosen to give a specified value of the Type I error probability. This is possible only in the case of the global Test. As explained before, since the GLR test performs multiple univariate tests, the test criterion has to be chosen by trial and error using simulation. The results obtained through such simulation can at best be used to make some broad conclusions and strictly cannot be generalized to all processes.

It is seen from the above discussion, that one test does not have uniformly higher power than the other. However, we recommend that for the purpose of *detecting whether one or more gross errors are present,* the global test (GT) should be used. This recommendation is based on the following considerations:

(1) The computation of the GT statistic is more efficient since a single test statistic is computed.
(2) In practice to instill confidence among process operators, it is necessary to keep the false alarm probability below a specified limit. The test criterion for GT can be chosen to precisely obtain this allowable limit. Note that any higher value of the test criterion can satisfy this limit but will result in a lower power. In the case of the GLR test, the lowest test criterion value that satisfies this limit can be chosen only through simulation.
(3) The GLR test requires knowledge about the gross error vectors for the different gross errors that can occur in the process for obtaining the test statistics. The global test does not require any information

regarding the gross errors for the purpose of detection. This may be an important practical consideration since complete knowledge regarding all possible gross errors that can occur in a process is not generally available.

It may be argued that GT is inferior since it can only detect the presence or absence of gross errors but for identifying the nature and location of gross error an identification strategy is required. (In the case of the GLR test, the test statistics can be directly used for identification as described in the next section.) This argument is shown to be without merit from two considerations:

(1) The use of GT for detection, does not preclude the use of the GLR test statistics for identifying the type and location of the gross error. In this case, it is necessary to construct the GLR test statistics only if a gross error is detected by the global test.
(2) It is demonstrated in the next section, that the identification strategy inherent in the GLR test is the standard serial elimination technique that was proposed and first used by Ripps [1] in combination with the global test for identifying measurements biases. Thus, the GLR test for detecting and identifying a single gross error can be viewed as a gross error detection strategy which has as one of its components the GT for detection and serial elimination for identification.

GROSS ERROR DETECTION USING PRINCIPAL COMPONENT (PC) TESTS

The variance-covariance matrices of constraint residuals $\mathbf{V}$ and of measurements adjustments ($\overline{\mathbf{W}}$ or $\mathbf{W}$) are always dense. This implies that, even if measurements are independent or weakly correlated, the reconciled data are always strongly correlated. The reconciled values and, hence, the measurement adjustments are correlated because they are related to each other via the process model. The same is true for the constraint residuals.

However, not all basic tests exploit the entire information contained in matrices $\mathbf{V}$, $\overline{\mathbf{W}}$ or $\mathbf{W}$. The non-MP constraint test and the univariate measurement test (MP or non-MP) described earlier in this chapter use only the diagonal terms of matrices $\mathbf{V}$, $\overline{\mathbf{W}}$, or $\mathbf{W}$, respectively. Alternatively, the principal component tests use the entire matrices. It is expected that such tests will be able to detect more subtle gross errors since they are multivariate tests. It was found that multivariate tests such as the global test often detect gross errors that are not detected by the univariate tests. This aspect is very important, because failure to detect all gross errors could result in an unsuccessful data reconciliation (the reconciled solution is infeasible or questionable).

Principal component tests are related to the univariate constraint and measurement tests, because they use a linear transformation of the constraint or measurement residual vectors. The following is a brief description of the principal component tests as given in Tong and Crowe [15]. As with the previous tests, we restrict this analysis to linear models with no unmeasured variables. The case with unmeasured variables can be handled with a projection matrix. Two basic types of principal component tests can be derived as follows:

Principal Component Tests for Residuals of Process Constraints

Let us consider a set of linear combinations of vector $\mathbf{r}$

$$\mathbf{p}_r = \mathbf{W}_r^T \mathbf{r} \tag{7-48}$$

where the columns of $\mathbf{W}_r$ are the eigenvectors of $\mathbf{V}$, satisfying

$$\mathbf{W}_r = \mathbf{U}_r \Lambda_r^{-1/2} \tag{7-49}$$

Matrix Λ_r is diagonal, consisting of the eigenvalues of $\mathbf{V}$, λ_{ri}, i = 1...q, on its diagonal and satisfies

$$\Lambda_r = \mathbf{U}_r^T \mathbf{V} \mathbf{U}_r \tag{7-50}$$

Matrix $\mathbf{U}_r$ consists of the orthonormalized eigenvectors of $\mathbf{V}$, so that

$$\mathbf{U}_r \mathbf{U}_r^T = \mathbf{I} \tag{7-51}$$

The vector $\mathbf{p}_r$ consists of *principal components of constraint residuals,* and its elements are *principal component scores.*

If gross errors are not present, then $\mathbf{r} \sim N(\mathbf{0}, \mathbf{V})$ and it can be shown that $\mathbf{p_r} \sim N(\mathbf{0}, \mathbf{I})$. Therefore, a set of correlated variables, $\mathbf{r}$, is transformed into a new set of uncorrelated variables, $\mathbf{p}_r$. The principal components are numbered in descending order of the magnitudes of the corresponding eigenvalues.

> ***Exercise 7-11.*** *If* $\mathbf{r} \sim N(0, V)$, *show that* $\mathbf{p_r} \sim N(0, I)$.

On the other hand, Equations 7-48 and 7-49 can be combined and rewritten as

$$\mathbf{r} = \mathbf{U}_r \Lambda_r^{1/2} \mathbf{p}_r \tag{7-52}$$

which means that the residual vector **r** can be uniquely reconstructed from its principal components if all of the principal components are retained, that is, $\mathbf{p}_r \in \Re^m$, where m is the number of equations (balance residuals). However, if fewer than m principal components are retained, we get

$$\mathbf{r} = \mathbf{U}_r \Lambda_r^{1/2} \mathbf{p}_r + (\mathbf{r} - \hat{\mathbf{r}}) \tag{7-53}$$

where

$$\hat{\mathbf{r}} = \mathbf{U}_r \Lambda_r^{1/2} \mathbf{p}_r \tag{7-54}$$

with $\mathbf{p}_r \in \Re^k$, and $k < m$. Equation 7-54 is referred to as the *principal component model* of vector **r**. Equation 7-53 indicates that the residuals in the vector **r** can be decomposed into the contributions from the principal components term, and the residuals of the principal component model, $\mathbf{r}-\hat{\mathbf{r}}$. This means that for gross error detection, instead of using statistical tests for **r**, we can perform hypothesis testing on $\mathbf{p}_r$ and $\mathbf{r}-\hat{\mathbf{r}}$.

Since each element of vector $\mathbf{p}_r$ is distributed as a standard normal variable, a detection rule similar to the univariate constraint test can be used and the test for constraint residual i is rejected if $p_{r,i}$ exceeds $Z_{1-\beta/2}$. Similar to the univariate tests, to limit the Type I error to level α, β can be chosen as in Equation 7-7 where the exponent m in this equation is replaced by the number of retained principal components, k.

Principal Component Tests on Measurement Adjustments

Similar to the principal component test statistics based on constraint residuals, *principal component measurement test statistics* can be defined as

$$p_{ai} = \left(\mathbf{W}_a^T \mathbf{a}\right)_i \qquad i = 1 \ldots k \qquad (7\text{-}55)$$

where the columns of $\mathbf{W}_a$ are the eigenvectors of $\overline{\mathbf{W}}$ and k is the number of retained principal components. In general, $k < n$, where n is the number of measurements.

> ***Exercise 7-12.*** *If* $\boldsymbol{a} \sim N\,(\boldsymbol{0}, \overline{\mathbf{W}})$, *show that* $\boldsymbol{p_a} \sim N\,(\boldsymbol{0}, \boldsymbol{I})$.

If gross errors are not present, then it can be shown that $\mathbf{p_a} \sim N\,(\mathbf{0}, \mathbf{I})$; therefore, the principal components on measurement adjustments are also not correlated. Similar to the measurement test, we can conduct a test on every p_{ai} by comparing it against a threshold $Z_{1-\beta/2}$.

Relationship between Principal Component Tests and Other Statistical Tests

The principal component tests are also based on a linear transformation of the constraint residuals as in Equation 7-35. It can be verified that the transformation matrix used for deriving principal component constraint test is $\mathbf{Y} = \mathbf{W}_r^T$ and for the principal component measurement test, the transformation matrix is $\mathbf{Y} = \mathbf{W}_a^T \mathbf{A}^T \mathbf{V}^{-1}$. Tong and Crowe [15] implied that, since the number of retained principal components is usually less than the number of principal components, the modified level of significance for principal component tests are smaller.

Therefore, we expect in general to reduce the overall Type I error in detecting gross errors by the principal component test. But this argument is without merit, because the Type I error probability for any test can always be reduced by simply choosing a smaller value of α. Furthermore, because principal component tests do not directly identify the gross error it is possible that the strategy used for identifying the gross error commits additional Type I errors. Some later examples in this chapter illustrate this problem.

Analogous to the global test, a collective global test based on principal components can also be proposed for which the test statistic is defined by

$$\gamma_k = \mathbf{p}_r^T \mathbf{p}_r \qquad (7\text{-}56)$$

The collective principal component test in Exercise 7-13 is called *truncated chi-square test* if all the principal components are not retained. Another important collective test statistic is defined by

$$Q_r = (\mathbf{r} - \hat{\mathbf{r}})^T (\mathbf{r} - \hat{\mathbf{r}}) \qquad (7\text{-}57)$$

known as the Q statistic or the squared prediction error, and sometimes, the Rao-statistic. It can be shown that Q_r is a weighted sum of squares of the last *m-k* principal components.

$$Q_r = \sum_{i=k+1}^{m} \lambda_i \mathbf{p}_i^2 \qquad (7\text{-}58)$$

The two quantities, γ_k and Q_r, are complementary. The former examines the retained and the latter examines the unretained principal component collectively; γ_k accounts for the amount of variance explained by the principal component model, while Q_r accounts for the amount of the variance unexplained. Tests based on these quantities can be conducted to examine whether a gross error is present in the retained or unretained principal components. For more information on collective principal component tests, see Tong and Crowe [15].

As previously stated, a major difference between the univariate tests and the multivariate chi-square tests is that the former does not take the correlation among the residuals into account and hence tends to be less reliable when correlation increases. However, the GLR (or MP measurement test) and MP constraint test, does incorporate the correlation by transforming the residuals using the inverse of the covariance matrix. This leads to a maximum power for correctly detecting a gross error over all other tests, but only when there is a single gross error.

When multiple gross errors are present, these tests no longer possess the maximum power. Tong and Crowe [15] indicated that the multivariate principal component tests not only provide better detection to subtle gross errors, but also have more power to correctly identify the variables in error over other tests. Again, this statement was not generally confirmed by an extensive comparison study [16].

Exercise 7-13. Collective Principal Component χ^2 Tests. *Similar to the global test (Equation 7-4) devise a test using the statistic defined by Equation 7-56, which includes the contribution of the k retained principal components of the balance residual vector* **r**.

Example 7-5

Let us apply the principal component tests based on measurement residuals to the process considered in the preceding examples. The nonzero eigenvalues of matrix **W** computed in Example 7-3 are all unity. The matrix $\mathbf{U}_a$ whose columns are the corresponding normalized eigenvectors is given by

$$\mathbf{U}_a = \begin{bmatrix} -0.7475 & 0.1152 & -0.0067 & 0.0287 \\ 0.4077 & -0.6881 & 0.0500 & -0.4232 \\ 0.3843 & -0.4449 & -0.5698 & 0.2554 \\ -0.0167 & 0.4956 & -0.6011 & -0.0597 \\ 0.0067 & 0.2523 & 0.0188 & -0.7382 \\ 0.3564 & 0.0773 & 0.5578 & 0.4542 \end{bmatrix}$$

Four principal components were retained in this example. The principal components are computed as [–0.5317, –2.1181, 0.2422, –3.0187]. At 5% level of significance, the tests for principal components 2 and 4 are rejected, while at the modified levels of significance only the test for the last principal component is rejected.

STATISTICAL TESTS FOR GENERAL STEADY-STATE MODELS

In the preceding sections, the different statistical tests for detecting gross errors were described for the simplest case when all the variables are measured directly. In general, unmeasured variables may be present, and the measurements may be indirectly related to the variables. Narasimhan and Mah [17] described simple transformations by which the general steady-state models can be converted to the above simple steady-state model.

Using these transformations all the statistical tests can be derived as described below.

If unmeasured variables exist, the constraint model is described by

$$\mathbf{A}_x\mathbf{x} + \mathbf{A}_u\mathbf{u} = \mathbf{c} \tag{7-59}$$

where $\mathbf{x}$: $n \times 1$ is the vector of measured variables and $\mathbf{u}$: $p \times 1$ is the vector of unmeasured variables and $\mathbf{A}_u$ is assumed to be of full column rank, p. As shown in Chapter 3, the unmeasured variables can be eliminated by pre-multiplying the constraints by a projection matrix $\mathbf{P}$: $(m\text{-}p) \times m$ of rank $m\text{-}p$, where m is the number of constraints, to give the reduced constraints:

$$\mathbf{PA}_x\mathbf{x} = \mathbf{Pc} \tag{7-60}$$

The constraint residuals for the reduced constraints can be defined exactly analogous to Equation 7-2:

$$\rho = \mathbf{P}(\mathbf{A}_x\mathbf{y} - \mathbf{c}) \tag{7-61}$$

One can show that the variance-covariance matrix of vector ρ is:

$$\mathbf{V}_\rho = \text{cov}(\rho) = \mathbf{PA}_x\Sigma(\mathbf{PA}_x)^T \tag{7-62}$$

> ***Exercise 7-14.*** *Using the rule of linear transformations in multivariate statistics, prove that the reduced constraint residuals follow a normal distribution with covariance matrix defined by Equations 7-62.*

The statistics of the global, constraint, and measurement tests can be obtained by using $\mathbf{PA}_x$, ρ, and $\mathbf{V}_\rho$, respectively, for $\mathbf{A}$, $\mathbf{r}$, and $\mathbf{V}$ in the appropriate equations. For deriving the GLR test statistics, we note that the gross error signature vectors for biases and leaks are also transformed due to the use of the projection matrix. These transformed signature vectors are given by

$$\mathbf{f}_{\rho k} = \mathbf{P}\mathbf{f}_k \tag{7-63}$$

where $\mathbf{f}_k$ are given by Equation 7-23. The GLR test statistics are now obtained using Equations 7-30 to 7-33 by substituting $\mathbf{f}_{\rho k}$, $\mathbf{V}_\rho$, and ρ, for $\mathbf{f}_k$, $\mathbf{V}$, and $\mathbf{r}$, respectively. It can be proved that the GLR test gives MP test statistics for detecting single gross errors even when unmeasured variables are present.

In some cases, the measurements may not be directly related to the variables as in Equation 3-1. An example of this was given in Chapter 2, where the pressure drop measurement is related to the square of the flow rate variable. Another example is the relationship between a *pH* measurement and the concentration of hydrogen ions and perhaps temperature of the process. These relationships are typically nonlinear, but for simplicity we represent them by the following linear equations.

$$\mathbf{y} = \mathbf{D}\mathbf{x} + \varepsilon \tag{7-64}$$

Let us assume that the constraints are given by

$$\mathbf{A}\mathbf{x} = \mathbf{c} \tag{7-65}$$

We define artificial variables $\mathbf{x}_a$ as

$$\mathbf{x}_a = \mathbf{D}\mathbf{x} \tag{7-66}$$

Then Equation 7-64 becomes

$$\mathbf{y} = \mathbf{x}_a + \varepsilon \tag{7-67}$$

Equations 7-65 and 7-66 can be jointly written as

$$\begin{bmatrix} \mathbf{0} \\ \mathbf{I} \end{bmatrix} \mathbf{x}_a + \begin{bmatrix} \mathbf{A} \\ -\mathbf{D} \end{bmatrix} \mathbf{x} = \begin{bmatrix} \mathbf{c} \\ \mathbf{0} \end{bmatrix} \tag{7-68}$$

Equations 7-67 and 7-68 represent an equivalent alternative model of the process in which the variables $\mathbf{x}$ are like “unmeasured variables” and variables $\mathbf{x}_a$ are like directly measured variables. Therefore, the method described for treating unmeasured variables can now be used to derive the statistics for all tests.

The technique described below can be applied even when the measurements are related to variables by nonlinear equations. However, the resulting modified constraint equations will be nonlinear and nonlinear data reconciliation and gross error detection techniques have to be used to solve this process.

> ***Exercise 7-15.*** *Derive MP measurement test statistics when unmeasured variables are present and all the measured variables are directly measured. Repeat the derivation for the case when some measured variables are indirectly measured.*

TECHNIQUES FOR SINGLE GROSS ERROR IDENTIFICATION

The second component of a gross error detection strategy deals with the problem of correctly identifying the type and location of a gross error which is detected by a test. It should be noted that the identification problem arises only if the detection test rejects the null hypothesis. Not all detection tests described in the preceding section are designed to distinguish between different gross error types. Only the GLR test is suitable for distinguishing between different types of gross errors, because it also uses information regarding the effect of each type of gross error on the process model.

In order to compare the different techniques developed in conjunction with the different tests for gross error identification and obtain a good understanding of the interrelationships, we initially restrict our consideration to gross errors caused by biases in measurements. In this section, we also consider only the problem of identifying a single gross error in the measurement. In this case, the identification problem reduces to simply identifying correctly the measurement which contains the gross error.

The techniques for identifying the measurement containing the gross error can be a simple rule or a complex strategy, depending on the test that is used. The measurement test and the GLR test, by virtue of the manner in which they derive the test statistics, use a simple rule to identify the gross error. It has already been pointed out in the preceding section that the GLR test and the MP measurement tests are identical if we restrict our consideration to gross errors caused by measurement biases only. In this case, there is a test statistic corresponding to each measurement. This identification rule used in these tests can be stated as follows:

Identify the gross error in the measurement that corresponds to the maximum test statistic exceeding the test criterion.

Because of the simplicity of the above rule, it is commonly stated that the measurement test or GLR test does not require a separate strategy for identifying gross errors. We have deliberately chosen to refer to the above rule as the identification component of these tests because we demonstrate later in this section that this rule is equivalent to the serial elimination strategy used in conjunction with the global test for identifying a gross error. It should also be noted that if we expect other types of gross errors to occur, such as leaks, then the GLR test which constructs a test statistic corresponding to each type and location of a gross error, simply extends the above rule to identify the gross error which corresponds to the maximum test statistic [11].

Serial Elimination Strategy for Identifying a Single Gross Error

If we use the global test for detecting the presence of gross errors, a comparatively more complex strategy has to be applied to identify the measurement containing the gross error. Ripps [1] first outlined a procedure which was later studied and refined by Serth and Heenan [18] and Rosenberg et al. [19]. This procedure is known as the *serial elimination procedure.*

In the serial elimination procedure, each measurement is deleted in turn and the global test statistic is recomputed. By eliminating a measurement, we make the corresponding variable unmeasured. Hence, the global test statistic has to be recomputed using the reduced constraint residuals as explained in the preceding section.

Due to the increase in the number of unmeasured variables, the objective function value, and thus the global test statistic, will decrease. Ripps [1] suggested that the gross error can be identified in that measurement whose deletion leads to the greatest reduction in the objective function value. Although, this strategy is called the serial elimination, it should strictly be called the measurement elimination.

Only in the context of multiple gross error detection described in the next chapter does the implication of serial elimination become clear. Instead of repeatedly solving the data reconciliation problem or computing the projection matrices for deleting each measurement in turn, Crowe [20] derived simplified expressions for the reduction in the data reconciliation objective function value due to the deletion of a measurement *i.* It is given by

$$\Delta J_i = J - J_i = \left(\mathbf{e}_i^T \mathbf{A}^T \mathbf{V}^{-1} \mathbf{r}\right)^2 / \mathbf{e}_i^T \mathbf{A}^T \mathbf{V}^{-1} \mathbf{e}_i \qquad (7\text{-}69)$$

It can readily be verified that the reduction in objective function value due to elimination of measurement *i* is equal to the the GLR test statistic (or square of the measurement test statistic) for variable *i*. This implies that if the rule used in conjunction with the global test is to identify the gross error in that measurement which gives maximum ΔJ_i, then this is precisely the same rule used in the GLR test (or MP measurement test) for identifying the gross error in the measurement corresponding to the maximum test statistic. In other words, the global test in combination with serial elimination strategy is equivalent to the GLR test.

> ***Exercise 7-16.*** *Show that the reduction in the global test statistic value obtained by eliminating a measurement is equal to the GLR test statistic value for that measurement.*

Another interesting and useful result is obtained by interpreting the principle involved in the GLR test from the viewpoint of data reconciliation objective function value. If we consider Equation 7-28, which defines the GLR test statistic, the two terms within parentheses on the RHS of this equation can be interpreted as the optimal objective value of data reconciliation problems. We have already noted that the first term in this expression is the optimal objective function value for the standard data reconciliation problem (see Exercise 7-1). The second term is the optimal objective function value of the following reconciliation problem in which the estimate of the gross error in measurement *i* is also obtained as part of the solution.

Problem P1:

$$\underset{x,b}{\text{Min}} \left(\mathbf{y} - \mathbf{x} - b\mathbf{e}_i\right)^T \Sigma^{-1} \left(\mathbf{y} - \mathbf{x} - b\mathbf{e}_i\right)$$

$$\text{subject to} \qquad \mathbf{A}\mathbf{x} = \mathbf{c}$$

Thus, the GLR test statistic can also be interpreted as the maximum difference between the optimal objective function values of data reconciliation under the assumption that there is no gross error, and the optimal objective function values obtained by solving Problem P1. It should be noted that in Problem P1, all the measurements are retained even if measurement i is assumed to contain a gross error; instead of eliminating measurement i, all measurements are used to obtain an estimate of the gross error. Based on this interpretation and the result given by Equation 7-69, it can be concluded that the optimal objective function values are equal, whether we choose to eliminate the measurement i (hypothesized to contain a gross error), or we choose to retain it and obtain an estimate of the gross error. This result is used in the following chapter to establish equivalence between different multiple gross error identification strategies.

Similar to the global test, the constraint test does not completely identify the type or location of the gross error. Although more informative, since it identifies the node (or equation) in gross error, the nodal test also requires additional identification. Mah et al. [21] developed an algorithm (for a mass flow network case) for identifying the measurements contributing to nodal imbalances. If no measurements are found in error, any significant node imbalance is attributed to a leak or a model error. A major problem with the nodal test is the possibility of error cancellations which makes it difficult to find the right location of the gross error. Such techniques are described in the next chapter.

> ***Exercise 7-17.*** *Prove that the optimal objective function value of Problem P1 is equal to the optimal objective function value obtained by solving the data reconciliation problem in which measurement* i *is eliminated and the corresponding variable is treated as unmeasured.*

Example 7-6

From the results of Examples 7-3 and 7-4, we observe that if we choose to identify a gross error in the measurement corresponding to the maximum test statistic, then the gross error in measurement 2 is correctly identified by both the MP measurement test and the GLR test. In fact, even if we consider gross errors due to leaks, the maximum GLR test statistic corresponds to a bias in measurement 2. The global test also rejects the null hypothesis and we can use the measurement elimination strategy to identify the location of the gross error. Table 7-2 shows the reduction in the global test statistic when different measurements are eliminated.

Table 7-2
Reduction in GT Statistic for Deletion of Different Measurement

Measurement Eliminated	Reduction in GT Statistic
1	1.571
2	10.27
3	0.244
4	4.002
5	2.884
6	6.040

Since the maximum reduction of the global test statistic is obtained when measurement 2 is eliminated, the measurement elimination procedure along with the GT identified the gross error correctly. This is not surprising because this procedure is similar to the use of the GLR test. It can be verified that the reduction in the GT statistic due to measurement elimination is identical to the GLR test statistics computed in Example 7-4.

Identifying a Single Gross Error by Principal Component Tests

We can identify the constraints in gross error by inspecting the contribution from the *j*th residual in $\mathbf{r}$, r_j, to a suspect principal component, say, $p_{r,i}$, which can be calculated by

$$g_j = \left(\mathbf{w}_{r,i}\right)_j r_j \qquad j = 1, \ldots, m \tag{7-70}$$

where $\mathbf{w}_{r,i}$ is the *i*th column vector of matrix $\mathbf{W}_r$.

Let us define $\mathbf{g} = (g_1, \ldots, g_m)^T$, and let $\mathbf{g}'$ be the same as $\mathbf{g}$ except that its elements are sorted in descending order of their absolute values. In general, the contributions of different residuals to the suspect principal component are different and are dominated by the first few elements. These are the major contributors to the suspect principal component. The major contributors are directly related to the constraints that should also be suspected. The number of major contributors, *k,* can be set so that

$$\left| \frac{\left(\sum_{j=1}^{k} g'_j\right) - p_{r,i}}{p_{r,i}} \right| < \varepsilon_1 \qquad (7\text{-}71)$$

where ε_1 is a prescribed tolerance such as 0.1.

Note that since the signs of these contributions can be either plus or minus, as can the signs of the elements of $\mathbf{w}_{r,i}$ and $\mathbf{r}$, the cancellation effect among the elements of $\mathbf{g}'$ should be taken into account in identifying the suspect constraints. This is done in Equation 7-71.

Similar to the nodal test, the principal component test on balance residuals only indicates which of the constraint residuals are major contributors to the suspect principal component. An additional strategy is required for identifying the source of the error (leak or measurement bias) and which of the measurements contains gross errors. We can, however, always use a principal component test on measurement adjustments in order to identify a measurement in gross error. That can be done by inspecting the contribution from the *j*th adjustment in $\mathbf{a}$, say, a_j, to a suspect principal component *i.*

The *j*th adjustment contribution can be calculated by

$$g_j = \left(\mathbf{w}_{a,i}\right)_j a_j \qquad j = 1,\ldots n \qquad (7\text{-}72)$$

where $\mathbf{w}_{a,i}$ is the *i*th eigenvector of $\mathbf{W}_a$ and *n* is the total number of measurements. We can study the contributions by checking the signs and magnitudes of the elements in $\mathbf{g}$. In general, as with the principal component test for balance residuals, the contributions vary and are dominated by a few elements. The identification rule for the principal component measurement test is the following:

Identify the gross error in the measurement that corresponds to the major contributor to the maximum principal component exceeding the test criterion.

Example 7-7

In order to identify the gross error using PC tests, we have to examine the contributions to the rejected principal component. Let us consider only the last principal component which is rejected at the modified level of significance (see Example 7-6). The contributors (measurement adjustments) to this principal component can be analyzed by computing the vector **g** (Equation 7-72). This vector is given by [0.0293, –1.1073, –0.1030, 0.0975, –1.0237, –0.9114]. The major contributor to the suspect principal component is measurement adjustment 2, and therefore a gross error is identified in this measurement.

Tong and Crowe [15] carried out extensive analysis of the principal component tests and outlined some practical guidelines for implementing a gross error detection and identification strategy for these tests. Most of their recommendations, such as making use of collective χ^2 tests first, using accurate variance-covariance matrix of measurement errors and the proper error distribution are valid for all strategies involving univariate tests.

In the end, they recommend that the PC tests should be used in combination with other statistical tests, since there is no guarantee that such tests will detect all gross errors. They also warn the user about the increased computational time in calculating the eigenvalues and eigenvectors, and also contribution analysis of the PC test statistics for gross error identification. The bottom line is that the PC tests are effective in certain situations, but they are not generally superior to the basic statistical tests described in this chapter.

DETECTABILITY AND IDENTIFIABILITY OF GROSS ERRORS

We close this chapter with the discussion of two important questions in gross error detection. The first is whether it is possible to detect gross errors in all measurements and whether gross errors in two or more measurements can be distinguished from each other. The concepts of detectability proposed by Madron [22], and identifiability discussed by different researchers [23, 24, 25], are used to answer these questions.

Detectability of Gross Errors

Similar to data reconciliation, an essential prerequisite for gross error detection is redundancy in measurements. Theoretically, it is possible to detect gross errors only in redundant measurements. *A gross error in a nonredundant measurement cannot be detected.* This is due to the fact that a nonredundant measurement is eliminated along with unmeasured variables and does not participate in the reduced reconciliation problem. Hence, no test statistic can be derived for a nonredundant measurement and a gross error in such a measurement cannot be detected.

In Chapter 3, methods were presented for observability and redundancy classification of process variables. Only redundant measurements are adjusted by reconciliation and observable unmeasured variables can be estimated. By adding sensors to measure new variables or by including additional constraints (if available), it is possible to eliminate unobservability and nonredundancy.

Both matrix and graphical approaches can be used for observability and redundancy classification. In practice, however, many redundant variables behave as nonredundant ones. We refer to such measurements as *practically nonredundant measurements.* Iordache et al. [22], Crowe [20], Madron [22], and Charpentier et al. [25] have reported difficulties in the reconciliation and gross error detection in such measurements. Similarly, even if some unmeasured variables are observable, their estimates may have such high standard deviations that they may be considered as *practically unobservable variables.*

If a measurement of a redundant variable contains a gross error, then data reconciliation should theoretically make a large adjustment to this measurement in order to obtain an estimate as close as possible to the true value of the variable. In some cases, however, due to the nature of the constraints and the standard deviations of variables, reconciliation may make an insignificant adjustment to the erroneous redundant measurement and instead make adjustments to other fault-free measurements to satisfy the constraints. Such a measurement is not truly redundant even if it is classified as redundant theoretically, and it is difficult to identify the gross error in such measurements.

Madron [22] defines a practically redundant measurement as one whose *adjustability* is greater than a selected threshold value. This condition is expressed analytically as

$$a_i = \left(1 - \sigma_{\hat{x}_i} / \sigma_{y_i}\right) > a_{cr} \tag{7-73}$$

where a_i is the adjustability, $\sigma_{\hat{x}_i}$ is the standard deviation of the reconciled value i, and σ_{y_i} is the standard deviation of the measurement error. The critical limit a_{cr} is a value from interval (0,1). For example, if a_{cr} is chosen as 0.1, all measurements i having $a_i < 0.1$ are considered practically nonredundant. For such measurements $\sigma_{\hat{x}_i}/\sigma_{y_i} > 0.9$, and, therefore, the adjustment made to the measured value is insignificant. The adjustability a_i is also a measure of the improvement in the accuracy of a measured value that can be achieved through data reconciliation.

Charpentier et al. [25] suggested using the ratio

$$d_i = \sqrt{\left(1 - \sigma_{\hat{x}_i}^2 / \sigma_{y_i}^2\right)} \tag{7-74}$$

for identifying the measurements with weak redundancy. This factor is a measure of *detectability* of an error. Since constraint imbalances indicate the existence of gross errors, the detectability of a gross error depends on its contribution to the imbalances of constraints. The contribution of a measurement in a constraint residual depends on the process constraint and relative accuracy of the measurements (relative standard deviations). The contribution of an error to the constraint imbalances is proportional to the detectability factor. The larger the detectability factor, the more likely is the gross error to be detected. This also implies that if the detectability factor d_i is large, then gross errors of small magnitudes in the corresponding measurements can be detected relatively easily.

A complete practical redundancy analysis is useful in identifying all measured variables with weak redundancy. For linear processes, the standard deviation of reconciled estimates can be computed analytically as described in Chapter 3 and adjustability or detectability measures can be computed. For nonlinear problems, however, these measures can be computed only after solving the reconciliation problem for a given set of measurements and by linearizing the constraints around the reconciled estimates. Equation 2-13 can be used to calculate the standard deviation of a reconciled value by a summation rule, as explained in Chapter 2.

Simulation studies have also been conducted and variables with the following characteristics have been identified as practically nonredundant:

- **Variables with relatively small standard deviations,** in comparison with the standard deviation of the other measurements belonging to the same balance. This is usually the case with the measurements whose order of magnitude is also small relative to the other variables in the same balance equation (for instance, flows of small streams that appear in balances with flows of large streams). The required ratio error/standard deviation for gross error detection is much larger for the variables with small standard deviation than for those with large standard deviation [23].
- **Parallel streams** (for instance, outlet flows from a splitter that are not constrained by any other balance [23]).
- **Flows that appear in the enthalpy balance, but not in mass balance** [25]. These are typically pumparound flows that are used in the main column enthalpy balance and associated heat exchanger balances, but they are not included in the tower mass balance. An overall mass balance using measured feed and product rates for the entire fractionation is usually chosen in order to avoid the large number of unmeasured flows around the column itself.
- **Temperatures of small streams** in the same balance with temperatures of large streams (even if the order of magnitude and standard deviation of such temperatures is similar).
- **Inlet temperature of the first heat exchanger** in the preheat train [25]. It usually appears in only one enthalpy balance, while the following temperatures enjoy extra redundancy by being part of at least two heat balances.
- **Measured variables that appear in only one equation with an unmeasured variable** which is not constrained by any other balance equation or bounded. A gross error in the measured variable is usually transferred to the unmeasured variable which has more freedom to adjust.

There is no simple solution for the data reconciliation and gross error detection in such variables. Extra constraints and extra instrumentation would certainly help, but that is not always possible. Sometimes artificial "measured" variables can be created from calculated values in order to enhance redundancy [25, 26]. The information about weak redundancy points can be provided to the users of a particular data reconciliation package in order to enable them to recognize the limitations in accuracy of the gross error detection methods and trigger decisions for improving their instrumentation.

Knowledge of practical variable classification is important information that can be included in the gross error detection algorithms. For instance, the detectability factor of a gross error can be used as a tie breaker when more than one measurement share the same value of the statistical test [16].

Example 7-8

For the process considered in the preceding examples, the covariance matrix of measurement errors is taken as the identity matrix. The covariance matrix for the estimates of all these variables can be computed using Equation 3-10. The diagonal elements of this matrix are the variances for the estimates. For this process, the variances for all the estimates turn out to be equal to 0.3333. Using these values in Equations 7-73 and 7-74, we obtain adjustability to be equal to 0.4226 and detectability to be 0.8165 for all variables. This implies that gross errors in all measurements have equal chance of being detected. On the other hand, if we take the true values of flow variables to be [100, 99, 1, 99, 1, 100] and assume the measurement error standard deviations to be 1% of the true values, then the adjustability and detectability values for different measurements are given in Table 7-3.

Table 7-3
Adjustability and Detectability Values for Process in Figure 1-2

Measurement	Error Variance	Adjustability	Detectability
1	1.0000	0.5025	0.8675
2	0.9801	0.4975	0.8646
3	0.0001	0.2929	0.7071
4	0.9801	0.4975	0.8646
5	0.0001	0.2929	0.7071
6	1.0000	0.5025	0.8675

From the results given in the above table, we can conclude that it is relatively more difficult to identify gross errors in the measurements of streams 3 and 5 compared to the others. Note that this can also be inferred from the first observation made in the discussion preceding this example about measurements with small standard deviations. In order to verify this observation, about 20 simulation trials were made in each of which a gross error of magnitude 5 to 15 times the standard deviation was simulated in flow measurement 1 and the GLR test was applied for identifying the gross error. Similarly, 20 trials were made with gross error in measurement 2, and so on for each position of the gross error. The results showed that while gross errors in streams 1, 2, and 6 were identified correctly in all trials, only 60% of the gross errors in stream 3 and 30% of the gross errors in stream 5 were identified correctly. Although the number of trials made are small, the trend of the results corroborates the observations.

Identifiability of Gross Errors

Even if a measurement has a high detectability, it is important to determine if a gross error in this measurement can be identified or distinguished from a gross error in any other measurement. For linear processes, this question can be answered in different ways, which we describe below.

Iordache et al. [23] pointed out that the test statistics of two different measurements are identical if the columns of matrix **A** corresponding to the two measured variables are proportional to each other. One special case of this occurs when two parallel streams link the same two nodes of a process. This implies that it is not possible to distinguish between gross errors that occur in these measurements. In the context of the GLR test, Narasimhan and Mah [13] indicated that if the signature vectors of two gross errors are proportional, then these cannot be distinguished from each other. If we restrict consideration to measurement biases only, then this observation is the same as the one made by Iordache et al. [23]. By using signature vectors, identifiability problems between different types of gross errors can be discovered.

Recently, Bagajewicz and Jiang [24] proposed the concept of *equivalent sets of gross errors.* A set of gross errors is equivalent to another set of gross errors if the two sets cannot be distinguished from each other. For the case of measurement biases, Bagajewicz and Jiang [24] proved that if a set of measurements of k variables forms a cycle of the process graph, then gross errors in any combination of $k–1$ measurements from this set cannot be distinguished from gross errors in any other combination. This can be easily verified, if we note that in serial elimination a measurement suspected to contain a gross error is eliminated, making the corresponding variable unmeasured.

Choosing to eliminate any set of *k–1* measurements from a cycle of *k* measurements will automatically make the remaining measurement non-redundant and will eliminate it from the reconciliation problem. This implies that the solution of the reduced reconciliation problem will be the same regardless of which combination of *k–1* measurements are eliminated. Thus, it is not possible to identify which set of *k–1* measurements from this cycle contains gross errors and all such sets are equivalent. For the same reason, it is not possible to distinguish gross errors in the measurements of all *k* variables of a cycle from any set of gross errors in the measurements of *k–1* variables of this cycle.

As a special case, if we consider a cycle formed by two streams (parallel streams), then is not possible to distinguish between a gross error in one stream from the other. It is also not possible to distinguish whether both the parallel stream measurements contain gross errors or if only one of them contains a gross error. Furthermore, if the number of independent constraints is equal to *m,* then all sets of *m* linearly independent gross errors are also equivalent. This is due to the fact that the reduced reconciliation problem will have no redundancy left and there is no information available to make any distinction between them.

We refer to equivalent sets of gross errors as belonging to an *equivalency class.* Equivalency classes can also be obtained in terms of the signature vectors of gross errors, which allow other types of gross errors such as leaks to be also considered. The following principle can be derived:

If the signature vectors for a set of k gross errors form a linearly dependent set of rank k−1, then it is not possible to theoretically distinguish between one combination of k−1 gross errors from any other combination of k−1 gross errors chosen from this set. It is also not possible to distinguish whether k gross errors or k−1 gross errors from this set are present in the process. (It is possible, however, to distinguish a combination of less than k−1 gross errors from other combinations.) As a special case, if the maximum number of independent signature vectors is m, then any set of m gross errors with linearly independent signature vectors is equivalent to any other such set.

Example 7-9

The process graph of the flow process considered in the preceding examples is shown in Figure 3-1. The following three cycles can be identified in this graph:

- Cycle 1 consisting of streams 2, 3, 4, and 5
- Cycle 2 consisting of streams 1, 2, 4, and 6
- Cycle 3 consisting of streams 1, 3, 5, and 6

Thus, the following equivalency classes of sets of biases are obtained:

Class 1. [2, 3, 4]; [2, 3, 5]; [2, 4, 5]; [3, 4, 5]; and [2, 3, 4, 5]
Class 2. [1, 2, 4]; [1, 2, 6]; [1, 4, 6]; [2, 4, 6]; and [1, 2, 4, 6]
Class 3. [1, 3, 5]; [1, 3, 6]; [1, 5, 6]; [3, 5, 6]; and [1, 3, 5, 6]

If we additionally consider, say, a process leak in node 1 (splitter), then we use the signature vectors to identify equivalent sets. The signature vectors for measurement biases are the columns of matrix **A.** The signature vector for a leak in node 1 is the first column of matrix **A,** which is identical to that for a bias in measurement 1. Thus, a leak in node 1 cannot be distinguished from a bias in measurement 1. In addition, we also obtain the same equivalent sets as obtained using cycles of a graph because the signature vectors for measurements biases in streams 2, 3, 4 and 5 are linearly dependent with rank 3 and so on.

It should also be kept in mind that if a set G of gross errors contains a subset of gross errors, say g_{Ci} belonging to an equivalency class C, then other equivalent sets of G can be obtained by replacing g_{Ci} with other sets which belong to C. Thus, for example, we can derive the equivalent sets for the combination [1, 2, 3, 4] by replacing the subset [2, 3, 4] by other sets of Class 1. Similarly, we can replace [1, 2, 4] by other sets of Class 2. Thus, we obtain another equivalency class given by

Class 4. [1, 2, 3, 4]; [1, 2, 3, 5]; [1, 2, 4, 5]; [1, 3, 4, 5]; [1, 2, 3, 6]; [1, 3, 4, 6]; [2, 3, 4, 6]; [1, 2, 5, 6]; [2, 3, 5, 6]; [1, 4, 5, 6]; [2, 4, 5, 6]; [3, 4, 5, 6]

The last 5 sets are added to Class 4, because they are equivalent to sets [1, 2, 3, 5]; [1, 2, 4, 5]; and [1, 3, 4, 5] which belong to Class 4. Equivalency Class 4 can also be generated using the fact that this process has 4 independent constraints and all sets of 4 gross errors with linearly independent signature vectors are equivalent.

Although identifiability problems can occur in linear processes, *in general, this is not a problem in nonlinear processes.* If nonlinear constraints are linearized around the reconciled estimates, it is highly unlikely that the columns of the linearized constraint matrix will become dependent. Even if this occurs, it has to be interpreted as a numerical problem rather than as an identifiability problem.

PROPOSED PROBLEMS

NOTE: The proposed problems that are included in this chapter require more extensive calculations. A computer program or a mathematical tool such as MATLAB is required in order to get solutions to these problems.

Problem 7-1. A mass flow network from Rosenberg et al. [19] is represented in Figure 7-1. The true mass flow rates (in lb/sec) in the stream order are given by the vector: [15 15 25 10 5 10 10 5 5 5 10 5 5 10 10 10]. All mass flow rates are considered measured. The standard deviation for each measurement is 2% of the measured value. Following the procedure explained at the end of Chapter 3, simulate random measured values and find the reconciliation solution. Next, simulate a single gross error (**bias** or **leak**) and

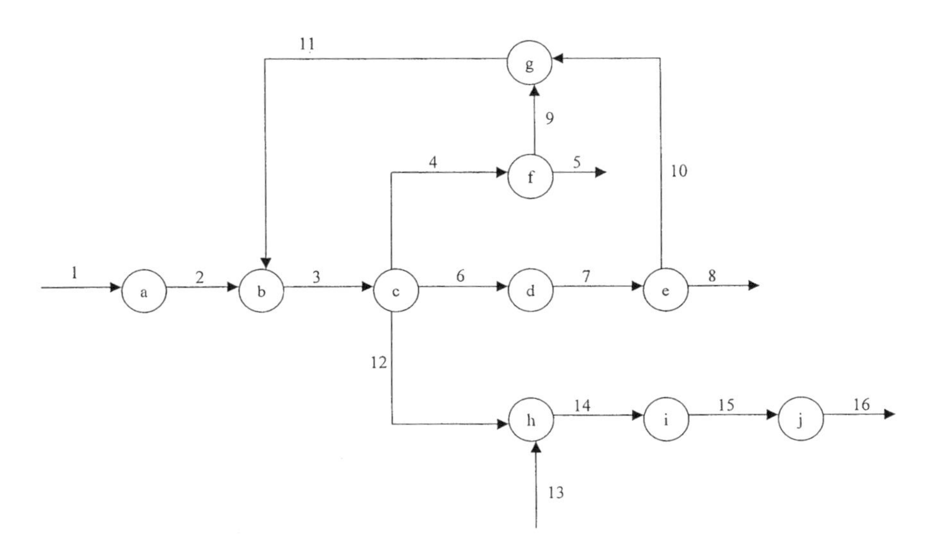

Figure 7-1. Mass flow network for Problem 7-1. *Reprinted with permission from [19]. Copyright ©1987 American Chemical Society.*

apply appropriate statistical tests (at α=0.05 level of significance) for gross error detection and identification. For a more extensive study, simulate gross errors of various magnitudes and in different locations. Also, calculate the detectability factors by Equation 7-74 and explain why some gross errors can be detected and correctly identified, while others cannot.

Problem 7-2. The steam-metering system for a methanol synthesis unit [18] is represented in Figure 7-2. The correct values of the steam flow rates are listed in Table 7-4. The values in Table 7-4 are 8-hour averages of the plant data except they have been adjusted to balance the system. All flow rates are considered measured. The standard deviation for each measurement is 2.5% of the measured value. Repeat the operations indicated in Problem 7-1, and explain the behavior of various statistical tests for gross errors simulated in streams with different detectability factors. Choose the level of significance α=0.05 for all statistical tests.

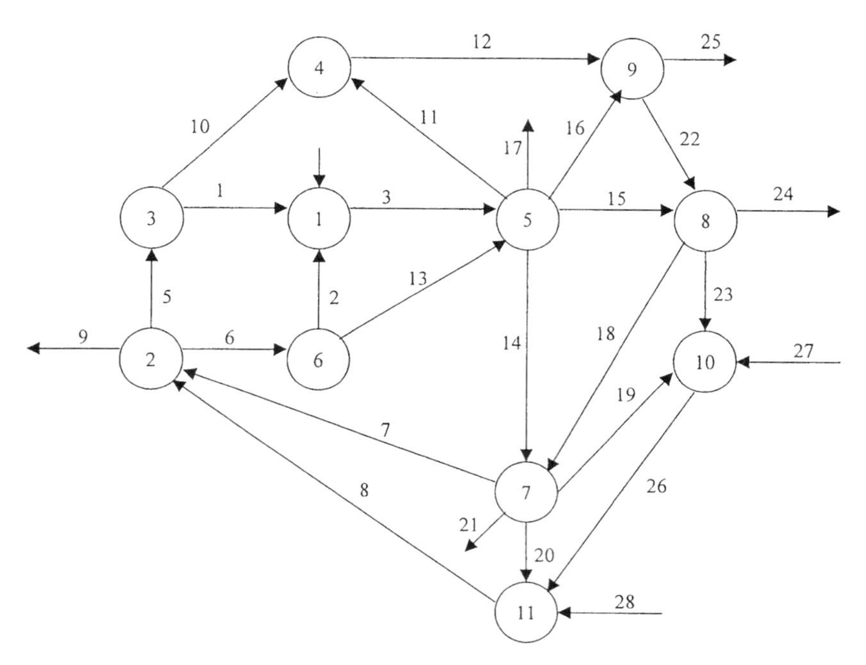

Figure 7-2. Steam metering system of a methanol synthesis plant [18]. *Reproduced with permission of the American Institute of Chemical Engineers. Copyright ©1986 AIChE. All rights reserved.*

Table 7-4
Correct Values of Flow Rates for the Steam System of Figure 7-2

Stream No.	Flow Rate 1,000 Kg/h	Stream No.	Flow Rate 1,000 Kg/h
1	0.86	15	60
2	1	16	23.64
3	111.82	17	32.73
4	109.95	18	16.23
5	53.27	19	7.95
6	112.27	20	10.5
7	2.32	21	87.27
8	164.05	22	5.45
9	0.86	23	2.59
10	52.41	24	46.64
11	14.86	25	85.45
12	67.27	26	81.32
13	111.27	27	70.77
14	91.86	28	72.23

Problem 7-3. A simplified diagram for an ammonia synthesis process from Crowe et al. [11] is represented in Figure 7-3. By using Crowe's projection matrix method, all unmeasured variables have been eliminated and a reduced model involving only measured variables is obtained. The constraint matrix for the reduced model is:

$$B = \begin{matrix} 0 & 0 & 0 & 1 & 0 & -1 & -0.5 & 0 \\ 1 & 0 & 0 & -1 & 0 & 0.98 & 0 & 0 \\ 0 & 1 & 0 & 0 & 0 & 0 & -1.5 & 0 \\ 0 & 0 & 1 & 0 & -0.02 & 0 & 0 & 0 \end{matrix}$$

and the measured values for the measured component flow rates are indicated in Table 7-5. A nondiagonal variance-covariance matrix for the measurement errors was used for this problem as follows:

$$\Sigma = \begin{matrix} 0.82 & 1.14 & 5.12E\text{-}3 & 0 & 0 & 0 & 0 & 0 \\ 1.14 & 6.34 & 0.0142 & 0 & 0 & 0 & 0 & 0 \\ 5.12E\text{-}3 & 0.0142 & 1.28E\text{-}4 & 0 & 0 & 0 & 0 & 0 \\ 0 & 0 & 0 & 8.16 & 0 & 0 & 0 & 0 \\ 0 & 0 & 0 & 0 & 0.326 & 0 & 0 & 0 \\ 0 & 0 & 0 & 0 & 0 & 3.81 & 0 & 0 \\ 0 & 0 & 0 & 0 & 0 & 0 & 3.08 & 0 \\ 0 & 0 & 0 & 0 & 0 & 0 & 0 & 32 \end{matrix}$$

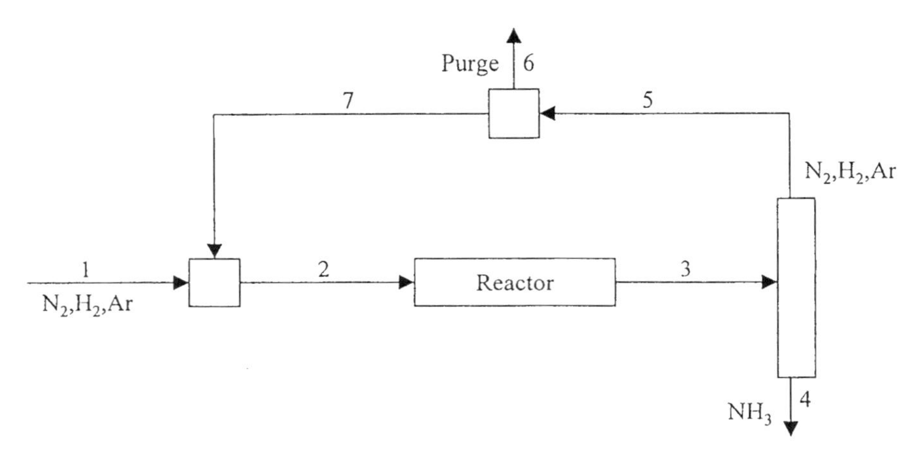

Figure 7-3. Flow diagram for a simplified ammonia plant [11]. *Reproduced with permission of the American Institute of Chemical Engineers. Copyright ©1983 AIChE. All rights reserved.*

At least one gross error exists in the given measured data. Apply various statistical tests at level α=0.05 to find the most likely location of the gross error.

Table 7-5
Measured Values for the Ammonia System in Figure 7-3

Species	Measured Flows (mol/s)
$N_2^{(1)}$	33
$H_2^{(1)}$	89
$Ar^{(1)}$	0.4
$N_2^{(2)}$	101
$Ar^{(2)}$	20.2
$N_2^{(3)}$	69
$NH_3^{(4)}$	62
$H_2^{(5)}$	205

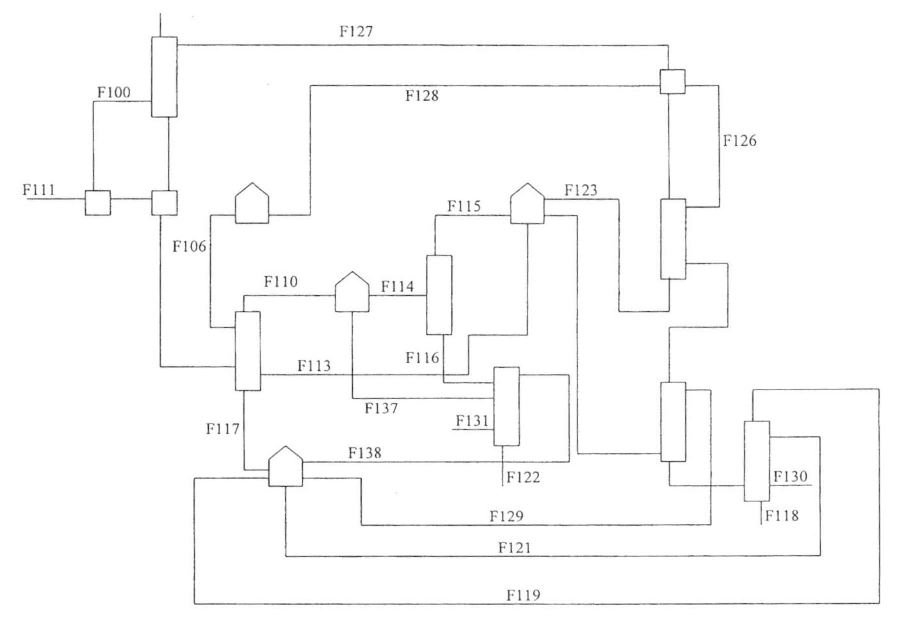

Figure 7-4. Flow diagram for a chemical extraction plant [27]. *Reproduced with permission of the Canadian Society for Chemical Engineering.*

Problem 7-4. A flowsheet for a chemical extraction plant from Holly et al. [27] is represented in Figure 7-4. The measured flow rates (in lb/hr), averaged over a 13-hour period, are given in Table 7-6. After eliminating the unmeasured flows by a projection matrix, the following reduced model is obtained:

$$F111 - F118 - F122 = 0$$
$$F114 - F115 - F116 = 0$$
$$F116 - F122 + F131 + F137 - F138 = 0$$
$$F110 - F114 - F137 = 0$$
$$F117 + F119 - F121 - F129 + F138 = 0$$
$$-F113 + F115 - F118 - F119 + F121 - F127 - F128 + F129 + F130 = 0$$

The variances for the measurement errors are also given in Table 7-6. Apply the global test and the measurement test (both at α=0.05), to identify the measurements suspected in gross error. Eliminate the suspected gross errors (one at a time) and recalculate the statistical tests. Which measurement is more likely to contain a gross error? Repeat the problem with the nondiagonal covariance matrix given by Holly et al. [27]. Explain any difference from the results with the diagonal variance-covariance matrix.

Table 7-6
Measured Values and Variances for the Chemical Extraction Process in Figure 7-4

Stream	Variance	Measured value (lb/hr)
F100	8.35E+05	18123
F110	1.07E+07	64557
F111	1.66E+07	79550
F113	4.95E+07	140130
F114	1.09E+07	65681
F115	6.3855	10.08
F116	6.77E+07	51411
F117	1.05E+08	180540
F118	4.70E+06	39573
F119	0.4303	2.624
F121	41096	3834.3
F122	8.27E+06	49312
F123	9.49E+06	60261
F126	23689	2962.7
F127	771.5	841.92
F128	6.547	10.235
F129	8.50E+07	181030
F130	7438	344.98
F131	8371	365.97
F137	4001	600.66
F138	8.27E+05	1762.4

SUMMARY

- Any gross error strategy needs to detect and also identify the location of gross errors.
- There are two types of errors associated with any statistical test: Type I error (when the test detects a nonexistent error) and Type II error (when the test fails to detect an existent error).
- Only the measurement test and the GLR test can directly identify the location of gross errors (by a simple identification rule).
- The GLR test is the only test which can identify both measurement biases and leaks by the same type of test.
- The gross error detection strategy by GLR test involves estimation of magnitudes of gross errors.
- Maximum power tests can be derived for the measurement test and for the nodal (constraint) tests. But the GLR test is more powerful than both of them for the single gross error case.
- The power of the GLR test is same as for the measurement test for a single measurement bias. Alternatively, the GLR test statistic is equivalent to the measurement test statistic for a single measurement bias.
- Principal component test is a linear combination of the eigenvectors of the variance-covariance matrix of constraining residuals or measurement adjustments.
- Principal component test cannot directly identify the location of gross error. It requires additional analysis in order to find the major constraints or measurements contributing to the principal components that failed the test.
- Serial elimination can be used to identify gross errors detected by the global test.
- The reduction in global test statistic after elimination of a measurement is equal to the GLR test statistic.
- The detectability of a gross error depends mainly on its magnitude and its location.
- Some gross errors can be detected, but not always properly identified.

REFERENCES

1. Ripps, D. L. "Adjustment of Experimental Data." *Chem. Eng. Progress Symp. Series* 61 (1965): 8–13 .
2. Almasy, G. A., and T. Sztano. "Checking and Correction of Measurements on the Basis of Linear System Model." *Problems of Control and Information Theory* 4 (1975): 57–69.
3. Madron, F. "A New Approach to the Identification of Gross Errors in Chemical Engineering Measurements." *Chem. Eng. Sci.* 40 (1985): 1855–1860.
4. Reilly, P. M., and R. E. Carpani. "Application of Statistical Theory of Adjustment to Material Balances," presented at the 13th Canadian Chem. Eng. Conference, Ottawa, Canada, 1963.
5. Mah, R.S.H., G. M. Stanley, and D. W. Downing. "Reconciliation and Rectification of Process Flow and Inventory Data." *Ind. & Eng. Chem. Proc. Des. Dev.* 15 (1976): 175–183.
6. Mah, R.S.H., and A. C. Tamhane. "Detection of Gross Errors in Process Data." *AIChE Journal* 28 (1982): 828–830.
7. Sidak, Z. "Rectangular Confidence Regions for the Means of Multivariate Normal Distribution." *Journal of Amer. Statis. Assoc.* 62 (1967): 626–633.
8. Rollins, D. K, and J. F. Davis. "Unbiased Estimation of Gross Errors in Process Measurements." *AIChE Journal* 38 (1992): 563–572.
9. Crowe, C. M. "Test of Maximum Power for Detection of Gross Errors in Process Constraints." *AIChE Journal* 35 (1989): 869–872.
10. Tamhane, A. C. "A Note on the Use of Residuals for Detecting an Outlier in Linear Regression." *Biometrika* 69 (1982): 488–499.
11. Crowe, C. M., Y.A.G. Campos, and A. Hrymak. "Reconciliation of Process Flow Rates by Matrix Projection. I: Linear Case." *AIChE Journal* 29 (1983): 881–888.
12. Tamhane, A. C., and R.S.H. Mah. "Data Reconciliation and Gross Error Detection in Chemical Process Networks." *Technometrics* 27 (1985): 409–422.
13. Narasimhan, S., and R.S.H. Mah. "Generalized Likelihood Ratio Method for Gross Error Identification." *AIChE Journal* 33 (1987): 1514–1521.
14. Narasimhan, S. "Maximum Power tests for Gross Error Detection Using Likelihood Rations." *AIChE Journal* 36 (1990): 1589–1591.
15. Tong, H., and C. M. Crowe. "Detection of Gross Errors in Data Reconciliation by Principal Component Analysis." *AIChE Journal* 41 (1995): 1712–1722.

16. Jordache, C., and B. Tilton. "Gross Error Detection by Serial Elimination: Principal Component Measurement Test versus Univariate Measurement Test," presented at the AIChE Spring National Meeting, Houston, Tex., March 1999.
17. Narasimhan, S., and R.S.H. Mah. "Treatment of General Steady State Models in Gross Error Detection." *Computers & Chem. Engng.* 13 (1989): 851–853.
18. Serth, R. W., and W. A. Heenan. "Gross Error Detecting and Data Reconciliation in Steam-Metering Systems." *AIChE Journal* 30 (1986): 243–247.
19. Rosenberg, J., R.S.H. Mah, and C. Iordache. "Evaluation of Schemes for Detecting and Identification of Gross Errors in Process Data." *Ind. & Eng. Chem. Proc. Des. Dev.* 26 (1987): 555–564.
20. Crowe, C. M. "Recursive Identification of Gross Errors in Linear Data Reconciliation." *AIChE Journal* 34 (1988): 541–550.
21. Mah, R.S.H., G. M. Stanley, and D. W. Downing. "Reconciliation and Rectification of Process Flow and Inventory Data." *Ind. & Eng. Chem. Proc. Des. Dev.* 15 (1976): 175–183.
22. Madron, F. *Process Plant Performance: Measurement and Data Processing for Optimization and Retrofits.* Chichester, West Sussex, England: Ellis Horwood Limited Co., 1992.
23. Iordache, C., R.S.H. Mah, and A. C. Tamhane. "Performance Studies of the Measurement Test for Detecting Gross Errors in Process Data." *AIChE Journal* 31 (1985): 1187–1201.
24. Bagajewicz, M. J., and Q. Jiang. "Gross Error Modeling and Detection in Plant Linear Dynamic Reconciliation." *Computers & Chem. Engng.* 22 (no. 12, 1998): 1789–1809.
25. Charpentier, V., L. J. Chang, G. M. Schwenzer, and M. C. Bardin. "An On-line Data Reconciliation System for Crude and Vacuum Units," presented at the NPRA Computer Conf., Houston, Tex., 1991.
26. Kneile, R. "Wring More Information out of Plant Data." *Chem. Engng* (Mar. 1995): 110–116.
27. Holly, W., R. Cook, and C. M. Crowe. "Reconciliation of Mass Flow Rate Measurements in a Chemical Extraction Plant." *Can. Jl. Of Chem. Eng.* 67 (Aug. 1989): 595–601.

8

Multiple Gross Error Identification Strategies for Steady-State Processes

In the preceding chapter, the different statistical tests for detecting the presence of gross errors and the methods for identifying a single gross error in the data were described. For a well-maintained plant, we should generally not expect more than one gross error to be present in the data. Therefore, a fundamental prerequisite of any gross error detection strategy is that it should have good ability to detect and identify correctly a single gross error.

If data reconciliation is applied to a large subsystem consisting of many measurements, however, or if the sensors are operating in a hostile environment, and/or plant maintenance procedures are inadequate, then it is possible for several gross errors to be simultaneously present in the data. Thus, there is a need to add a third component to the gross error detection strategy which provides the capability to detect and identify multiple gross errors.

Generally in the research literature, a gross error detection strategy is presented as a single entity without clearly distinguishing the different components of detection, identification of a single gross error, and identification of multiple gross errors. As mentioned in the preceding chapter, we have chosen to separately analyze the three different components used in a gross detection strategy to gain a better insight into the similarities and differences between the various methods proposed. Our main focus in this chapter is the component of the gross error detection strategy that deals with multiple error identification.

Multiple gross error identification strategies have been proposed by different researchers over the past four decades and it is not our aim to describe in detail every strategy in this chapter. In order to gain some perspective, we have attempted to classify these strategies into different classes depending on the core principle on which they are based. Within each of these categories we have chosen to describe one of the strategies in detail, depending on ease of description, and then indicate the different variants of this strategy proposed by other researchers.

All the techniques developed for multiple gross error identification can be broadly classified as either *simultaneous* strategies or *serial* strategies. They may also differ in the type of information exploited for identification. For example, some of the strategies also make use of information about lower and upper bounds on the variables to enhance the identification process. Lastly, as we have noted in the methods used for single gross error identification, not all strategies are designed to distinguish between different types of faults. For ease of comparison and description, we will restrict our considerations to the identification of gross errors caused by sensor biases, and indicate wherever pertinent the extension that can be made to include other types of gross errors. Linear systems are initially considered followed by the treatment of nonlinear processes.

STRATEGIES FOR MULTIPLE GROSS ERROR IDENTIFICATION IN LINEAR PROCESSES

Simultaneous Strategies

Identification Using Single Gross Error Test Statistics

Simultaneous strategies for multiple gross error identification attempt to identify all gross errors present in the data simultaneously or in a single iteration. In the case of the measurement test or GLR test, this strategy is a simple extension of the identification rule used for identifying a single gross error and can be stated as follows:

Identify gross errors in all measurements whose corresponding test statistics exceed the test criterion.

In the case of the GLR test, the above rule can be easily extended to identify other types of gross errors by making use of the corresponding test statistics. The effectiveness of this rule was investigated by Serth and Heenan [1] and was generally found to result in too many mispredictions. The main reason why the above rule does not work well is due to what has been referred to as the *smearing effect.*

Since the variables are all related through the constraints and the constraint residuals are used in deriving the test statistics, a gross error in one measurement may cause the test statistics of good measurements to exceed the test criterion but not the one which contains the gross error. In other cases, the test statistic of the measurement containing the gross error and those of other measurements can simultaneously exceed the test criterion. The degree of smearing depends on many factors, such as the level of redundancy, differences in standard deviations of measurement errors, and magnitude of the gross error [2].

Identification Using Combinatorial Hypotheses

Another simultaneous gross error identification strategy is based on explicitly postulating all the alternative hypotheses not only for a single gross error but also for the different combinations of two or more gross errors in the data. A test statistic can be derived for each of these alternatives and the most probable one chosen. This strategy is especially suited for application along with the GLR test. It can be recalled from the preceding chapter that the GLR test considers all possible single gross error hypotheses as part of the alternative hypothesis. We can extend this to include hypothesis of all possible combinations of two or more gross errors. In other words, the hypotheses can be formulated as follows:

H_0 (no gross error in the data):

$$E[\mathbf{r}] = 0 \tag{8-1a}$$

H_1 (composed of alternatives H_1^i, $H_1^{i1,i2}$, . . . $H_1^{i1,i2,...ik}$)

where

H_1^i (single gross error alternatives):

$$E[\mathbf{r}] = b\mathbf{f}_i \tag{8-1b}$$

$H_1^{i1,i2}$ (two gross error combination alternatives):

$$E[\mathbf{r}] = b_1\mathbf{f}_{i1} + b_2\mathbf{f}_{i2} \tag{8-1c}$$

⋮

$H_1^{i1,i2,...ik}$ (k gross error combination alternatives):

$$E[\mathbf{r}] = b_1\mathbf{f}_{i1} + b_2\mathbf{f}_{i2} + \dots + b_k\mathbf{f}_{ik} \tag{8-1d}$$

where the indices i_1, i_2 and so on are chosen to exhaustively consider all possible combinations of gross errors. Thus, if t_{max} is the maximum number of gross errors considered to be simultaneously present in the data, then $2^{t_{max}}$ alternatives are present in the composite alternative hypotheses. Corresponding to each of these alternatives, the GLR test statistic can be derived. In general, let us consider the alternative hypothesis of *k* gross errors in the data corresponding to the gross error vectors $\mathbf{f}_{i1}$, $\mathbf{f}_{i2}$, . . . $\mathbf{f}_{ik}$. The expected values of the constraint residuals under this alternative hypothesis can be written as

$$E[\mathbf{r}] = \mathbf{F}_k\mathbf{b} \tag{8-2}$$

where **b** is a column vector of unknown magnitudes of the gross errors and $\mathbf{F}_k$ is a matrix whose columns are the gross error vectors $\mathbf{f}_{i1}$, $\mathbf{f}_{i2}$, . . . $\mathbf{f}_{ik}$ corresponding to the *k* gross errors hypothesized (see Chapter 7). We can obtain the likelihood ratio for this hypothesis and following the same procedure as in deriving the test statistic T_k in Equation 7-30 we can obtain the test statistic for this hypothesis as

$$T^{i_1,i_2,\cdots i_k} = (\mathbf{F}_k^T\mathbf{V}^{-1}\mathbf{F}_k)^{-1}(\mathbf{F}_k^T\mathbf{V}^{-1}\mathbf{r})^2 \tag{8-3}$$

The maximum likelihood estimates of the corresponding gross error magnitudes are given by

$$\hat{\mathbf{b}} = (\mathbf{F}_k^{T'}\mathbf{V}^{-1}\mathbf{F}_k)^{-1}(\mathbf{F}_k^T\mathbf{V}^{-1}\mathbf{r}) \tag{8-4}$$

In order to determine the number and type of gross errors, however, we cannot apply the simple rule of choosing the maximum test statistic among all alternative hypotheses. This is due to the fact that all the test statistics do not follow the same distribution because of the differences in the number of degrees of freedom. The test statistics for *k* gross errors given by Equation 8-3 can be shown to follow a central chi-square distribution with *k* degrees of freedom under the null hypothesis. In order to choose the most probable hypothesis, we can compute the Type I error probabilities for each of the test statistics given by

$$\alpha^{i_1,i_2,\cdots i_k} = \Pr\left(\chi_k^2 \geq T^{i_1,i_2,\cdots i_k}\right) \qquad k = 1,2,\ldots t_{max} \tag{8-5}$$

where χ_k^2 is the random variable which follows a central chi-square distribution with *k* degrees of freedom. We can now choose the minimum among the Type I error probabilities given by Equation 8-5. If this exceeds the modified level of significance β for a chosen allowable probability of Type I error α, then we can conclude that gross errors are present and the hypothesis corresponding to the minimum Type I error probability gives the number and type of gross errors present. Furthermore, the estimates of the gross errors are given by Equation 8-4. It can be easily verified that this method is equivalent to the choice of the maximum likelihood ratio test statistic for the case of identifying a single gross error described in the preceding chapter.

> ***Exercise 8-1.*** *Derive the GLR test statistic (Equation 8-3) and maximum likelihood estimates (Equation 8-4) corresponding to the alternative hypothesis of k gross errors.*

It must be noted that in any case it is not possible to detect and identify more gross errors than the number of constraint equations, that is, t_{max} can at most be equal to *m*. This is because for every gross error hypothesized either the corresponding measurement is eliminated or an extra unknown parameter corresponding to the magnitude of the gross error has to be estimated, which reduces the degree of redundancy by one.

Furthermore, due to gross error identifiability problems as discussed in the preceding chapter, only one combination of gross errors belonging to each equivalency class needs to be considered in the alternative hypothesis. For example, in the case of two parallel streams, only one of the hypotheses for a gross error in one of these stream flow measurements has to be considered. Similarly, it is not possible to simultaneously identify gross errors in flow measurements of streams forming a cycle of the process. In other words, only those combinations of gross errors can be considered whose gross error signature vectors are linearly independent, so that the matrix $\mathbf{F}_k$ will be of full column rank and the inverse of the matrix $\mathbf{F}_k^T\mathbf{V}^{-1}\mathbf{F}_k$ in Equation 8-4 is guaranteed to exist.

It was proved in the preceding chapter that for a single gross error the GLR test is equivalent to the use of global test combined with measurement elimination strategy. This result is valid even when multiple measurements are simultaneously eliminated. In other words, the GLR test statistic given by Equation 8-3 for the hypothesis of gross errors being simultaneously present in *k* measurements is identical to the reduction in the global test statistic due to elimination of the k measurements suspected of containing gross errors. We describe below a strategy proposed by Rosenberg et al. [3] based on the global test which is essentially based on this principle.

The strategy proposed by Rosenberg et al. [3] makes use of the global test along with elimination of measurements. Corresponding to each alternative hypothesis of 8-1, the measurements which are suspected to contain gross errors are eliminated and the global test statistic is computed. Using these statistics, gross errors in measurements corresponding to the most probable hypothesis (the lowest Type I error probability) are identified. Instead of comparing all the alternative hypotheses simultaneously, however, Rosenberg et al. [3] considered in sequence single gross error alternatives, followed by two gross error alternatives and so on in increasing order of the number of gross errors hypothesized.

If at any stage in this sequence the global test statistic computed after eliminating the suspect measurements is found to be less than the critical values at the α level of significance drawn from the appropriate chi-square distribution, then the procedure is terminated. This approach therefore attempts to identify as few gross errors as necessary to accept the null hypothesis that no more gross errors are present in the remaining measurements. Rosenberg et al. [3] performed simulation studies to compare the performance of this strategy with other strategies. In particular, they found that this simultaneous strategy performs better than the simple simultaneous strategy based on measurement tests described in the preceding section. However, their comparison was limited to cases when only one gross error was present in the measurements.

The simultaneous strategy described above has not been used much due to the combinatorial increase in the number of alternative hypotheses to be tested for each of which a statistic has to be computed, leading to excessive computational burden. However, the speed and power of computers has been increasing rapidly, and it is worthwhile to study whether the simultaneous strategy described above gives better performance than serial strategies and has acceptable computational requirements.

Identification Using Simultaneous Estimation of Gross Error Magnitudes

The method due to Rosenberg et al. [3] implicitly assumes that a *minimal number of gross errors* are present in the system, because it considers hypotheses of 2 or more gross errors only if the hypothesis of one gross error is not statistically acceptable. This is also true of all serial strategies that we will discuss later in this chapter. Such methods generally do not perform well when many gross errors are present. An alternative strategy is to initially presume as many gross errors to be present in the data as can be identified and then to discard some of these possibilities based on additional criteria. Rollins and Davis [4] developed a simultaneous strategy called the *unbiased estimated technique* (UBET) which is based on this principle.

It has been pointed out earlier that the maximum number of gross errors that can be identified is equal to the number of process constraints, *m*. Moreover, the signature vectors for these gross errors must be linearly independent in order that their magnitudes can be uniquely estimated. In the UBET method, it is initially assumed that *m* gross errors (whose signature vectors are linearly independent) are present. Furthermore, it is assumed that the types and locations of these candidate gross errors are specified. (We will describe later methods that can be used to choose this initial candidate list of gross errors using the basic statistical tests).

Let **F** be an $n \times m$ matrix whose columns are the signature vectors of the assumed gross errors. The magnitudes of these gross errors can be simultaneously estimated using **F** instead of $\mathbf{F}_k$ in Equation 8-4. The estimated magnitudes of gross errors will be *unbiased* if all these gross errors are actually present; hence, the name UBET for this method.

In order to decide whether an assumed gross error is actually present, we can test the hypotheses that the magnitude of the assumed gross error is equal to zero or not. The estimated magnitudes of gross errors can be used for this purpose. If no gross errors are present, then it can be proved that $\hat{b}_1$ is normally distributed with mean zero and variance d_{ii}, which are the diagonal elements of matrix **D**, where $\mathbf{D} = \mathbf{F}^T\mathbf{V}^{-1}\mathbf{F}$. We can conclude that the magnitude of gross error *i* is non-zero if $|\hat{b}_i/d_{ii}|$ exceeds $Z_{1-\alpha/2}$, where α is a chosen level of significance; otherwise, we can conclude that gross error *i* is not present.

In order to select an initial candidate set of *m* gross errors, we can make use of the basic statistical tests described in Chapter 7. For example, one simple method is to compute the GLR test statistics for each gross error using Equation 7-30 and to pick the first *m* gross errors with the largest statistics, after taking into account equivalency considerations to ensure that the signature vectors of selected gross errors are linearly independent. Rollins and Davis [4] made use of the nodal test in order to select an initial candidate set of gross errors, but this requires nodal tests to be performed on combinations of nodes as well.

Jiang et al. [5] were the first to point out the need for taking into account equivalency of gross errors when choosing the candidate set and they appropriately modified the UBET strategy. They also made use of principal component tests instead of nodal tests to choose the candidate set of gross errors. Their simulation results showed that the overall performance of the gross error detection strategy did not significantly depend on whether the nodal or principal component tests were used to choose the initial candidate set of gross errors.

Since the UBET assumes that a maximum identifiable number of gross errors may be present, it can be expected to perform well when many gross errors are present and poorly when only a few gross errors are actually present. This was also confirmed through simulation studies [4].

Other simultaneous strategies were proposed by Jiang et al. [5], which have similarities with the method of Rosenberg et al. [3], and also by Sanchez et al. [6], who designed various strategies for simultaneous identification and estimation of measurement biases and leaks.

The above three simultaneous strategies are illustrated using the simple flow process example considered in Chapter 1.

Example 8-1

We consider the simple heat exchanger with bypass process shown in Figure 1-2 for the case when all the flow variables are measured. We assume that three gross errors of +5, +4 and –3 units in measurements of streams 1, 2, and 5, respectively, are present. The true, measured, and reconciled values of all flows for this case are shown in Table 8-1.

Table 8-1
Data for Heat Exchanger with Bypass Process Containing Three Gross Errors

Stream Number	True Flow Values	Measured Flow Values	Reconciled Flow Values
1	100	106.91	102.0533
2	64	68.45	67.1667
3	36	34.65	34.8867
4	64	64.20	67.1667
5	36	33.44	34.8867
6	100	98.88	102.0533

Let us compute the GLR test statistics for all allowable gross error combination hypotheses using Equation 8-3. The maximum number of gross errors that can be identified for this process is equal to 4, since there are only 4 constraints. Table 8-2 shows the GLR test statistics and the corresponding α levels for different combinations up to a maximum of 4 gross errors. Since streams 2, 3, 4, and 5 form a cycle, it is not possible to distinguish gross errors in all four of these measurements from a combination of gross errors in any three of these four stream measurements (see Example 7-9). For the same reason, it is also not possible to distinguish gross errors in any 3 of these streams from any other combination of 3 streams chosen out of these four streams. Thus, in Table 8-2 the test statistics for all these equivalent combinations are listed in the same row. Similarly, other equivalent combinations are listed together in the same row of Table 8-2.

Consider the GLR test statistics for a single gross error given in Table 8-2. The GLR test criterion at 5% level of significance is 3.84. If we apply the simple identification strategy based on the GLR test, then gross errors in measurements of streams 1, 4, and 6 are identified. Thus, Type I errors for measurements 4 and 6 are committed. Furthermore, the gross error in stream 2 is not identified. These identification errors are caused by the smearing effect of the gross errors.

We will now apply the simultaneous strategy of testing all possible hypotheses of one or more gross errors. From the results of Table 8-2, it is observed that a gross error in stream 1 has the highest test statistic among all single gross error hypotheses. Similarly, the combination [1, 2] has the largest test statistic among all 2 gross errors hypotheses, combination [1, 2, 5] among 3 gross error hypotheses and all four gross error hypotheses have the same test statistic (since the reduced problem does not have any redundancy for any of these hypotheses).

Among these different combinations, the hypothesis of gross errors in measurements 1 and 2 gives the least Type I error probability of 1.4126E-10. Hence, the simultaneous strategy based on testing all possible hypotheses identifies gross errors in measurements 1 and 2. Thus, this strategy identifies the gross errors in measurements 1 and 2 correctly, although it does not identify the gross error in stream 5 (in this example, it is tacitly assumed that the Type I error probabilities for different hypotheses are computed accurately, even though they are very small). The gross error in measurement 5 is not identified; therefore, a Type II error with respect to that measurement is committed.

Table 8-2
GLR Test Statistics for all Hypotheses of One or More Gross Errors

Measurement Combination	GLR Test Statistic	Type I Error Probability
[1]	35.381	2.7114E-09
[2]	2.470	0.1160
[3]	0.084	0.7719
[4]	13.202	2.7970E-04
[5]	3.139	0.0764
[6]	15.105	1.0169E-04
[1, 2]	45.361	1.4126E-10
[1, 3]	36.910	9.6644E-09
[1, 4]	40.295	1.7786E-09
[1, 5]	35.467	1.9878E-08
[1, 6]	36.491	1.1915E-08
[2, 3]	2.968	0.0227
[2, 4]	13.282	0.0013
[2, 5]	7.469	0.0239
[2, 6]	15.489	4.3307E-04
[3, 4]	13.61	0.0011
[3, 5]	4.982	0.0828
[3, 6]	16.803	2.2459E-04
[4, 5]	13.997	9.1329E-04
[4, 6]	37.725	6.4279E-09
[5, 6]	23.133	9.4773E-06
[1, 2, 3]	45.742	6.4362E-10
[1, 2, 4]; [1, 2, 6]; [1, 4, 6]; [2, 4, 6]; [1, 2, 4, 6]	45.522	7.1663E-10
[1, 2, 5]	46.254	5.0086E-10
[1, 3, 4]	43.234	2.1948E-09
[1, 3, 5]; [1, 3, 6]; [1, 5, 6]; [3, 5, 6]; [1, 3, 5, 6]	37.223	4.1278E-08
[1, 4, 5]	40.318	9.1230E-09
[2, 3, 4]; [2, 3, 5]; [2, 4, 5]; [3, 4, 5]; [2, 3, 4, 5]	14.014	0.00289
[2, 3, 6]	17.610	5.2933E-04
[2, 5, 6]	24.600	1.8722E-05
[3, 4, 6]	37.854	3.0349E-08
[4, 5, 6]	41.415	5.3382E-09
All combinations of 4 measurements out of 6 except [1, 2, 4, 6]; [1, 3, 5, 6] and [2, 3, 4, 5]	46.254	2.1804E-09

If we apply the technique of Rosenberg et al. [3], we would first compute the GT and check whether a gross error exists. For these process data, the GT statistic is 46.254 while the test criterion at 5% level of significance is 9.488. Since the GT is rejected, we compare all single gross error hypotheses and find that a gross error in measurement 1 has the least Type I error probability. If we eliminate this measurement and recompute the GT we find it is equal to 10.873 while the test criterion at 5% level of significance is 7.815.

We consider all two gross error hypotheses and find that the combination [1, 2] has the least Type I error probability. We therefore eliminate these two measurements and recompute the GT statistic to be 0.893. Since the test criterion is now 5.991, the GT is not rejected and we terminate the procedure. In this case also, two of the gross errors are identified correctly without committing a Type I error. As with the GLR test, however, a Type II error with respect to measurement 5 was committed.

In order to apply the UBET method, we have to first choose a candidate set of 4 gross errors. Using the GLR test statistics for single gross errors listed in Table 8-2, we choose biases in measurements 1, 6, 4, and 5 as candidates. Note that signature vectors for these gross errors are linearly independent. The estimates of these gross errors are obtained as [3.81, –4.25, –1.25, –4.22] and the variances of these estimates are computed as [3, 2, 2, 3]. Hence, the test statistics for the magnitudes of these gross errors are computed as [2.2, 3.0, 0.86, 2.44]. For $\alpha = 0.05$, the critical value is 1.96. Based on these test statistics we, therefore, conclude that the magnitudes of gross errors in measurements 1, 6 and 5 are nonzero. Thus, UBET commits a Type I error in 6 and a Type II error in 2.

It should also be noted that, although we have restricted our considerations to identification of measurement biases, we can use the above simultaneous strategies for identifying leaks if we combine these component strategies with statistical tests such as the GLR or nodal tests that have the capability to detect and identify leaks.

Serial Strategies

As opposed to simultaneous strategies, serial techniques identify gross errors serially, one by one. Many serial strategies in combination with different statistical tests have been proposed and their performances studied [1, 3, 7]. They differ by the type of statistical tests that are used, the manner in which the gross errors are identified, the tie-breaking rules used when the criteria for gross error identification are identical for different gross errors, and the criteria used in the algorithm for terminating the serial strategy.

Furthermore, some serial strategies also make use of information regarding upper and lower bounds on variables. A better understanding of these techniques can be obtained by focusing on the core principle used in the strategies ignoring the processing details and the statistical test used for gross error detection. We first describe the two main principles exploited in serial strategies before describing the different algorithms developed using these principles.

Principle of Serial Elimination

The principle of serial elimination first suggested by Ripps [8] has been described in Chapter 7 for identifying a single gross error. This principle is useful in identifying gross errors caused by measurement biases only, because it relies on eliminating measurements suspected of containing a bias. This basic principle can as well be utilized for multiple gross error identification by identifying gross errors serially. At each stage of the serial procedure, a gross error is identified in one measurement (based on some criteria) and the corresponding measurement is eliminated before proceeding to the next stage. The major advantage of serial elimination is that it does not require any prior knowledge about the existence or location of gross errors.

Principle of Serial Compensation

The principle of serial compensation was first suggested by Narasimhan and Mah [9] in conjunction with the use of the GLR test. Unlike the serial elimination technique, this principle can be used to identify other types of gross errors, besides measurement biases. At each stage of the serial procedure, a gross error is identified (based on some criteria) and the measurements or model are compensated using the identified type and location of the gross error and the estimated magnitude of the gross error, before proceeding to the next stage.

Equivalence Between Different Strategies

The criteria used for identifying a single gross error in each stage of the serial procedure can be based on one of the statistical tests. Historically, the serial elimination principle has been used in conjunction with the global test or measurement test, while the serial compensation principle has been used with the GLR test. However, it was established in the preceding chapter that the measurement test and GLR test are identical tests for identifying measurement biases. Thus the use of the serial elimination principle in conjunction with the GLR test is the same as using it in conjunction with the measurement test.

Furthermore, it was also proved in Chapter 7 that if the global test is used in conjunction with serial elimination strategy, then this is precisely the same as using the GLR test. The former identifies a gross error in the measurement whose elimination gives the maximum reduction in the global test statistic, while the latter identifies a gross error in the measurement corresponding to the maximum test statistic. Thus, identical results are obtained by using the serial elimination strategy in conjunction with the global test, measurement test or GLR test provided they use the same principle for single gross error identification at each stage of the serial procedure.

The serial elimination and serial compensation principles were historically derived from different viewpoints. However, it is proved subsequently that if a modified serial compensation strategy as developed by Keller et al. [10] is used, then this is exactly identical to the serial elimination strategy. This result unifies all the different approaches. The reader is urged to keep these results in mind when going through the description of the some of the most efficient strategies described below.

Serial Elimination Strategies That Do Not Use Bounds

Among the serial elimination strategies which do not use bound information, the version proposed by Serth and Heenan [1] known as the *iterative measurement test* (IMT) provides the basic structure. This strategy makes use of the measurement test for gross error detection, and the rule for identifying a single gross error in the measurement corresponding to the maximum test statistic at each stage of the iterative procedure. For multiple gross error detection, it uses the serial elimination strategy. The algorithm terminates when the maximum of the test statistics for all remaining measurements does not exceed the test criterion. The details of the algorithm are as follows:

Algorithm 1. Iterative Measurement Test (IMT) Method [1]

For ease of understanding the algorithm, the following sets are defined: Let S be the set of all original measured variables. Let C be the set of measurements which are identified as containing gross errors (initially set C is empty). At each stage of the iterative procedure, the measurements in set C are eliminated and the variables corresponding to these measurements are treated as unmeasured and projected out of the reconciliation problem as described in Chapter 3. Let T be the set of measured variables in the reduced reconciliation problem after projection of all unmeasured variables. Initially, T is the set of measurements which occur in the reduced reconciliation problem after projecting out all unmeasured variables which are present.

Step 1. Solve the initial reconciliation problem. Compute the vectors $\hat{\mathbf{x}}$ (Equation 3-8), **a** (Equation 7-11), and **d** (Equation 7-15).

Step 2. Compute the measurement test statistics $z_{d,j}$ (Equation 7-17) for all measurements in set T.

Step 3. Find z_{max} the maximum absolute value among all $z_{d,j}$ from Step 2 and compare it with the test criterion $Z_c = Z_{1-\beta/2}$. If $z_{max} \leq Z_c$ proceed to Step 5. Otherwise, select the measurement corresponding to z_{max} and add it to set C. If two or more measurement test statistics are equal to z_{max}, select the measurement with the lowest index *j* to add to set C.

Step 4. Remove the measurements contained in C from set S. Solve the data reconciliation problem treating the variables corresponding to set C also as unmeasured. Obtain T, the set of measurements in the reduced data reconciliation problem, and the vectors **a** and **d** corresponding to these measurements. (Note that Serth and Heenan [1] designed this elimination scheme for a mass flow balance problem; therefore, in their algorithm, removing measurements is equivalent to removing streams from the network by nodal aggregation. In general, this can be achieved by a projection matrix, as described in Chapter 3.) Return to Step 2.

Step 5. The measurements y_j, $j \in C$ are suspected of containing gross errors. The reconciled estimates after removal of these measurements are those obtained in Step 4 of the last iteration.

Note that the test criterion $Z_{1-\beta/2}$ should strictly be recalculated at each Step 3, since β depends on the number of measured variables in the model (Equation 7-7, where *m* is replaced by the number of measurements in the model). The more measured variables are eliminated, the fewer the number of simultaneous multiple tests, therefore the lower the probability of Type I error overall. It is possible, however, to simply use a global test at each stage to first detect whether any additional gross errors are present and use the measurement test statistics for identifying the gross error location. The level of significance at each stage k can be maintained at α and the critical value can be chosen from the chi-square distribution with degrees of freedom $m-k+1$ where $k-1$ is the number of gross errors identified so far. This ensures the Type I error probability to be maintained at α.

The measurement test used in Steps 2 and 3 can be replaced by a different statistical test. Principal component measurement test [11] can be used instead, but that requires more computational time. The principal component analysis should be used only if that provides superior power and lower Type I error. But that is not easily achieved without additional identification [5, 12, 13].

Example 8-2

The serial elimination strategy is illustrated using the same process as in Example 8-1. We will use the global test at each stage for detecting whether gross errors are present, and if so use the GLR test statistics (which is equivalent to the measurement test) for identifying the gross error. The level of significance for the global test is chosen as 0.95 for all stages of the serial strategy.

For the same measured data as in Example 8-1, the global test statistics for each stage of the serial elimination procedure and the corresponding test criteria are listed in Table 8-3. It is observed that the global test rejects the null hypothesis in stages 1 and 2, but not in stage 3. Thus, two gross errors are identified by this algorithm. In the first stage, since the maximum GLR test statistic is attained for measurement 1 (refer to Table 8-2), a gross error in this measurement is identified. In the second stage after eliminating measurement 1, the GLR test statistic is maximum for stream 2 among all remaining measurements. This can be verified by comparing the test statistics for all combinations of two gross errors which contain stream 1. Thus, two of the three gross errors are correctly identified without a Type I error being committed.

Table 8-3
Gross Error Identification Using Serial Elimination Strategy

Stage k	Global Test Statistic	Degrees of Freedom $\nu = m-k+1$	Test Criterion $\chi^2_{\nu,0.05}$	Gross Error Identified in Measurement
1	46.254	4	9.488	1
2	10.873	3	7.815	2
3	0.893	2	5.991	-

Serial Compensation Strategies That Do Not Use Bounds

The serial compensation strategy was developed for use with the GLR test [9] for multiple gross error identification. It exploits the capability of the GLR test to detect different types of gross errors and also makes use of the estimates of the gross errors obtained as part of this method. Again, in this method gross errors are detected serially one at each stage. The component method used in this strategy to identify a gross error at each stage is the simple rule of identifying the gross error corresponding to the maximum GLR test statistic.

Since the gross error could either be associated with a measurement or the model (in case of leaks), elimination of measurements is not appropriate. Instead, the estimated magnitude is used to compensate the corresponding measurement or model constraints. The GLR test is applied to the compensated constraint residuals to detect and identify any other gross error present. The procedure stops when the maximum of the test statistics among all remaining gross error possibilities does not exceed the test criterion. We refer to this algorithm as the *simple serial compensation strategy* (SSCS) since a modified version is described later. The details of the algorithm are as follows:

Algorithm 2. Simple Serial Compensation Strategy (SSCS) [9]

Step 1. Compute the constraint residuals $\mathbf{r}$ and covariance matrix of constraint residuals $\mathbf{V}$ using Equations 7-2 and 7-3 if no unmeasured variables are present. (Otherwise, the method for treating unmeasured variables described in Chapter 7 can be used to obtain the projected constraint residuals and its covariance matrix.)

Step 2. Compute the GLR test statistics T_k (Equation 7-30) for all gross errors.

Step 3. Find T, the maximum value among all T_k from Step 2, and compare it with the test criterion $\chi^2_{1-\beta,1}$. If $T \le \chi^2_{1-\beta,1}$ proceed to Step 5. Otherwise, identify the gross error corresponding to T. If two or more gross error test statistics are equal to T, arbitrarily select one. Let f^* be the gross error vector corresponding to T and b^* be the estimated magnitude (Equation 7-29).

Step 4. If the gross error identified in Step 3 is a bias in measurement *j,* then compensate the measurements using the estimated magnitude of the bias. The compensated measurements are given by

$$\mathbf{y}_c = \mathbf{y} - b^* \mathbf{e}_j \tag{8-6}$$

On the other hand, if the gross error identified is associated with the model constraints, for example leak *i* corresponding to leak vector f^*, then the constraint model is compensated using the estimated magnitude. The compensated model is given by

$$\mathbf{Ax} = \mathbf{c} - b^* \mathbf{f}^* \tag{8-7}$$

In either case, the compensated constraint residuals are given by

$$\mathbf{r}_c = \mathbf{r} - b^* \mathbf{f}^* \tag{8-8}$$

Return to Step 2 replacing the constraint residuals with the compensated constraint residuals.

Step 5. Compute the reconciled estimates for **x** using the compensated measurements and compensated model. Equivalently, the estimates can be obtained using the compensated constraint residuals instead of the original constraint residuals in Equation 3-8.

The principle of compensating the measurements or model at each stage of the above strategy implicitly assumes that the gross errors identified in the preceding stages and their estimated magnitudes are correct. In order to understand this clearly, it is instructive to consider the hypotheses that is tested by this algorithm at each stage of the serial procedure.

Let us assume that at the beginning of stage *k+1,* we have already identified *k* gross errors corresponding to gross error vectors $\mathbf{f}^*_1, \mathbf{f}^*_2, \ldots, \mathbf{f}^*_k$

and their estimated magnitudes are given by $b^*_1, b^*_2, \ldots, b^*_k$. The hypothesis for stage $k+1$ under the assumption that all gross errors identified in the previous stages as well as their estimated magnitudes are correct can be stated as

H_0^{k+1} (only gross errors identified in previous stages are present):

$$E[\mathbf{r}] = \sum_{j=1}^{k} b^*_j \mathbf{f}^*_j \quad \text{or} \quad E[\mathbf{r}_c] = 0 \tag{8-9a}$$

H_1^{k+1} (one additional gross error is present):

$$E[\mathbf{r}] = \sum_{j=1}^{k} b^*_j \mathbf{f}_j + b\mathbf{f}_i \quad \text{or} \quad E[\mathbf{r}_c] = b\mathbf{f}_i \tag{8-9b}$$

where $\mathbf{f}_i$ in the alternative hypothesis can be any one of the gross error vectors corresponding to remaining gross error possibilities not identified in the preceding k stages. It can be noted that in terms of compensated constraints residuals, the hypotheses formulation is similar to hypotheses for detecting and identifying a single gross error and thus the GLR test statistics for the remaining gross error possibilities can be computed by using the compensated residuals in Equations 7-30 to 7-32.

If multiple gross errors are present in the data, the serial procedure of trying to identify them results in mispredicting the type or location of the gross error. Even if the gross error is correctly identified, the estimate of its magnitude may be grossly incorrect. Thus, the compensated residuals can contain spurious large errors not present in the original residuals, which can impair the accuracy of identification of remaining gross errors. The serial compensation strategy may thus give rise to a large number of mispredictions, especially when many gross errors are present and when the magnitudes of gross errors are large in comparison with the standard deviations of the random errors in measurements. This was demonstrated through simulation studies by Rollins and Davis [4].

On the other hand, if only a few gross errors are present in the data, then the SSCS strategy performs as well as the IMT as shown through simulation studies [9] and typically requires about one-fifth to one-half the computing time required by the serial elimination methods. This is due to the fact that that serial elimination requires recomputation of the test statistics at each elimination step, which in turn involves recalculation of all matrices involved in order to get a new solution. Construction of a projection matrix is required after each deletion of a measurement from the network. Only for diagonal Σ and mass flow balance equations certain efficient elimination procedures can be developed [14].

Algorithm 3. Modified Serial Compensation Strategy (MSCS) [10]

In order to obviate the problem of too many mispredictions by the SSCS strategy due to incorrect compensation, a modified procedure was proposed by Keller et al. [10]. In their modified strategy, only the types and locations of gross errors identified in the previous iterations are assumed to be correct and the estimates of the gross errors are not used in compensation. The modified strategy still uses the GLR test for detecting gross errors and the rule for identifying a gross error corresponding to the maximum test statistic. However, the hypotheses that are tested at each stage of the serial procedure are different from 8-9a and 8-9b and can be stated as follows using the same notation as in 8-9a and 8-9b:

H_0^{k+1} (only gross errors identified in previous stages are present):

$$E[\mathbf{r}] = \sum_{j=1}^{k} b_j \mathbf{f}_j^* \tag{8-10a}$$

H_1^{k+1} (one additional gross error is present):

$$E[\mathbf{r}] = \sum_{j=1}^{k} b_j \mathbf{f}_j^* + b\mathbf{f}_i \tag{8-10b}$$

It should be noted that in the null and alternative hypotheses, the magnitudes of the gross errors are assumed to be unknown and their maximum likelihood estimates are computed as part of the computation of the GLR test statistics. However, it can be observed that in the hypotheses formulation the locations of the gross errors identified in the preceding *k* stages are assumed to be correct. The GLR test statistic at stage $k+1$ for a gross error corresponding to vector $\mathbf{f}_i$ (not identified in the preceding stages) is given by

$$T_i^{k+1} = \underset{\mathbf{b}_k}{\text{Max}}\left\{(\mathbf{r}-\mathbf{F}_k\mathbf{b}_k)^T\mathbf{V}^{-1}(\mathbf{r}-\mathbf{F}_k\mathbf{b}_k)\right\} -$$

$$\underset{\mathbf{b}_k, b}{\text{Max}}\left\{(\mathbf{r}-\mathbf{F}_k\mathbf{b}_k - b\mathbf{f}_i)^T\mathbf{V}^{-1}(\mathbf{r}-\mathbf{F}_k\mathbf{b}_k - b\mathbf{f}_i)\right\} \tag{8-11}$$

Keller et al. [10] demonstrated through simulation that the MSCS strategy commits less mispredictions as compared to SSCS, especially when a large number of gross errors are present.

Although the MSCS strategy was devised as a modification of the SSCS strategy, a look at hypotheses 8-10a and 8-10b shows that in reality no compensation is being applied, and the original constraint residuals are utilized for gross error detection in all stages. If we restrict our consideration to gross errors caused by sensor biases, then this strategy is in fact exactly equivalent to the serial elimination strategy. The proof of this follows from the interpretation of the GLR test statistic as the difference between the optimal objective function values of two data reconciliation problems (see Chapter 7).

The GLR test statistics for stage *k+1* of the MSCS strategy, given by Equation 8-11, can also be interpreted in a similar manner. The first term in the RHS of 8-11 is the optimal objective function of the data reconciliation problem in which the magnitudes of the gross errors identified in the preceding *k* stages are simultaneously estimated as part of the data reconciliation problem. The formulation of this problem is an extension of Problem P1 of Chapter 7.

Similarly, the second term in the RHS of 8-11 is equal to the optimal objective function of the data reconciliation problem in which the magnitudes of gross errors identified in preceding *k* stages and the magnitude of the gross error hypothesized in stage *k+1* are simultaneously estimated. In addition, it was also pointed out in Chapter 7 that the optimal objective function value is the same whether we choose to retain the measurement containing a gross error and estimate the magnitude of the gross error, or we choose to eliminate the measurement containing a gross error and treat it as an unmeasured variable in the data reconciliation problem.

This result is also true when there are several measurements containing gross errors. In other words, we can also interpret the RHS of 8-11 as the difference in the optimal objective function values of two data reconciliation problems, one in which the measurements identified as containing gross errors in the preceding *k* stages are eliminated, and the other in which an additional measurement in which a gross error is hypothesized at stage $k+1$ is also eliminated. But this is precisely the strategy used in IMT. Thus, the MSCS and IMT are identical strategies for identifying gross errors in measurements. However, the MSCS has the additional capability of identifying other types of gross errors. We can therefore regard MSCS as a *serial elimination* strategy that can handle different types of gross errors.

Example 8-3

The serial compensation strategy is applied for gross error detection on the same measured data for the flow process considered in the preceding examples. Again, the global test is used for detecting whether gross errors are present and the GLR test statistics are used for identifying the gross error. The main difference between this and the preceding example is that the residuals at each stage are compensated using the magnitude of the estimated bias and Equation 8-8. The results are presented in Table 8-4, which show that two gross errors are detected in measurements 1 and 2.

Table 8-4
Gross Error Identification Using Modified Serial Compensation Strategy

Stage k	Global Test Statistic	Degrees of Freedom $\nu = m-k+1$	Test Criterion $\chi^2_{\nu,0.05}$	Gross Error Identified in Measurement	Estimated Magnitude of Gross Error
1	46.254	4	9.488	1	7.285
2	10.870	3	7.815	2	3.748
3	1.51	2	5.991	-	-

Serial Strategies That Use Bounds

The algorithms described in the preceding section do not ensure the feasibility of the reconciled solution. Some of the flow rates in the final solution may be negative or may have large unreasonable values. Furthermore, if reliable upper and lower bounds of both measured and unmeasured variables are known but are not imposed, the reconciliation can provide a solution that violates certain bounds (infeasible solution). This very fact may be indicative of other undetected gross errors in the data or that the identified gross errors may be incorrect.

The information about upper and lower bounds on variables can be exploited in the gross error detection strategy. Serth and Heenan [1] proposed a heuristic method of utilizing bound information in gross error detection by modifying the IMT method. The *modified IMT method* (MIMT) still uses the measurement test for detection and the serial elimination strategy for multiple gross error detection. However, the component strategy used for identifying a gross error at each stage is not a simple rule of choosing the measurement with the maximum test statistic. Instead, the MIMT method identifies gross errors in measurements only if their deletion from the original set S gives a data reconciliation solution that satisfies the bounds for all measured variables.

Algorithm 4. Modified Iterative Measurement Test (MIMT) Method [1]

Step 1. Solve the initial reconciliation problem. Compute the vectors **x, a,** and **d**.

Step 2. Compute the measurement test statistics $z_{d,j}$ for all measurements in set T.

Step 3. Find z_{max} the maximum absolute value among all $z_{d,j}$ from Step 2 and compare it with the test criterion $Z_c = Z_{1-\beta/2}$. If $z_{max} \leq Z_c$, proceed to Step 6. Otherwise, select the measurement corresponding to z_{max} and *temporarily* add it to set C. If two or more measurement test statistics are equal to z_{max}, select the measurement with the lowest index *j* to add to set C.

Step 4. Remove the measurements contained in C from set S. Solve the data reconciliation problem treating the variables corresponding to set C also as unmeasured. Obtain T, the set of measurements in the reduced data reconciliation problem, and the vectors **a** and **d** corresponding to these measurements.

Step 5. Determine if the reconciled values for all variables in set T and set C are within their prescribed lower and upper bounds. If all reconciled values are within the bounds, store the current solution and return to Step 2. Otherwise, delete the last entry in C, replace it by the measured variable corresponding to the next largest value of $|z_{d,j}| > Z_c$, and return to Step 4. If $|z_{d,j}| \leq Z_c$ for all remaining variables, delete the last entry in set C and go to Step 6.

Step 6. The measurements y_j, $j \in C$ are suspected of containing gross errors. The reconciled estimates after removal of these measurements are those obtained in Step 4 of the last iteration.

> ***Exercise 8-2.*** *Rewrite Algorithm 4 for a problem with both measured and unmeasured variables solved by QR decomposition as described in Chapter 3.*

There are several limitations in the manner in which bounds are treated in the MIMT algorithm as listed below:

(1) The algorithm only checks for bound violations in the measured variables (more specifically only in the measurements of sets T and C). Bounds on unmeasured variables are ignored in the algorithm.
(2) The algorithm may terminate even if the test statistics for some of the measurements exceed the test criterion, due to the modifications in Step 5 of the algorithm. In essence, the method relies on bound information at the expense of the information provided by the statistical test.
(3) As an extreme case, the algorithm can also terminate with all the test statistics below the test criterion, but the reconciled solution violating the bounds for some of the measured variables. This can happen if the test statistics of the initial reconciled solution do not exceed the test criterion in Step 2 of the algorithm.
(4) Since the method is based on serial elimination of measurements, it has the same limitation as IMT of being applicable only for identifying biases in measurements and not other types of gross errors.

Rosenberg et al. [3] proposed the *extended measurement test* (EMT) and *dynamic measurement test* (DMT) strategies that also exploit bound information in gross error detection. Strictly, EMT cannot be classified as a serial strategy because it involves the simultaneous elimination of two or more measurements similar to the simultaneous strategy using the global test described earlier. The EMT strategy initially creates a candidate set of measurements suspected of containing gross errors by using the measurement test and selecting those for which the test is rejected.

From this candidate set, combinations of one or more measurements are eliminated in order. Gross errors in the eliminated set of measurements are identified provided the following two conditions are met:

(a) The measurement test for all remaining measurements are not rejected.
(b) Reconciled estimates of all variables (including those that are eliminated) satisfy the bounds.

In extreme cases, if it is not possible to identify a set of measurements which when deleted satisfies the above two conditions, then all measurements are suspected of containing gross errors.

The DMT algorithm is very similar to MIMT except that it checks for bound violations in the estimates of both measured and unmeasured variables (including variables whose measurements are eliminated). Moreover, DMT initializes the set C with the measurement having the largest MT statistic and enlarges the set C at each iteration by the measurement with the largest rejected MT statistic. For details of EMT and DMT the reader is referred to the paper by Rosenberg et al. [3]. These algorithms also have the limitation that they can be used for identifying measurement biases only.

Algorithm 5. Bounded GLR (BGLR) Method [15]

The above serial strategies suffer from some limitations described in the preceding section due to the heuristic manner in which bound information is utilized in gross error detection. If reliable bounds on variables are available and the reconciled estimates of variables are expected to satisfy these bounds, then it seems more appropriate to include them as inequality constraints in the data reconciliation problem. Due to these inequality constraints, the solution to the data reconciliation problem cannot be obtained analytically, even if the model constraints are linear.

A quadratic programming optimization technique has to be used as described in Chapter 5 for solving the resulting data reconciliation problem. (In general, if the model constraints are nonlinear, then a successive quadratic programming technique as described in Chapter 5 has to be employed [15].) Despite the complexity, this method offers an elegant and theoretically more rigorous approach for including bounds in data reconciliation and gross error detection. Such an approach is used in the BGLR method as described in the following steps:

Step 1. Solve the bounded reconciliation problem including the bounds on both measured and unmeasured variables in the reconciliation problem. (Use a nonlinear optimization technique described in Chapter 5.)

Step 2. Identify the active constraints at the solution of the reconciliation problem. (The active constraints include all the conservation constraints as well as any bound constraints that are satisfied as equalities at the optimal solution.) Denote the matrix formed by the active constraints by $\bar{\mathbf{A}}$, and the RHS of the constraints by $\bar{\mathbf{c}}$.

Step 3. Compute the constraint residuals, $\bar{\mathbf{r}}$, and covariance matrix of these residuals, $\bar{\mathbf{V}}$, using all the active constraints identified in Step 2 (use Equations 7-2 and 7-3 with $\mathbf{A}$ replaced by $\bar{\mathbf{A}}$ and $\mathbf{c}$ replaced by $\bar{\mathbf{c}}$).

Step 4. Compute the GLR test statistics using Equation 7-30 to 7-32 with $\mathbf{r}$ and $\mathbf{V}$ replaced by $\bar{\mathbf{r}}$ and $\bar{\mathbf{V}}$, respectively.

Step 5. If the maximum among the GLR test statistics $T \leq \chi^2_{1-\beta}$ go to Step 6. Otherwise apply the simple serial compensation strategy using $\bar{\mathbf{A}}$, $\bar{\mathbf{r}}$, and $\bar{\mathbf{V}}$, instead of $\mathbf{A}$, $\mathbf{r}$, and $\mathbf{V}$.

Step 6. Solve the bounded data reconciliation problem using the compensated measurements and the compensated model to obtain the reconciled estimates of all variables.

The strategy described above is a simpler and generalized version of the original method developed by Narasimhan and Harikumar [15] in which separate tests are applied to the variables which are at their bounds (also referred to as the *restricted variables*) and the other unrestricted variables.

A few points are worth mentioning with respect to the above strategy. In Step 5 of the above strategy, it is implicitly assumed that the same set of constraints will be active in all stages of the serial compensation strategy. The theoretically correct procedure is to solve the bounded data reconciliation problem using compensated measurements and compensated model constraints at each stage after a gross error is identified, in order to determine the new set of active constraints. Since this will increase the computational burden significantly, it has not been used in the above strategy. If there is no limitation of computing power, however, then it is advisable to solve the bounded data reconciliation problem at each stage of the serial procedure.

Secondly, in the BGLR method described above the simple serial compensation strategy is used for multiple gross error detection. Instead, the serial elimination strategy or modified serial compensation strategy can be used as well. Lastly, the above strategy does not have the limitations of the MIMT and other methods. The estimates that are obtained at the end of the procedure will satisfy the bounds, and the test statistics of all measurements will be below the threshold value (provided the bounded data reconciliation problem is solved at each stage of the serial procedure).

In the extreme case, if an infeasible solution is obtained in Step 1, this indicates that the bounds imposed on variables are too restrictive and have to be relaxed. One important advantage that this method offers is that it can be used even when more complex inequality constraints (such as thermodynamic feasibility restrictions imposed on the temperatures of heat exchangers) are used. If the inequality constraint is active at the optimal data reconciliation solution, then this is simply included as an equality constraint in the constraint set when gross error identification is applied.

> ***Exercise 8-3.*** *Develop the BGLR algorithm for the case when (a) the bounded data reconciliation problem is solved after each stage of the serial compensation procedure, (b) modified serial compensation is used as the strategy for multiple gross error detection instead of serial compensation, and (c) when both changes (a) and (b) are made.*

Simulation studies conducted by different researchers [1, 3, 15] show that bound information enhances the performance of gross error detection strategies only if the measured values are close to the bounds. Both simultaneous and serial algorithms described above are not limited to linear flow processes. If nonlinear constraints are involved, they can be linearized before applying any of these detection schemes. Gross error detection in nonlinear processes are discussed in greater detail later in this chapter.

Example 8-4

We apply the BGLR method to the process considered in the preceding examples. For this purpose we will assume that lower bounds and upper bounds on variables are specified as in Table 8-5.

Tight bounds on variables 1 and 2 are specified to illustrate the effect of bounds on gross error detection. The data reconciliation solution obtained without using these bounds are shown in Table 8-1.

Table 8-5
Lower and Upper Bounds on Flows of Heat Exchanger with Bypass Process

Measurement	Lower Bound	Upper Bound
1	99	102
2	60	67
3	30	40
4	55	75
5	30	40
6	90	110

The reconciled estimates violate the upper bounds on streams 1 and 2. Therefore, the data reconciliation problem is solved using the bounds and the reconciled estimates are obtained as [101.97, 67.00, 34.97, 67.00, 34.97, 101.97]. It is observed that the upper bound on stream 2 flow is active at the optimal solution. By including this constraint along with the flow balances we get the expanded constraint matrix and RHS of constraints

$$\overline{\mathbf{A}} = \begin{bmatrix} 1 & -1 & -1 & 0 & 0 & 0 \\ 0 & 1 & 0 & -1 & 0 & 0 \\ 0 & 0 & 1 & 0 & -1 & 0 \\ 0 & 0 & 0 & 1 & 1 & -1 \\ 0 & 1 & 0 & 0 & 0 & 0 \end{bmatrix} \qquad \overline{\mathbf{c}} = \begin{bmatrix} 0 \\ 0 \\ 0 \\ 0 \\ 67 \end{bmatrix}$$

The last of the above constraints is the active upper bound constraint on the flow of stream 2. Using the enlarged constraint set, we compute the global test statistic which is equal to 46.338. This is greater than the test criterion of 11.07 ($\chi^2_{5,0.05}$) and the GT is rejected. The maximum GLR test statistic is obtained for measurement 1 and so a gross error is identified in this measurement in stage 1. We continue this procedure by solving the bounded data reconciliation problem at each stage to determine the active constraints and the results are given in Table 8-6. Again only two of the 3 gross errors are correctly identified. The final reconciled values are [99.0, 64.7033, 34.2967, 64.7033, 34.2967, 99.0], which satisfy the bounds.

Table 8-6
Gross Error Identification Using Bounded GLR Method

Stage k	Active Constraints s	Global Test Statistic	DOF ν = $(m-k+1+s)$	Test Criterion $\chi^2_{\nu,0.05}$	Gross Error Identified
1	1 (UB on 2)	46.338	5	11.075	1
2	None	10.873	3	7.815	2
3	1 (LB on 1)	2.61	3	7.815	-

Combinatorial Strategies

There are several gross error identification strategies which make use of the nodal test. As pointed out in the previous chapter, the use of the nodal test requires a strategy even for identifying a single gross error. Since these strategies cannot be easily classified as either serial or simultaneous, we have chosen to categorize them separately. Most of these strategies are specifically tailored to flow processes and cannot be applied to nonlinear processes.

The basic principle used in the gross error identification strategy based on the nodal test was first proposed by Mah et al. [16]. If a gross error is present in any flow measurement, then this affects the constraint residuals in which the measurement occurs. Thus, we can expect the nodal test for the two nodes on which the corresponding stream is incident to be rejected. If these two nodes are merged, however, then the corresponding interconnecting stream is eliminated and the nodal test for this combination node (also called a pseudo-node) will most probably not be rejected. In order to exploit this principle, nodal tests are conducted on the residuals around single nodes as well as combinations of two or more nodes which are connected by streams. If a nodal test for any combination is not rejected, then the flow measurements of all streams which are incident on the pseudo-node may be assumed to be free of gross errors.

It should be noted that no decisions can be made concerning the flow measurements of streams interconnecting any two nodes forming the pseudo-node. On the other hand, if the nodal test is rejected, then one or more measurements incident on the pseudo-node can contain gross errors. No direct statement, however, can be made regarding any of these measurements. By selecting suitable combinations of nodes on which nodal tests are performed, it is possible to identify a set of measurements which are likely to free of gross errors. The remaining measurements are suspected of containing gross errors.

Further screening of this candidate set can be performed using other serial or simultaneous strategies to identify the measurements containing gross errors. These strategies can also be used to identify leaks in nodes along with measurement biases [1, 7]. We describe below the *linear combination technique* (LCT) algorithm by Rollins et al. [18] which uses the above strategy.

Algorithm 6. The Linear Combination Technique [18]

In the LCT method proposed by Rollins et al. [18], nodal tests are conducted on constraint residuals around single nodes as well as certain combinations of nodes. In general, for any of these nodal tests, the hypotheses can be expressed as follows:

$$H_{oi}\text{: } \mathbf{l}_{\mathrm{i}}^{\mathrm{T}}\mu_{\mathrm{r}} = 0 \quad \text{versus} \quad H_{li}\text{: } \mathbf{l}_{\mathrm{i}}^{\mathrm{T}}\mu_{\mathrm{r}} \neq 0 \tag{8-12}$$

where μ_r is the expected value of the constraint residuals and $\mathbf{l}_i$ is a vector of zeros and ones representing the linear combination in the *i*th test. At α level of significance, H_{oi} is rejected if:

$$\frac{\mathbf{l}_{\mathrm{i}}^{\mathrm{T}}\mathbf{r}}{\sqrt{\mathbf{l}_{\mathrm{i}}^{\mathrm{T}}\mathbf{V}\mathbf{l}_{\mathrm{i}}}} \geq Z_{1-\alpha/2} \tag{8-13}$$

If H_{oi} is not rejected, all measurements of streams incident on the pseudo-node are considered to be free of gross errors. After all linear combination tests are performed, two sets of measured variables are obtained. One set, SET1, contains variables whose measurements are not suspected to contain gross errors. The complementary set, SET2, consists of variables whose measurements are suspected of containing gross errors. Of course the algorithm may result in incorrectly classifying the measurements, giving rise to Type I errors (when good measurements are placed in SET2), or Type II errors (when faulty measurements are placed in SET1). The chosen level of significance α for the tests plays an important role in balancing the two types of errors.

In order to reduce the number of linear combinations for hypothesis testing, Rollins et al. [18] adopted the following rules:

a. Conduct *m* nodal (constraint) tests on individual nodes. If H_{ok} for node (balance) *k* (k=1, . . ., m) is not rejected, no nodal test on combination nodes containing node *k* is conducted.

b. A gross error in a small flow is generally difficult to detect and if it is implicitly assumed that such stream measurements do not contain gross errors, then no nodal test on combinations of nodes connected by a low flow rate stream is conducted.
c. No nodal test is performed on node combinations which are not connected.
d. No nodal test is performed on a nodal combination containing two nodes which are connected by a stream whose measurement has been classified to be free of a gross error.

The above rules are used to essentially avoid conducting nodal tests on pseudo-node combinations which do not provide any additional information for identifying the good measurements. Mah et al. [16] used a similar strategy and made use of the above rules except rule (b). In addition, their procedure also attempted to identify leaks in nodes as a last resort, when the nodal test for a node is rejected, but all the streams incident on this node are classified as good (placed in SET1). Serth and Heenan [1] proposed three different variants of the above strategy in one of which information about bounds on variables was also exploited. Yang et al. [7] used a combination of measurement and nodal tests in which the measurement test was used to identify an initial candidate list of suspect measurements while the nodal tests were used to counter-check whether the suspect measurement does contain a gross error.

Although strategies based on nodal tests reduce the number of mispredictions (Type I errors) as compared to serial strategies [1], they suffer from the following drawbacks:

(1) If multiple gross errors are present in the data, due to partial or complete cancellation of these errors, nodal tests for node combinations on which these streams are incident may not be rejected resulting in incorrect classification (Type II errors).
(2) They are designed for linear flow processes and it is difficult to extend them to nonlinear processes.

Example 8-5

The LCT algorithm is applied on the data for the heat exchanger with bypass process considered in the preceding examples. Nodal tests (using $\alpha = 0.05$) are performed on single and combination of appropriate nodes and the streams are classified as shown in Table 8-7.

Table 8-7
Gross Error Detection Using LCT Algorithm

Node Combination	NT Statistic	Status	Measurements Classified Free of Gross Errors
1	2.200	Rejected	–
2	3.005	Rejected	–
3	0.856	Not Rejected	3 and 5
4	0.716	Not Rejected	4 and 6
1+2	4.653	Rejected	–
1+3	–	Not performed – Stream 3 is good	
1+4	–	Not performed – disconnected	
2+3	–	Not performed – disconnected	
2+4	–	Not performed – Stream 4 is good	
3+4	–	Not performed – Stream 5 is good	

Other than measurements 1 and 2, the rest are classified as good and thus two of the three gross errors are correctly identified by LCT.

Other collective methods used to simultaneously identify leaks and measurement biases and estimate their error magnitudes have been recently proposed by Jiang and Bagajewicz [19].

PERFORMANCE MEASURES FOR EVALUATING GROSS ERROR IDENTIFICATION STRATEGIES

The performance of the serial elimination and compensation methods described above can be compared by means of computer simulation experiments in which known errors are introduced into the data and the ability of the schemes to identify and correct the errors is evaluated. Such comparisons have been frequently reported [3, 5, 6, 9, 10, 15, 18, 19]. Since the error detection procedure is stochastic in nature, the performance is averaged over a suitably large number of trials. A minimum of 1,000 simulation trials is recommended.

Given the set of true values for all process variables, for each simulation trial, a measurement vector is generated as

$$\mathbf{y} = \mathbf{x} + \varepsilon + \delta \tag{8-14}$$

where δ is the vector of gross errors. Using random numbers from the standard normal distribution, a vector ε of random errors is first added to the true values. For this, the standard deviations are usually taken as some fixed percentage of the true values. A specified number of gross errors are added to obtain the measurement vector. The location of the gross errors are uniformly and randomly selected over the set of measured variables, while the magnitudes of the gross errors are uniformly and randomly chosen between specified upper and lower bounds. The sign of the gross error is also chosen randomly. The magnitudes of the gross errors are constrained by

$$l(x_i + \varepsilon_i) \leq \delta_i \leq u(x_i + \varepsilon_i) \tag{8-15}$$

where l is a lower fraction and u is an upper fraction of the random value (not including the gross error). For instance, $l = 0.05$ and $u = 0.50$.

For the purpose of gross error detection, the level of significance α for the statistical test is also required. This value is frequently chosen so that the average Type I error when no gross errors are present is 0.1.

Different performance measures have been used to evaluate gross error detection performance [3, 9] as follows:

1. The **overall power** (OP) of the method to identify gross errors correctly is given by

$$\text{Overall power} = \frac{\text{Number of gross errors correctly identified}}{\text{Number of gross errors simulated}} \tag{8-16}$$

   The overall power is computed only for simulation trials in which gross errors are simulated.

   Rollins and Davis [4] defined an *overall power function* (OPF) as follows, which is a more conservative measure.

$$\text{OPF} = \frac{\text{Number of simulation trials with perfect identification}}{\text{Number of simulation trials}} \tag{8-16a}$$

   By perfect identification, it is implied that all gross errors and their locations are correctly identified and no mispredictions are made. Obviously, an ideal strategy is one which results in an OPF of unity. Note that it is always possible to get an OP value of 1 by predicting all measurements to contain gross errors. This is not satisfactory because too many mispredictions are also made in the bargain. In this case, however, the OPF value will be zero if some of the measurements do not contain gross errors.

Sanchez et al. [6] have modified the definition of OPF, taking into account *equivalency* of gross errors, and have denoted their performance measure as OPFE. This measure is also computed like OPF except that "perfect identification" is interpreted to account for equivalency of gross errors. Perfect identification of gross errors is achieved, if the set of gross errors identified in a simulation trial belongs to the same equivalency class as the set of gross errors actually present in the measured data. This also implies that no mispredictions are made in the simulation trial.

2. The **average number of Type I errors** (AVTI), which defines the number of incorrect identifications made by a method, is given by

$$\text{AVTI} = \frac{\text{Number of gross errors wrongly identified}}{\text{Number of simulation trials made}} \tag{8-17}$$

The measure AVTI is computed separately for each simulation run, whether or not gross errors are simulated. The interpretation of "wrong identifications" can be suitably made after taking equivalency of gross errors into consideration.

3. Another measure of performance is the **selectivity,** defined as

$$\text{Selectivity} = \frac{\text{Number of gross errors correctly identified}}{\text{Total number of gross errors detected}} \tag{8-18}$$

It may be noted that the denominator includes only those simulation trials where a gross error is simulated.

4. **Average error of estimation** (AEE) is the fourth type of performance measure. It gives the accuracy of estimation of the bias magnitude on the average. It is used to compare serial compensation methods, where estimates of the gross error magnitudes are also provided.

$$\text{AEE} = \frac{1}{\text{N}} \sum_{1}^{\text{N}} \left| \frac{\text{estimated value} - \text{actual value}}{\text{actual value}} \right| \tag{8-19}$$

COMPARISON OF MULTIPLE GROSS ERROR IDENTIFICATION STRATEGIES

Different simulation studies have been conducted comparing the performance of gross error detection strategies. Among the different strategies proposed for multiple gross error identification we would like to determine the best, in terms of the measures described above. Unfortunately, the performance studies conducted so far do not provide a definite answer. Nevertheless, some conclusions can be drawn from these studies.

Before making a comparison it is important to ensure that different strategies are compared on the same basis. Since it is always possible to improve the power of a strategy at the expense of a greater Type I error probability, it is important to ensure that the different strategies have the same Type I error probability, so that a judgment can be made based on their power and selectivity measures, etc.

Secondly, it does not matter whether the GT, MT or GLR test is used for detection because the same performance can be obtained with all these tests by using the same component strategies for single and multiple gross error identification. This follows from the equivalency results between these three tests proved in the previous chapter. However, strategies that make use of the nodal tests or principal component tests are distinct and cannot be combined with other tests. Our comparison is focused on the component strategy used for multiple gross error detection.

Since modified serial compensation is identical to serial elimination and also has the ability to handle gross errors other than biases, among these two only MSCS needs to be considered. Keller et al. [10] showed that MSCS performs better than SSCS because it commits fewer mispredictions when multiple gross errors are present in the data. The reason for this is that SSCS uses the estimated magnitudes of gross errors identified in preceding stages to compensate the measurements, which can itself introduce errors due to incorrect estimates. MSCS avoids this by collectively estimating the magnitudes of gross errors hypothesized. Rollins and Davis [4] also showed that SSCS commits an unacceptably high number of Type I errors when the standard deviations are very small or when many gross errors are present in the data. Based on these studies, it appears that MSCS is the best among the serial strategies.

Studies to determine the best simultaneous strategy for multiple gross error identification have not yet been performed. The simultaneous strategy proposed by Rosenberg et al. [3] was compared with serial elimination and was found to perform equally well. However, in this comparison the equivalency between different gross errors has not been taken into consideration properly, since the effect of equivalency of gross errors was not completely known at the time of their study. The simultaneous strategy based on testing all possible gross error combination hypotheses has not been evaluated in any study so far, since it was considered to be computationally intensive.

A comparison between LCT and SSCS was made by Rollins and Davis [4] and they showed that LCT performs much better especially when standard deviations of measurement errors are small and when a large number of gross errors are present in the data. However, since MSCS is better than SSCS a comparison between LCT and MSCS is more appropriate. A major problem with LCT is that at present it is applicable only to linear flow processes.

The strategy based on principal component tests was recently compared with MSCS and other strategies by Jiang et al. [5]. Their study is the only one where equivalency between different gross error sets is properly taken into account before computing the performance measures. Their study did not demonstrate superior performance of PC test strategy. This is also confirmed by recent studies by Jordache and Tilton [12] and Bagajewicz et al. [13].

GROSS ERROR DETECTION IN NONLINEAR PROCESSES

All statistical tests and serial elimination or serial compensation techniques presented in this chapter can be used for detection and identification of gross errors in processes described by nonlinear models. The usual procedure is first to perform a linearization of the process model followed by an identification method designed for linear equations. Typically, the measured data are reconciled under the assumption that no gross errors are present and the constraint equations are linearized around the reconciled estimates.

This strategy, although very popular, may not be suitable for highly nonlinear processes with data corrupted by significant gross errors. The reconciled estimates which are obtained based on the assumption that no gross errors exist in the data may be far from the true values and the resulting linearized model may not be a good approximation. Consequently, this approach may fail to identify the true gross errors. There is no guarantee of a successful gross error detection and identification even for pure linear models. Nonlinear processes are much more complex, and they require special methods of gross error detection.

One step forward was provided by Kim et al. [20]. They tailored the MIMT serial elimination algorithm to fit the nonlinear data reconciliation problem. Their enhanced algorithm differs from the MIMT algorithm in two ways. First, in Step 1, the data reconciliation problem is solved using nonlinear programming (NLP) techniques. Second, in Step 5, the reconciled values $\hat{\mathbf{x}}$ and the measurement adjustments are also calculated based on the nonlinear solution.

The variance-covariance matrix of the adjustment vector, however, is calculated from a linearized model as proposed by Serth and Heenan [21]. Their method, tested on a CSTR reactor model, showed superior performance in comparison with the MIMT algorithm used with successive linearization, especially when the number of gross errors increases. The NLP solver is more robust and provides more reliable estimates for reconciled values and gross errors which enhances the performance of gross error detection and identification. If large gross errors exist in the data, however, they need to be screened out prior to application of this technique. Moreover, the computational time can be very high for large-scale industrial problems.

A new strategy for detection of gross errors in nonlinear processes was recently proposed by Renganathan and Narasimhan [22]. In their approach, a test strategy analogous to the GLR method was proposed that does not require linearization of the constraints. A brief description of this method follows.

Gross Error Identification in Nonlinear Processes Using Nonlinear GLR Method

It was proved in the preceding chapter that the GLR test statistic for a measurement i is identical to the difference in the *objective function* (OF) values of two data reconciliation problems, one of which assumes that no gross errors are present, whereas the other assumes that a gross error is present in measurement i. The formulation of the data reconciliation problem when a gross error is assumed in measurement i, was also described in the preceding chapter (Problem P1 in Chapter 7).

For identifying a single gross error in the GLR method, the maximum difference between the OF values over all the gross errors hypothesized is obtained. If this difference exceeds a critical value then a gross error is detected and is identified in the variable which gives the maximum OF difference. In other words, the gross error model that gives the minimum least squares OF value is selected, which means that the gross error model that best fits the observed data is selected as the most likely possibility.

For **nonlinear processes,** a gross error detection test can be obtained by applying the above principle of GLR test, that is, the test statistic is obtained as the maximum difference in OF values between the no gross error model and the gross error model for variable *i*. The test statistic T is given by

$$T = \max_{i} T_i \tag{8-20}$$

where

$$T_i = \text{OF for no gross error model} \\ \quad - \text{OF for } i\text{th gross error model} \tag{8-21}$$

The OF for no gross error model is obtained by solving the standard nonlinear DR problem as formulated in Chapter 5. The OF for *i*th gross error model is obtained by the solving the nonlinear DR problem analogous to Problem P1 described in the preceding chapter, obtained by simply replacing the linear constraints (Equation 7-1) with the nonlinear constraints of the process (Equations 5-8 and 5-9). These nonlinear DR problems are solved using nonlinear programming techniques as described in Chapter 5. The test statistic is compared with a prespecified threshold (critical value) T_{cr} and a gross error is detected if T exceeds T_{cr}. This means that the corresponding gross error model best fits the data and so the variable corresponding to that gross error model is identified to be biased.

The magnitude of bias *b* is also obtained as part of the solution of Problem P1. Although no statistical reasoning can be given for the choice of the critical value, the same test criterion as in the case of linear processes can be used. This may be adjusted by trial and error using simulation if it is desired to obtain a specified Type I error probability. It should be noted that, in this approach, the nonlinear constraints are treated as such and not approximated by a linear form. Furthermore, bounds and other inequality constraints can be included as part of the constraints, and the gross error detection test can still be applied as described above since it uses only the optimal objective function values. For ease of reference, we denote this test as a *nonlinear GLR test* (NGLR).

The NGLR test described above can be applied to detect at most one gross error. However, this test can also be combined with any simultaneous or serial strategy for multiple gross error detection described in this text. As a particular case, we describe the use of NGLR test along with MSCS strategy for multiple gross error identification caused by measurement biases.

NGLR Test with Modified Serial Compensation Strategy (MSCS)

In MSCS, at each stage of application of the test, only the locations of previously detected gross errors are assumed to be correct, but the estimates of all the gross error magnitudes are assumed to be unknown and are therefore estimated simultaneously. This process is repeated until no further gross errors are detected. Applying this strategy along with the NGLR test, we obtain the test statistic at stage *k+1* as in Equations 8-21 and 8-22, but the OF values at stage *k+1* for the no gross error model and gross error model for variable *i* are, respectively, obtained by minimizing the following objective functions, subject to the nonlinear constraints given by Equations 5-8 and 5-9.

Problem P2

$$\underset{\mathbf{x},\mathbf{b}_k}{\text{Min}} \left(\mathbf{y} - \mathbf{E}_k^* \mathbf{b}_k - \mathbf{x}\right)^T \mathbf{Q}^{-1} \left(\mathbf{y} - \mathbf{E}_k^* \mathbf{b}_k - \mathbf{x}\right) \tag{8-22}$$

Problem P3

$$\underset{\mathbf{x},\mathbf{b}_k,b_i}{\text{Min}} \left(\mathbf{y} - \mathbf{E}_k^* \mathbf{b}_k - b_i \mathbf{e}_i - \mathbf{x}\right)^T \mathbf{Q}^{-1} \left(\mathbf{y} - \mathbf{E}_k^* \mathbf{b}_k - b_i \mathbf{e}_i - \mathbf{x}\right) \tag{8-23}$$

where $\mathbf{b}_k$ is a *k* vector of unknown biases and $\mathbf{E}_k$ is a matrix whose columns are the unit vectors. The *j*th column vector of $\mathbf{E}_k$ has a unit value in the position corresponding to the measurement in which a gross error was identified in stage *j* for j = 1, . . . k. The minimization with respect to the unknown magnitudes of gross error magnitudes, $\mathbf{b}_k$, for computing the objective function values implies that only the locations of gross errors identified in the previous stages are assumed to be correct. Their magnitudes, however, have to be estimated simultaneously along with the gross error hypothesized in the present stage, which are actually the premises for the hypotheses in MSCS. For highly nonlinear processes such as reactors, Renganathan and Narasimhan [22] demonstrated that the NGLR method gives better performance than methods which rely on linearization of the constraints.

BAYESIAN APPROACH TO MULTIPLE GROSS ERROR DETECTION

The major problem in gross error detection is how to enhance the power and selectivity of the test without increasing the frequency of false detections (Type I errors). One way to enhance the power and selectivity of gross error tests is to use the information from the past data and, particularly, the frequency of past failures on measuring instruments.

To incorporate historical information on measuring instrumentation, we can make use of the Bayes theorem in statistics [23]. A gross error detection and identification procedure for measurement biases based on this approach has been developed by Tamhane et al. [24, 25] for steady-state processes.

The usual detection techniques for steady state processes are applied to the snapshots of data or averages of data collected within a time period. But if a significant instrument failure occurs within the data collection period, the statistical tests using averages of data may not be able to capture that failure. Even if eventually they will capture that failure, it might take a long time until an instrument problem is detected by a statistical test.

The new Bayesian approach is not a one-time application, as with the previous steady-state tests or combination of tests, but rather a *sequential application* that incorporates historical data collected and updated over time. The sequential approach raises more questions than the statistical tests using data averages. For instance, how often are the instruments inspected to confirm the occurrence of gross errors? If a gross error is confirmed, how soon is the instrument repaired? Is it reasonable to assume that the repair is perfect, that is, the instrument will be as good as new? Should the model include a factor for the aging of instruments? Incorporating all these issues in the sequential application framework makes the Bayesian algorithm a much more challenging task than the previous gross error detection and identification algorithms.

To simplify the description of the Bayes test for gross error detection we will first assume a **one-time application of this test.** We consider a relatively short measurement period of N consecutive observations. The average vector **y** of the *N* observations satisfies the following measurement model:

$$\mathbf{y} = \mathbf{x} + \varepsilon + \delta \otimes \mathbf{e} \tag{8-24}$$

where: **x** is the $n \times 1$ vector of true values, ε is the $n \times 1$ vector of random errors and $\delta \otimes \mathbf{e}$ represent a vector formed by products $\delta_i \mathbf{e}_i$ ($i = 1 \ldots n$); δ_i is the magnitude of a gross error in measurement *i,* and $\mathbf{e}_i$ is unity if a gross error is present in measurement *i* or zero otherwise. We assume that a gross error can occur in any measurement *i* only at the beginning of the measurement period. The vector **x** is assumed to satisfy the following linear constraints:

$$\mathbf{Ax} = 0 \tag{8-25}$$

Note that if a nonlinear model is used, a linearization of the original model should be first performed. The following assumptions will also be made for the one-time application of the Bayes test:

a. Vector ε follows a multivariate normal distribution $N(\mathbf{0},\mathbf{Q})$ with known covariance matrix $\mathbf{Q} = (1/N)\Sigma$, where Σ is the covariance matrix of the individual data vectors.
b. A gross error in any measurement *i* is a known constant magnitude, say δ_i^*.
c. The values of prior probabilities of instrument failures are known. We assume that each element δ_i of δ is modeled as a Bernoulli random variable taking on values δ_i^* and 0 with probabilities p_i and $1 - p_i$, respectively ($i = 1 \ldots n$). If $\delta_i = \delta_i^*$, $i \in I$ and $\delta_i = 0$, $i \notin I$, then the corresponding gross error vector is denoted by δ_I. There are 2^n possible states of nature δ_I where δ_I ranges from $(0, 0, \ldots, 0)$ for the state with no gross error, to $(\delta_i, \delta_2, \ldots, \delta_n)$ for the state with gross errors occurring in every measurement.
d. Gross errors (or instrument failures) in different instruments occur independently of each other. Then the prior probability that $\delta = \delta_I$ is given by

$$\pi_I = \prod_{i \in I} p_i \prod_{i \notin I} (1 - p_i) \tag{8-26}$$

We will refer to the π_I's as the *group prior probabilities.* The Bayes theorem is applied to compute the *group posterior probability* $\tilde{\pi}_I$ of δ_I, given the group prior probability π_I of δ_I and the measured data. A general Bayes formula for posteriors $\tilde{\pi}_I$ is given by

$$\tilde{\pi}_I = \frac{\pi_I f(\text{data} \mid \delta_I)}{\sum_J \pi_J f(\text{data} \mid \delta_J)} \tag{8-27}$$

where $f(\text{data}|\delta_I)$ denotes the conditional probability density function (p.d.f.) of the given data given that the true state of nature is δ_I, and the summation in the denominator is over all 2^n subsets *J*. The *Bayes decision rule for identification* of the most likely state of nature is the following:

The most likely state of nature δ_I corresponds to the maximum posterior in Equation 8-27.

Therefore, if $\tilde{\pi}_I^* = \max \tilde{\pi}_I$, then the measurements $i \in I^*$ are declared in gross error. If $I^* = \varnothing$, then all measurements are declared free of gross errors. From the identification rule above, we can see that the Bayesian approach falls into the category of *simultaneous strategies for gross error detection and identification.*

Since it also assumes knowledge about the magnitude of gross errors, the Bayesian strategy is closely related to the serial compensation strategies based on GLR test. In fact, for equal prior probabilities of gross error occurrence p_i for all measurements $i = 1 \ldots n$, Formula 8-27 becomes similar to Equation 7-27 used to derive the GLR test. There are two differences though: (i) instead of direct measured data $\mathbf{y}$, a linear transformation of vector $\mathbf{y}$ is used for the GLR test (namely the residual vector $\mathbf{r} = \mathbf{Ay}$) and (ii) the denominator in Equation 8-27 is a summation over all possible states of nature δ_J rather than the state of nature δ_Φ corresponding to the case of no gross error as in Equation 7-27.

Note that in Equation 8-27 we cannot use $\mathbf{y}$ directly for *data* because its p.d.f. involves the true vector $\mathbf{x}$ which is unknown. What is required is a transformed vector $\mathbf{Cy}$ such that

(i) the p.d.f. of $\mathbf{Cy}$ is free of $\mathbf{x}$ and depends only on δ;
(ii) the covariance matrix $\mathbf{CQC}$ is nonsingular.

Equation 8-25 indicates that matrix $\mathbf{C}$ must satisify the condition, $\mathbf{C} = \mathbf{MA}$, for some $m \times m$ matrix $\mathbf{M}$. But $\mathbf{C}$ is not unique. It can be shown

[24, 26] that the choice which leads to a maximal dimensional transformation gives rise to the following Bayesian formula:

$$\tilde{\pi}_I = \frac{\pi_I \mathbf{exp} - \left\{0.5(\mathbf{y} - \delta_I)^T \mathbf{W}(\mathbf{y} - \delta_I)\right\}}{\sum_J \pi_J \mathbf{exp} - \left\{0.5(\mathbf{y} - \delta_J)^T \mathbf{W}(\mathbf{y} - \delta_J)\right\}} \tag{8-28}$$

where **W** is the covariance matrix of the vector **d** of modified measurement adjustments (see Equation 7-15 in Chapter 7).

Next, we will present the **sequential application of the Bayesian test,** which is the desired implementation of the Bayesian strategy for gross error detection. The sequential application of the Bayesian test enables continuous updating of the prior probabilities of instrument failures for better gross error detection and identification. The measurement model is now time-dependent even though the underlying process is steady state:

$$\mathbf{y}(t) = \mathbf{x} + \boldsymbol{\varepsilon}(t) + \delta \otimes \mathbf{e}(t-1) \tag{8-29}$$

where t is the index for time periods; $\mathbf{y}(t)$ is the average of $N \geq 1$ data vectors observed during time period t; $\varepsilon(t)$ is the vector of random errors assumed to follow a multivariate normal distribution $N(0,\mathbf{Q})$; $\delta \otimes \mathbf{e}\,(t-1)$ represent the vector of gross errors present at time $(t-1)$, i.e., at the beginning of measurement time period t.

Initially we assume that the occurrence of gross errors are independent Bernoulli random variables with a constant (with respect to time) failure rate θ_i for the ith instrument. In other words, the probability that the ith instrument fails (in the sense of a gross error occurring in the measured value) at the start of any given period t is the same, namely, θ_i, and the failure of a given instrument in different time periods are independent. For a given θ_i, the conditional probability that instrument i is in a failed state at time $t-1$ is given by

$$p_i(t-1 \mid \theta_i) = 1 - (1-\theta_i)^{\tau_i(t-1)} \qquad i = 1 \ldots n \tag{8-30}$$

where $\tau_i(t-1)$ is the time since the last check on instrument i. Note that θ_i is quite different from $p_i(t-1)$, the probability that the ith instrument is in a failed state at time $t-1$. The instrument may be in a failed state at time $t-1$

if it failed in any of the previous time periods and has not been repaired. To compute $p_i(t-1)$ required in Equation 8-26 which is used to calculate the Bayesian test (Equation 8-28) we assume a prior distribution on each θ_i and compute $p_i(t-1|\theta_i)$ with respect to this prior distribution. This approach has the capability of using the past instrument failure to update the prior distributions on the θ_i's. Independent beta distributions [27],

$$f_i(\theta_i) = \frac{\Gamma(l_i + m_i)}{\Gamma(l_i)\Gamma(m_i)} \theta_i^{l_i - 1}\left(1 - \theta_i^{m_i - 1}\right) \qquad (8\text{-}31)$$

can be used conveniently for this purpose, because they are conjugate priors with respect to the geometric distributions which are followed by the instrument lifetimes (see Equation 8-35). In Equation 8-31 above, $\Gamma(.)$ denote the gamma function, and l_i and m_i are two parameters with the following interpretation: l_i is the number of previous failures for instrument i and m_i is the sum of previous lifetimes for instrument i; the ratio $l_i / (l_i+m_i)$ is the mean of the beta prior distribution, denoted by $\hat{\theta}_i$. The parameters l_i and m_i are updated using data on past failures of instruments [25] as follows:

$$l_i^{(n_f)} = l_i^{(0)} + n_f \quad \text{and}$$

$$m_i^{(n_f)} = m_i^{(0)} + \sum_{j=1}^{n_f} t_i^{(j)} - n_f \qquad i = 1 \ldots n \qquad (8\text{-}32)$$

where n_f is the number of past failures for instrument i and $t_i^{(j)}$ is the lifetime of instrument i for the jth failure, that is, the number of time periods between its jth failure and $(j-1)$th failures (or $t_i^{(0)}$ for j=1). A method of choosing initial values $l_i^{(0)}$ and $m_i^{(0)}$ was proposed by Colombo and Constantini [28]:

$$l_i^{(0)} = \frac{1 - s_i\hat{\theta}_i^{(0)}}{s_i - 1} \quad \text{and} \quad m_i^{(0)} = \frac{(1-\hat{\theta}_i^{(0)})l_i^{(0)}}{\hat{\theta}_i^{(0)}} \qquad i = 1 \ldots n \qquad (8\text{-}33)$$

Parameter $1 < s_i < 2$ enables the user to select which factor is more important in the estimation of the prior $\theta_i^{(0)}$; a large value (close to 2) yields small values of $l_i^{(0)}$ and $m_i^{(0)}$, which means that more weight is given to the current data than the prior information and vice versa. By choosing $\hat{\theta}_i^{(0)}$ and s_i, $l_i^{(0)}$ and $m_i^{(0)}$ can be estimated from Equation 8-33 above.

The Bayesian formula can be used to compute the posterior p.d.f. of θ_i, given its prior p.d.f. and the conditional p.d.f. of the failure data.

$$f_i(\theta_i \mid \text{failure data}) = \frac{g_i(\text{failure data} \mid \theta_i) f_i(\theta_i)}{\int_0^1 g_i(\text{failure data} \mid \theta_i) f_i(\theta_i) d\theta_i} \quad (8\text{-}34)$$

where

$$g_i(\text{failure data} \mid \theta_i) = \theta_i (1 - \theta_i)^{\tau_i - 1} \quad (8\text{-}35)$$

In Equation 8-35, g_i is the probability that instrument *i* lasts (was not in failed state) for exactly τ_i time periods. The $p_i(t-1)$ (required in Equation 8-26) is computed by

$$p_i(t-1) = \int_0^1 p_i(t-1 \mid \theta_i) d\theta_i$$
$$= 1 - \frac{\Gamma(l_i + m_i)\Gamma[m_i + \tau_i(t-1)]}{\Gamma(l_i)\Gamma[l_i + m_i + \tau_i(t-1)]} \quad (8\text{-}36)$$

Although not explicitly stated, the following assumptions have been also made so far in the Bayesian model:

(i) The magnitudes of gross errors are known and constant values δ_i^*
(ii) The instrument failure probabilities are independent of instrument ages
(iii) Checking and corrective actions are immediate and perfect

These assumptions, however, usually do not hold in real life. In the next section we will show how these assumptions can be relaxed in the practical implementation of the Bayesian strategy.

First, *the magnitudes δ_i's of the gross errors can be sequentially updated* based on past data rather than considering them known constants. There are various ways of estimating the magnitudes of gross errors. One method was presented in Chapters 7 and 8 in connection with the GLR test. Another procedure was suggested by Romagnoli [29]. Iordache [26] proposed a simplified method which makes use of the modified adjustment vector **d** defined by Equation 7-15. This method, though, is suitable only for linear constraints as given by Equation 8-25. The expected value of vector **d** can be obtained by combining Equation 7-15 with Equation 3-8:

$$E(\mathbf{d}) = \mathbf{Q}^{-1}E(a) = \mathbf{Q}^{-1}E(\mathbf{y}-\mathbf{x}) = \mathbf{A}^T(\mathbf{AQA}^T)^{-1}\mathbf{A}\delta = \mathbf{W}\delta \qquad (8\text{-}37)$$

If we approximate E(**d**) in Equation 8-37 by the observed vector **d** = **Wy**, then the vector δ of gross errors can be estimated from the equation

$$\mathbf{W}\delta = \mathbf{d} \qquad (8\text{-}38)$$

Note that in Equations 8-37 and 8-38 the vector δ is actually $\delta \otimes \mathbf{e}\,(t-1)$ described in Equation 8-29. Furthermore, matrix **W** is generally singular, therefore a least-squares solution should be obtained. One way is to use a Moore-Penrose pseudo inverse of **W**. This solution, which also involves a singular value decomposition of matrix **W**, provides a minimum Euclidean norm of vector δ, that is, it minimizes $\delta^T\delta$. The solution is unique [30, 31] and can be written as:

$$\delta^* = W^+\mathbf{d} \qquad (8\text{-}39)$$

Note that, in order to obtain a meaningful solution, only the estimates for the δ_i's associated with the measurements declared in gross error by the Bayes test are updated. Therefore, all $\mathbf{e}_i$'s except those corresponding to measurements suspected in gross error in the previous step (t–1) are zero.

Secondly, *the failure rates θ_i's will not be constant,* but will increase with the ages of the instruments. Let $\theta_i(T_i)$ be the failure probability for instrument *i* when its actual age is T_i. The following model can be used for $\theta_i(T_i)$ [27]:

$$\theta_i(T_i) = \begin{cases} 0 & T_i = 0 \\ 1 - [1-\theta_1(1)]\exp\{\beta_i(T_i - 1)\} & T_i = 1, 2 \end{cases}$$

where $0 < \theta_i(1) < 1$ and $\beta_i \geq 0$ are given constants. If $\beta_i = 0$, a constant failure rate model is obtained. For $\beta_i > 0$, the failure rate increases with age (as $T_i \rightarrow \infty$, $\theta_i(T_i) \rightarrow 1$). Note that the model described by Equation 8-40 has not been implemented in the Bayes test yet, because it is rather complicated.

It was only used to simulate gross errors based on the aging function for θ_i [23, 24]. In the Bayes test, Equation 8-28 constant θ_i is still assumed.

Thirdly, *delays in checking* and *imperfect corrective actions* can also be taken into account. Immediate instrument checking after gross error detection, followed by a corrective action is not usually feasible in practice. First, because of inherent Type I errors, we may want to verify over a longer period of time the consistency in the gross error detection. A simple rule such as 2/3 (2 *out of* 3) can be adopted; that means that a gross error should be detected by the Bayes test in some particular measurement *i* at least twice out of three consecutive time periods.

Second, even with sustained evidence of gross errors, the operators may want to postpone the instrument checking and correction at a more convenient time (for instance, at the end of the shift or at the scheduled maintenance time). Until then, a gross error is assumed detected, but not corrected in the instrument *i*. Therefore, the parameters l_i, m_i, and θ_i are not updated until the instrument was checked and found to cause a gross error. But the magnitude of the gross error δ_i will be updated continuously until the instrument has been repaired.

Based on the above assumptions, a **Bayesian gross error detection and identification algorithm** can be implemented as follows:

Step 0. *Initialization.* At the beginning input the following information:

(i) Constraint matrix **A** (or the model to be linearized), covariance matrix Σ, and the number of data vectors per sampling period, N. If average data of N data vectors is used, the covariance matrix $\mathbf{Q}=(1/N)\Sigma$ will be used instead of Σ.

(ii) For each measurement $i = 1 \ldots n$, enter the following initial estimates: $\hat{\delta}_i^{(0)}$, $\hat{\theta}_i^{(0)}$, $l_i^{(0)}$, and $m_i^{(0)}$. Note that $l_i^{(0)}$ and $m_i^{(0)}$ can be initialized from $\hat{\theta}_i^{(0)}$ and a parameter $1 < s_i < 2$, Equation 8-33; $\hat{\delta}_i^{(0)}$ can be initialized as constant number of standard deviation, i.e., $c_i\sigma_i^{(0)}$.

(iii) Set the ages $\tau_i(0)$ to all instruments *i* equal to 1 (fresh instruments). Set the time period *t* equal to 1 also.

Step 1. *Read in the N data vectors* for period *t* and compute their average vector $\mathbf{y}(t)$.

Step 2. *Calculate the prior probabilities* $p_i(t-1)$, Equation 8-30 and the group priors π_I, Equation 8-26.

Step 3. *Calculate the posterior probabilities* π_I, Equation 8-28. Note that, if all possible states of nature for the vector $\delta \otimes \mathbf{e}$ are considered, the computational time for Steps 2 and 3 is exceedingly large. There are two ways to reduce the computational time:

(i) Since the denominator for all posteriors is the same, only the numerators should be computed; we can even calculate the natural logarithm of the numerators, thus avoiding the computation of the exponential functions.
(ii) The number of states of nature for the vector $\delta \otimes \mathbf{e}$ can be reduced to a much smaller subset, by the following strategy: Calculate first the posteriors associated with the states of nature involving only one e_i equal to 1 at a time (single gross error case) and the δ_ϕ case (no gross error). Select the measurements corresponding, say, to the top 25% of posterior values and calculate the posteriors of combinations of at most three of those measurements (we assume that at most three gross errors can simultaneously exist). Other strategies for reducing the number of hypotheses to be tested [18] can also be adopted.

Step 4. *Decision.* Find the maximum group posterior, $\tilde{\pi}_1^* = \max \tilde{\pi}_1$, among all selected combinations I. If the corresponding set I^* is empty, no gross error is detected. Otherwise, the measurements in set I^* are suspected in gross error. However, the final decision is delayed. For instance, if rule 2/3 is used, a measurement is declared in gross error when detected at least twice out of three consecutive time periods. The same is true for the no gross error case.

Step 5. *Action.* If the current time *t* corresponds to the scheduled inspection time for the instrumentation, the following actions are assumed:

(i) Check the instruments in set I^* detected at Step 4. Note that instrument checking may be delayed. In that case, I^* is the set of all instruments declared in gross error between two consecutive inspections.
(ii) Decide which ones are actually faulty, say subset I'.
(iii) Repair or replace the instruments in set I'.
(iv) Update the age of instruments I', i.e., $\tau_i(t)=1$ for a true failure followed by a corrective action.

Step 6. *Reestimate the magnitudes* δ_i^* *of the gross errors,* Equation 8-39. Only the magnitudes of the gross errors for the measurements in the detected set I^* are reestimated. Note that, due to inherent Type I errors, the magnitudes of some falsely detected errors are also reestimated. But, if averaged data is used, their estimates should be much smaller than those for the true gross errors.

Step 7. *Reestimate the parameters* $l_i^{(t)}$ *and* $m_i^{(t)}$ for beta prior distribution of θ_i (i=1 . . . n). The associated parameter $\theta_i^{(t)}$ is also reestimated, by the ratio $l_i / (l_i+m_i)$.

Step 8. *Set time period* $t = t + 1$ and return to Step 1.

A sequence of failures (gross error occurences) for a certain instrument *i*, followed by detection and correction actions according to the Bayesian algorithm described above is shown in Figure 8-1.

More details about the sequential Bayesian algorithm can be found in Tamhane et al. [24, 25]. Iordache [26] performed a comparative evaluation of the Bayesian algorithm with a similar strategy based on the measurement test, using simulation runs with 10,000 time periods. The performance criteria used were the probability of Type I errors, the power of correct detection and the average delay time before a correct detection for the same average number of Type I errors.

In general, the Bayesian method outperforms the measurement test in the following situations: high frequencies of gross error occurrence (multiple gross errors), large spread in the magnitudes and frequencies of gross errors, and long delays in confirmation and repairs. On the other hand, the Bayesian method converges very slowly. Starting with initial guess of θ_i equal to 33% or 300% of the true value, a large number of observed failures (on the order of 100) is needed before θ_i converges. Therefore, accurate initial estimates of θ_i's are needed before the Bayesian method may be put to practical use.

If there is uncertainity about the prior estimates of $\theta_{i'}$'s, one strategy is to place more weight on the current data until more historical data is obtained. The performance of the Bayesian method is much less dependent on the estimates of δ_i's. More work is required to make the Bayesian approach really competitive with other gross error identification strategies. If all implementation details are clarified, it will become an appropriate strategy for online applications.

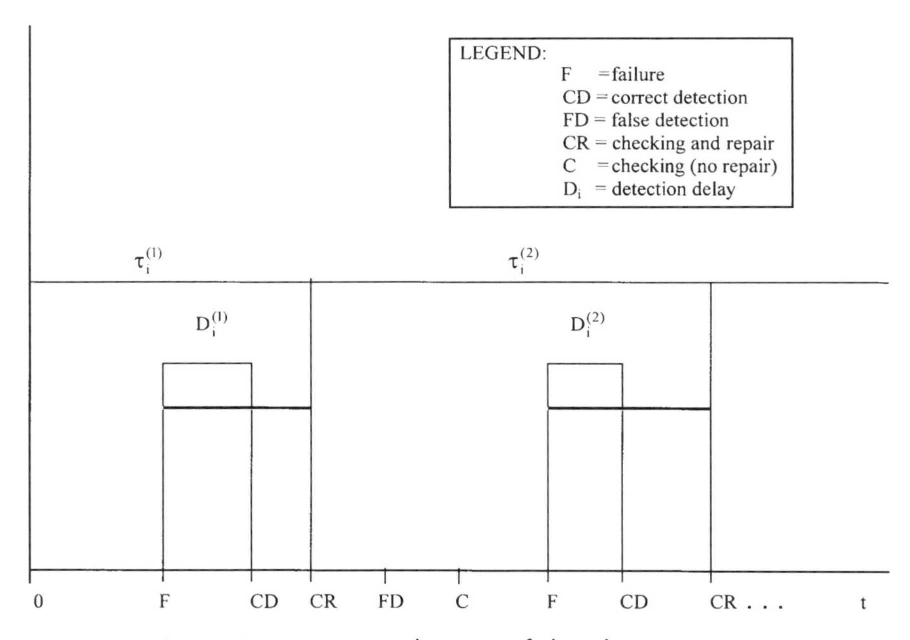

Figure 8-1. A sequential Bayesian failure detection process.

PROPOSED PROBLEMS

NOTE: The proposed problems that are included in this chapter require more extensive calculations. A computer program or a mathematical tool such as MATLAB is required in order to get solutions to these problems.

Problem 8-1. Use the the steam-metering system for a methanol synthesis unit represented in Figure 7-2 and the data indicated in Problem 7-2 to simulate multiple gross errors. For faster solving and a more clear analysis, at most three simultaneous gross errors are recommended. If possible, apply both serial elimination and serial compensation strategies (with appropriate statistical tests at α=0.05 level of significance) and perform a comparison of results. Gross errors of various magnitudes and in locations with different detectability factors are recommended.

Problem 8-2. A section of a heat exchanger train from reference [32] is shown in Figure 8-2. A *heat transfer fluid* (HTF) is used to heat two hydrocarbon streams (A and B). Five heat balance equations can be employed in order to describe the system. The first four are obtained by equating the shell and the tube side duties. The fifth is an HTF energy balance.

The process operating values and the measured variables are given in Table 8-8. Standard deviations of the measurement errors are also listed in Table 8-8. The units for flows are bbl/hr and for the temperatures degrees F. Densities, in lb/bbl, are: HTF: 290, A: 300, B: 320. The enthalpies, in Btu/lb, are related to temperature by the following equations:

$$\text{HTF: } H = 0.323T + 1.14E\text{-}4T^2$$
$$\text{A/B: } H = 0.424T + 2.80E\text{-}4T^2$$

Solve first the nonlinear data reconciliation to obtained the reconciled values for the case with no gross errors (the data in Table 8-8 is free of

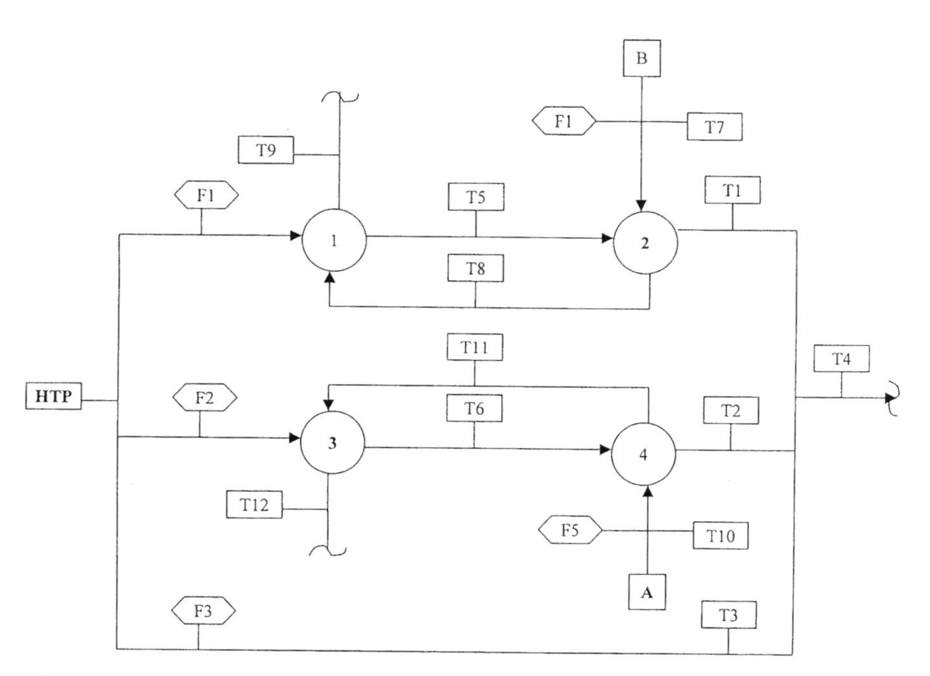

Figure 8-2. Heat exchanger network. *Reproduced from reference [30] with permission of Gulf Publishing Co.*

gross errors). To speed up the calculations, successive linearization is recommended, but any NLP solver can also be used, if available.

Next, change the measured value for flow F5 from 312.9 to 362.5 and for the temperature T8 from 662.0 to 669.1. Apply appropriate gross error detection and identification strategies to find the location of gross errors and their corrected values. To compare the results, see reference [32].

Table 8-8
Data for the Heat Exchanger Network in Figure 8-2

Tag	PV	MV	SD
F1	191.8	182.5	16.2
F2	477.7	462.4	15.4
F3	557.7	562	20.3
F4	315.9	318.3	10.8
F5	298.2	312.9	13.2
T1	674.4	675.5	1.9
T2	603.8	604.8	3.2
T3	750.2	749	2.3
T4	682.5	680.2	3.5
T5	705.7	705.2	2.1
T6	662.1	663.7	2.7
T7	650.7	650.1	1.1
T8	661	662	1.7
T9	676	676.5	2.1
T10	470.1	470	1.4
T11	530	529	2.1
T12	618.6	620	2.3
Node No.	**Q(MMBtu/hr)**	**UA(Btu/hr deg F)**	
1	1.209	20769	
2	0.825	24710	
3	5.908	44814	
4	3.774	28397	

Notation:
PV = operating process value
MV = measured value
SD = standard deviation

SUMMARY

- For detection and identification of multiple gross errors, simultaneous or serial strategies have been designed. Two major types of serial strategies exist: serial elimination and serial compensation.
- The simultaneous strategy for mutiple gross errors based on measurement tests is simple, but it usually detects too many nonexistent errors.
- The simultaneous strategies based on a GLR test require testing of multiple hypotheses and take significant computational time. A method of selecting the most likely hypothesis is required for these strategies.
- The simultaneous strategy based on a GLR test is equivalent to serial elimination strategy based on a global test (elimination of one, two, three, and so on measurements).
- The major advantage of serial elimination is that it does not require prior knowledge about the location and magnitude of gross errors. It might take significant computational time, however, and could suffer a decrease in the solution accuracy due to the reduction in redundancy if too many measurements are eliminated.
- A certain modified serial compensation strategy is equivalent to serial elimination.
- The maximum number of measurements that can be eliminated is equal to the number of constraints (or reduced constraints, when unmeasured variables exist).
- For all gross error detection strategies, the global test should be applied first (to avoid unnecessary calculation of statistical tests, when no gross error exists).
- The SSCS algorithm may detect many nonexistent gross errors, because of possible wrong compensation.
- The MSCS algorithm is equivalent to the serial elimination strategy using the iterative measurement test (IMT) for detection and identification of measurement biases. The MSCS, however, can detect leaks as well.

SUMMARY *(continued)*

- Bounds on variables enhance the performance of gross error detection strategies only if the measured variables are close to either bound.
- Combinations of tests (e.g., nodal and measurement tests) can be successfully implemented but require a strategy for reducing the number of test hypotheses.
- Strategies based on combinations of tests cannot be easily extended to nonlinear models.
- The MIMT and MSCS strategies can be applied to nonlinear models (with some modifications).
- One way to enhance the power of detection and identification of a gross error is to include prior probability of instrument failure and a prior estimate of the magnitude of the gross error and apply a Bayesian type test. The prior infomation can be continuously updated by a sequential application of the Bayesian algorithm.

REFERENCES

1. Serth R. W., and W. A. Heenan. "Gross Error Detection and Reconciliation in Steam-metering Systems." *AIChE Journal* 32 (1986): 733–747.

2. Iordache, C., R.S.H. Mah, and A. C.Tamhane. "Performance Studies of the Measurement Test for Detecting Gross Errors in Process Data." *AIChE Journal* 31 (no. 7, 1985): 1187–1201.

3. Rosenberg, J., R.S.H. Mah, and C. Iordache. "Evaluation of Schemes for Detecting and Identifying Gross Errors in Process Data." *Ind. & Eng. Chem. Proc. Des. Dev.* 26 (1987): 555–564.

4. Rollins, D. K., and J. F. Davis. "Unbiased Estimation of Gross Errors in Process Measurements." *AIChE Journal* 38 (1992): 563–572.

5. Jiang, Q., M. Sanchez, and M. J. Bagajewicz. "On the Performance of Principal Component Analysis in Multiple Gross Error Identification." *Ind. & Eng. Chem. Research* 38 (no. 5, 1999): 2005–2012.

6. Sanchez, M., J . Romagnoli, Q. Jiang, and M. Bagajewicz. "Simultaneous Estimation of Biases and Leaks in Process Plants." *Computers & Chem. Engng.* 23 (no. 7, 1999): 841–858.

7. Yang, Y., R.Ten, and L. Jao. "A Study of Gross Error Detection and data Reconciliation in Process Industries." *Computers Chem. Engng.* 19 (Suppl., 1995): S217–S222.
8. Ripps, D. L. "Adjustment of Experimental Data." *Chem. Eng. Progr. Symp. Series* 61 (no. 55, 1965): 8–13.
9. Narasimhan, S., and R.S.H. Mah. "Generalized Likelihood Ratio Method for Gross Error Identification." *AIChE Journal* 33 (1987): 1514–1521.
10. Keeler, J. Y., M. Darouach, and G. Krzakala "Fault detection of Multiple Biases or Process Leaks in Linear Steady State Systems." *Computers & Chem. Engng.* 18 (1994): 1001–1004.
11. Tong, H., and C. M. Crowe. "Detection of Gross Errors in DATA Reconciliation by Principal Component Analysis." *AIChE Journal* 41 (no. 7, 1995): 1712–1722.
12. Jordache, C., and B. Tilton. "Gross Error Detection by Serial Elimination: Principal Component Measurement Test versus Univariate Measurement Test," presented at the AIChE Spring National Meeting, Houston, Tex., March 1999.
13. Bagajewicz, M., Q. Jiang, and M. Sanchez. "Performance Evaluation of PCA Tests for Multiple Gross Error Identification." *Computers & Chem. Engng.* 3 (Suppl., 1999): S589–S592.
14. Romagnoli, J. A., and G. Stephanopolous. "Rectification of Process Measurement Data in the Presence of Gross Errors." *Chem. Eng. Science* 36 (1981): 1849–1863.
15. Harikumar, P., and S. Narasimhan. "A Method to Incorporate Bounds in Data Reconciliation and Gross Error Detection—II. Gross Error Detection Strategies." *Computers & Chem. Engng.* 17 (no. 11, 1993): 1121–1128.
16. Mah, R.S.H., G. M. Stanley, and D. W. Downing. "Reconciliation and Rectification of Process Flow and Inventory Data." *Ind. & Eng. Chem. Proc. Des. Dev.* 15 (1976): 175–183.
17. Rosenberg, J. "Evaluation of Schemes for Detecting and Identifying Gross Errors in Process Data." M.S. Thesis, Northwestern University, Evanston, Ill.,1985.
18. Rollins, D. K., Y. Cheng, and S. Devanathan. "Intelligent Selection of Hypothesis Tests to Enhance Gross Error Identification." *Computers & Chem. Engng.* 20 (1996): 517–530.
19. Jiang, Q., and M. Bagajewicz. "On a Strategy of Serial Identification with Collective Compensation for Multiple Gross Error Estimation in Linear Data Reconciliation." *Ind. & End. Chem. Research* 38 (no. 5, 1999): 2119–2128.

20. Kim, I. W., M. S. Kang, S. Park, and T. F. Edgar. "Robust Data Reconciliation and Gross Error Detection: The Modified MIMT using NLP." *Computers & Chem. Engng.* 21 (no. 7, 1997): 775–782.

21. Serth R. W., C. M. Valero, and W. A. Heenan "Detection of Gross Errors in Nonlinearly Constrained Data: A Case Study." *Chem. Eng. Comm.* 51 (1987): 89–104.

22. Renganathan, T., and S. Narasimhan. "A Strategy for Detection of Gross Errors in Nonlinear Processes." *Ind. & Eng. Chem. Res.* 38 (1999): 2391–2399.

23. Box, G.E.P, and G. C. Tiao. *Bayesian Inference in Statistical Analysis.* Reading, Mass.: Addison-Wesley, 1973.

24. Tamhane, A. C., C. Iordache, and R.S.H. Mah, "A Bayesian Approach to Gross Error Detection in Chemical Process Data. Part I: Model Development." *Chemometrics and Intel. Lab. Sys.* 4 (1988): 33–45.

25. Tamhane, A. C., C. Iordache, and R.S.H. Mah. "A Bayesian Approach to Gross Error Detection in Chemical Process Data. Part II: Simulation Results." *Chemometrics and Intel. Lab. Sys.* 4 (1988): 131–146.

26. Iordache, C. *A Bayesian Approach to Gross Error Detection in Process Data.* Ph.D. Dissertation, Northwestern University, Evanston, Ill., 1987.

27. Mann, N. R., R. E. Schafer, and N. D. Singpurwalla. *Methods for Statistical Analysis of Reliability and Life Data.* New York: Wiley, 1974.

28. Colombo, A. G., and D. Constantini. "Ground-hypotheses for Beta Distribution as Bayesian Prior." *IEEE Trans. on Reliability* R-29 (no. 1, 1980): 17–20.

29. Romagnoli, J. A. "On Data Reconciliation: Constraint Processing and Treatment of Bias." *Chem. Eng. Science* 38 (1983): 1107–1117.

30. Golub, G. H., and C. Reinsch. "Singular Value Decomposition and Least-squares Solutions." *Numer. Math.* 14 (1970): 403–420.

31. Seber, G.A.F. *Linear Regression Analysis* New York: Wiley, 1977.

32. Albers, J. E.," Data Reconciliation with Unmeasured Variables." *Hydroc. Proc.* (March 1994): 65–66.

9

Gross Error Detection in Linear Dynamic Systems

As it currently stands, industrial applications of dynamic data reconciliation have not been attempted. It is therefore not surprising that only a few attempts have been made to address problems in the subject of gross error detection in dynamic systems even in the research literature. Several developments, however, have occurred in the closely related topic of model-based fault diagnosis which can be gainfully exploited for gross error identification in chemical processes also. The purpose of this chapter is to expose the reader to the issues involved in this problem area and to also provide an introduction to the problem of model-based fault diagnosis.

It should be pointed out that, typically, gross error detection is more concerned with the problem of detecting biases in measured data or process leaks, whereas fault diagnosis treats a wider class of problems associated with sensors, actuators, and the process model. The introduction to fault diagnosis that we provide is only brief in as much as it pertains to the problem of detecting biases in measurements. The interested reader can refer to the book by Patton et al. [1] and the more recent book by Gertler [2] for a more comprehensive treatment of this subject. For the purposes of this chapter, we use the terms *fault* and *gross error* interchangeably.

In Chapter 7, we listed the basic requirements that any gross error detection strategy for steady-state processes should fulfill. These are the abilities to (i) detect the presence of one or more gross errors, (ii) identify the type and location of the gross error, (iii) identify multiple gross errors, and (iv) provide estimates of the gross errors. These requirements carry over to gross error detection techniques for dynamic systems also. In addition, these techniques should consider the following issues:

(i) In steady-state processes, gross error detection strategies exploit only spatial redundancy in the data for the purpose of detecting and identifying gross errors. Similar to dynamic data reconciliation, however, which exploits temporal redundancy in the data for improving the accuracy of estimation, gross error detection and identification can also exploit temporal redundancy for improving diagnostic performance. This is typically achieved by applying gross error detection techniques to a window of measurements made within a chosen time period.
(ii) Since a gross error has an effect only on those measurements that are made after its occurrence, it is also important to estimate as part of the overall gross error detection strategy the time instant at which a gross error has occurred. In the following section, we describe a procedure to meet these requirements.

It can be noted from the preceding chapters that for steady-state processes the techniques of data reconciliation and gross error detection go hand in hand. This is also valid for dynamic systems and the type of state estimator used also has an impact on the gross error detection strategy. For the sake of simplicity we consider only gross errors caused by biases in sensors, although, in principle, the method we describe can be applied for identifying other types of faults. Furthermore, we restrict our consideration to linear dynamic systems for which a Kalman filter estimator is used for state estimation.

PROBLEM FORMULATION FOR DETECTION OF MEASUREMENT BIASES

We consider a linear dynamic system for which Equation 6-1 describes the dynamic evolution of the state variables. If biases in measurements are not present, then Equation 6-2 can be used to describe the relation between measurements and state variables. We assume that the statistical properties of the state and measurement noises given by Equations 6-3 through 6-7 are obeyed for the process. We further assume that a Kalman filter given by Equations 6-13 through 6-17 is used to estimate the state variables at each sampling time.

Consider the case when a bias of magnitude b in measurement i occurs at a time $t = t_0T$. One theoretical model for the bias is to assume that it occurs instantaneously at some time instant and once it occurs its magnitude remains constant for all subsequent times (until the sensor is recalibrated). This is also referred to as a step jump in the measurements [3]. In this case, the measurement model may be written as

$$\mathbf{y}_k = \mathbf{H}_k \mathbf{x}_k + b u_{k-t_0} \mathbf{e}_i + \mathbf{v}_k \tag{9-1}$$

where $\mathbf{e}_i$ is the ith unit vector, is the unit step function which is 1 for all times $k \geq t_0$ and is 0 otherwise (as in Chapter 6, the subscripts t_0 and k are used to represent time instants t_0T and kT).

Equations 6-1 and 9-1 together represent the gross error (or fault) model caused by a measurement bias in sensor i. Although a bias is modeled as a step change, it can also be modeled as a drift with constant rate of change as follows.

$$\mathbf{y}_k = \mathbf{H}_k \mathbf{x}_k + [s(k - t_0)] u_{k-t_0} \mathbf{e}_i + \mathbf{v}_k \tag{9-2}$$

where s is the rate at which the bias changes.

The objectives of a gross error detection strategy are to (i) detect whether a gross error has occurred, (ii) determine the time t_0 at which a gross error has occurred and the measurement i that contains the bias, and (iii) estimate the magnitude b or the rate of change s of the bias as the case may be. For the sake of definiteness, in the subsequent sections we model sensor biases as step changes.

STATISTICAL PROPERTY OF INNOVATIONS AND THE GLOBAL TEST

The principal quantities which are used in gross error detection strategies are the *innovations* defined in Chapter 6, which are computed as part of the Kalman filter estimator at each time.

$$\nu_k = \mathbf{y}_k - \mathbf{H}_k \hat{\mathbf{x}}_{k|k-1} \tag{9-3}$$

The innovations are analogous to measurement residuals which have been used by the measurement test for detecting gross errors in steady state processes. Under the null hypothesis that no biases are present in the measurements made from initial time to current time k, it can be proved [4] that the innovations are normally distributed with expected values and covariance matrix given by

$$E[\nu_k] = \mathbf{0} \tag{9-4}$$

$$\mathrm{Cov}[\nu_k] = \mathbf{V}_k = \mathbf{H}_k \mathbf{P}_{k|k-1} \mathbf{H}_k^T + \mathbf{Q}_k \tag{9-5}$$

Moreover the innovations at time *k* and time *j* are not correlated, that is,

$$\mathrm{Cov}[\nu_k, \nu_j] = \mathbf{0} \qquad j \neq k \tag{9-6}$$

> ***Exercise 9-1.*** *Prove that the innovations follow a Gaussian distribution with statistical properties given by Equations 9-4 through 9-6, when no gross errors are present in the measurements.*

Utilizing the properties of the innovations it is possible to construct a statistical test analogous to the global test defined in Chapter 7 for detecting the presence of a gross error. The test statistic is given by

$$\gamma = \nu_k^T \mathbf{V}_k^{-1} \nu_k \tag{9-7}$$

Under the null hypothesis that no gross errors are present in the measurements, it can be proved that γ follows a chi-square distribution with n degrees of freedom where n is the number of measurements. For a chosen level of significance α, we can choose as the test criterion $\chi^2_{1-\alpha, n}$ and reject the null hypothesis if γ exceeds the criterion. This test can be applied at every sampling time instant to detect when a gross error has occurred. If the test rejects the null hypothesis for the first time at some time instant, say $\hat{t}_0$, then it may be concluded that a gross error has occurred at time $\hat{t}_0$.

Of course, this conclusion is subject to the Type I and Type II error probabilities of the test. In order to protect against these errors, one possibility is to conclude that a gross error has occurred at time t_0, if the test rejects the null hypothesis not only at time $\hat{t}_0$, but also for M out of the next N time instants. Such a simple voting system has been proposed by Rollins and Devanathan [5]. A more elegant approach is to use the *sequential probability ratio test* (SPRT) first proposed by Wald [6] and used in fault diagnosis by Montgomery and Williams [7], among others. This approach, however, has so far not been used in gross error detection.

Instead of constructing the global test using only the innovations at every sampling instant, the innovations obtained over a time window of, say, *N* sampling instants may be jointly used because they have a joint Gaussian distribution by virtue of property, Equation 9-6. The global test statistic in such a case is given by

$$\bar{\gamma} = \sum_{i=k}^{k+N} v_i^T V_i^{-1} v_i \tag{9-8}$$

This test was proposed by Mehra and Peschon [8] and it can be proved that under the null hypothesis, $\bar{\gamma}$ has a chi-square distribution with *Nn* degrees of freedom. For a given level of significance, the test criterion can be chosen from this distribution to detect whether a gross error is present among the *N* measurements within the time window. Using this test, however, it is difficult to estimate the time of occurrence of a gross error.

Example 9-1

We consider a level control process for which a linear discrete model was derived in Example 6-1. Based on the data given in that example, measurements corresponding to the closed loop behavior of the process were simulated without adding any biases to the measurements and the Kalman filter is used to obtain estimates of level and valve position. The steady state value of the covariance matrix of estimation errors for this process can be computed as

$$P_{ss} = \begin{bmatrix} 0.0053 & -0.0005 \\ -0.0005 & 0.0047 \end{bmatrix}$$

and hence the steady state covariance matrix of innovations can be computed as

$$V_{ss} = \begin{bmatrix} 82.5944 & 1.8335 \\ 1.8335 & 46.5330 \end{bmatrix}$$

The global test statistic $\bar{\gamma}$ is computed at each time as well as the cumulative global test statistic $\bar{\delta}$. These are shown in Figures 9-1 and 9-2, respectively, along with the chi-square test criterion at 5% level of significance. It is observed from these plots that the GT rejects the null hypothesis 2 out of 100 sampling instants. The cumulative global test, however, does not reject the null hypothesis at any time within this set of 100 samples. It should be noted that the test criterion for the simple GT is constant at 5.99 while the test criterion for the cumulative GT increase with time, since the number of degrees of freedom increases.

Measurements were also simulated for a bias in valve position of 0.5 volts occurring at initial time and the corresponding GT and cumulative GT statistics for 100 samples are shown in Figures 9-3 and 9-4, respectively. While the GT rejects the null hypothesis for 26 out of 100 samples, the cumulative GT rejects the null hypothesis for all samples. The results indicate that the cumulative GT does not commit Type I errors and is able to detect the presence of the bias for all sampling times. This is expected because the cumulative GT also exploits temporal redundancy. It should be cautioned, however, that, in this simulation, the time at which a gross error occurs is known exactly and, hence, the cumulative GT commits no Type I or Type II errors.

In a sequential application of the cumulative GT, the time of occurrence of the gross error is not known precisely. Therefore, all the measurements used in computing the cumulative GT statistic will not contain the effect of the gross error and the test may not have perfect detection capability. Nevertheless, the results indicate that sequential tests such as SPRT [6], which is also a cumulative test, should be used for the purposes of gross error detection in dynamic processes.

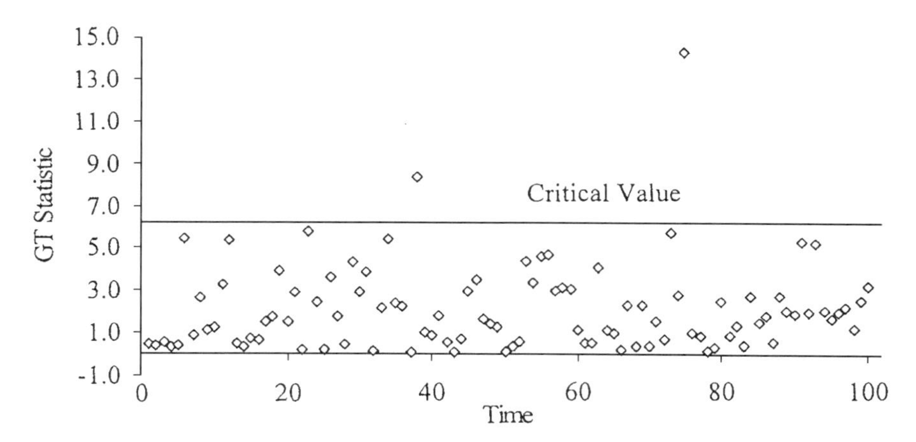

Figure 9-1. Global test statistic for measurements without gross errors.

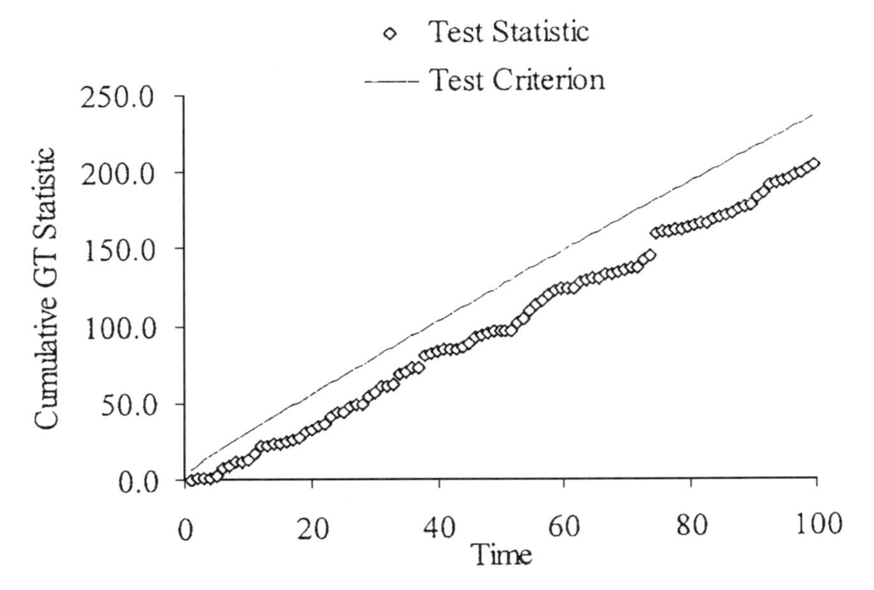

Figure 9-2. Cumulative global test statistic for measurements without gross errors.

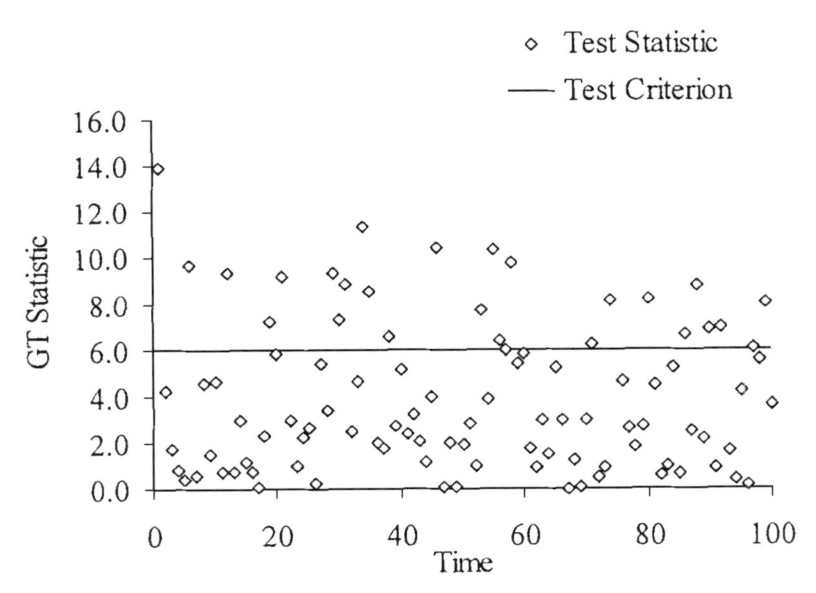

Figure 9-3. Global test statistic for measurements with bias in valve position.

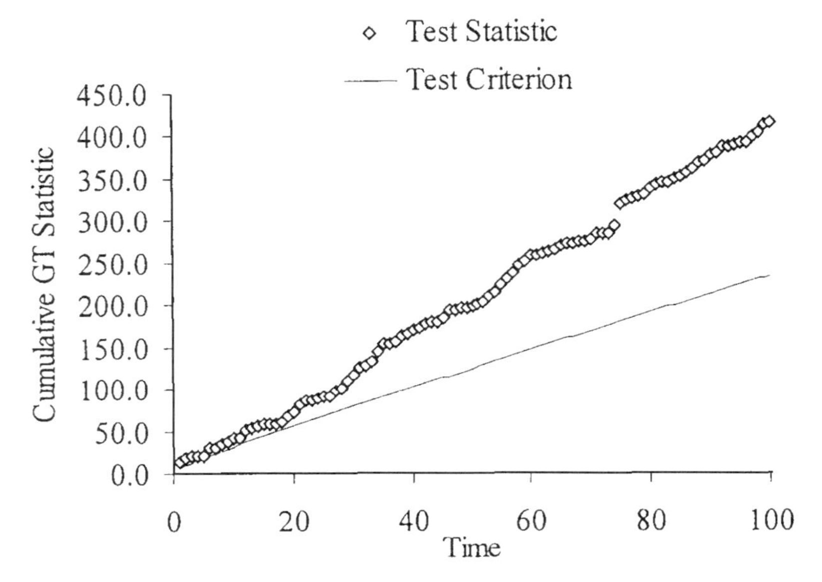

Figure 9-4. Cumulative global test statistic for measurements with bias in valve position.

The global test can detect whether a gross error has occurred and can also be appropriately used to estimate the time of occurrence of a gross error. It requires an identification strategy, however, to determine the type and location of the gross error (as in the case of steady-state processes described in Chapter 7).

Serial elimination strategies, such as those developed for steady-state systems, have not been adapted for dynamic systems as yet—though, in principle, it is possible to devise such strategies in combination with the global test. The generalized likelihood ratio (GLR) test, however, which was used in steady state processes was in fact based on the GLR test proposed by Willsky and Jones [3] for fault diagnosis in dynamic processes. Thus, this test can be used for identifying the type and location of the gross error. In fact, this technique was applied by Narasimhan and Mah [9] for identifying different types of faults including sensor biases for dynamic chemical processes. We discuss the features of the GLR technique for detection, identification, and estimation of sensor biases.

GENERALIZED LIKELIHOOD RATIO METHOD

The GLR test for steady state processes, which is described in Chapter 7, was shown to be capable of identifying different types of gross errors, provided a model for the effect of the gross error on the process (also known as the gross error model) is given. In the case of a dynamic process, the effect of a sensor bias of magnitude b in measurement i that occurs at time t_0 is given by Equation 9-1. The evolution of the state variables is still described by Equation 6-1. Without the knowledge that this gross error has occurred, the Kalman filter estimates will continue to be obtained using Equations 6-13 through 6-17. Therefore, until time $t_0 - 1$ when there is no bias in the measurements, the expected values of the innovations at each time will still be zero. At subsequent times, however, the expected values of the innovations at any time $k \geq t_0$ are given by

$$E[\nu_k] = b\mathbf{G}_{k,t_0}\mathbf{e}_i \qquad k \geq t_0 \tag{9-9}$$

The matrix $\mathbf{G}_{k,t0}$ is referred to as the *signature matrix* and depends on the time k at which the innovations are computed and the time t_0 at which a gross error has occurred. It depends on the system matrices and the type of control law used. For a control law based on the estimates as given by Equation 6-9 we can recursively compute the signature matrix using the following equations.

$$\mathbf{T}_{k,t_0} = \mathbf{A}_k\mathbf{T}_{k-1,t_0} + \mathbf{B}_k\mathbf{C}_{k-1}\mathbf{J}_{k-1,t_0} \tag{9-10}$$

$$\mathbf{G}_{k,t_0} = \mathbf{H}_k\left[\mathbf{T}_{k,t_0} - (\mathbf{A}_k + \mathbf{B}_k\mathbf{C}_{k-1})\mathbf{J}_{k-1,t_0}\right] + \mathbf{I} \tag{9-11}$$

$$\mathbf{J}_{k,t_0} = (\mathbf{A}_k + \mathbf{B}_k\mathbf{C}_{k-1})\mathbf{J}_{k-1,t_0} + \mathbf{K}_k\mathbf{G}_{k,t_0} \tag{9-12}$$

with all the above matrices initialized to the $\mathbf{0}$ matrices when $k < t_0$. It can also be proved that even if a sensor bias is present in the measurements, the innovations follow a Gaussian distribution with covariance matrix given by Equation 9-5. Moreover, the innovations at different times are not correlated. This result is valid in general for other types of *additive* faults, that is, faults whose effect on the process can be modeled as an additive term to the normal process model (compare Equations 6-2 and 9-1 in the case of a sensor bias).

> ***Exercise 9-2.*** *Prove that when a gross error caused by a bias in measurement i occurs at time t_0, the expected values of the innovations are given by Equations 9-9, where the signature matrix is computed recursively using Equations 9-10 through 9-12. Also show that the innovations follow a Gaussian distribution with statistical properties given by Equations 9-4 through 9-6.*

Since the innovations at different times are uncorrelated, they jointly follow a Gaussian distribution. both under the null hypothesis and under the alternative hypothesis that a sensor bias is present. Based on the statistical properties of the innovations established in Exercises 9-1 and 9-2, the GLR test can be applied to a window of *N* innovations computed from time t_0 to t_0+N. The GLR test statistic (which is equal to twice the natural logarithm of the maximum likelihood ratio as in Equation 7-28) can be obtained in this case using

$$T = \underset{i}{\text{Max}}\ T_i \tag{9-13}$$

where T_i is the maximum likelihood test statistic for a bias in measurement *i* given by

$$T_i = \frac{\left(\mathbf{e}_i^T \sum_{j=t_0}^{t_0+N} \mathbf{G}_{j,t_0}^T \mathbf{V}_j^{-1} \nu_j \right)^2}{\mathbf{e}_i^T \left(\sum_{j=t_0}^{t_0+N} \mathbf{G}_{j,t_0}^T \mathbf{V}_j^{-1} \mathbf{G}_{j,t_0}^T \right) \mathbf{e}_i} \tag{9-14}$$

A sensor bias is identified in the measurement i^* which has the maximum test statistic among all measurements. The maximum likelihood estimate of the magnitude of the bias is given by

$$\hat{b} = \frac{\mathbf{e}_i^T \sum_{j=t_0}^{t_0+N} \mathbf{G}_{j,t_0}^T \mathbf{V}_j^{-1} \nu_j}{\mathbf{e}_i^T \left(\sum_{j=t_0}^{t_0+N} \mathbf{G}_{j,t_0}^T \mathbf{V}_j^{-1} \mathbf{G}_{j,t_0}^T \right) \mathbf{e}_i} \tag{9-15}$$

Although it is possible to apply the GLR test using only the innovations at time t_0 to identify the gross error that has occurred at this time, we have exploited temporal redundancy in the data by exploiting all the innovations from time t_0 to t_0+N in the GLR test.

Exercise 9-3. *Using the joint distribution of innovations obtained in Exercises 9-1 and 9-2, derive the GLR test statistic for identifying the location of a sensor bias. Also derive the maximum likelihood estimate of the sensor bias magnitude given by Equation 9-15. Follow a similar procedure as outlined in Chapter 7.*

Example 9-2

The GLR test is applied to the level control process studied in Example 9-1. Using the system model developed in Example 6-1, the expected values of the innovations for biases of specified magnitudes in either the level or valve position measurements can be computed using Equations 9-9 through 9-12. The expected values of innovations (measurement residuals) in level at different sampling instants, after a sensor bias of 0.2 volts (0.317 cm) in level measurement or a sensor bias of 0.5 volts (0.3185 cm) in valve position measurement has occurred at start time, are plotted in Figure 9-5.

Similarly, the expected values of valve position innovations for the same sensor biases are plotted in Figure 9-6. These plots essentially describe the expected evolution of the innovations after the particular sensor bias of specified magnitude has occurred, assuming that the time at which the bias occurs is known precisely. For a bias of a different magnitude, these curves will be shifted up or down. From these figures, it is clear that the expected trend in the two innovations are different for the two sensor biases and it should, in principle, be possible to distinguish between these biases.

For a given set of measurements, the GLR method essentially determines the best fit of the pattern of measurements to these expected trends (shifted appropriately to determine the best estimate of the magnitude) in order to identify the bias that has occurred. The method also accounts for correlation among these trends and the relative magnitude of errors in the innovations. As a particular case, measurements corresponding to a bias

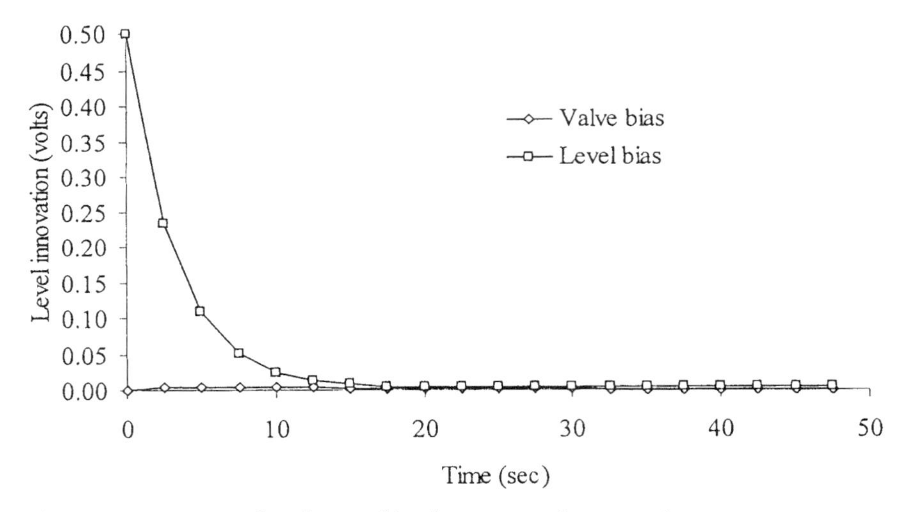

Figure 9-5. Expected evolution of level innovation for sensor biases.

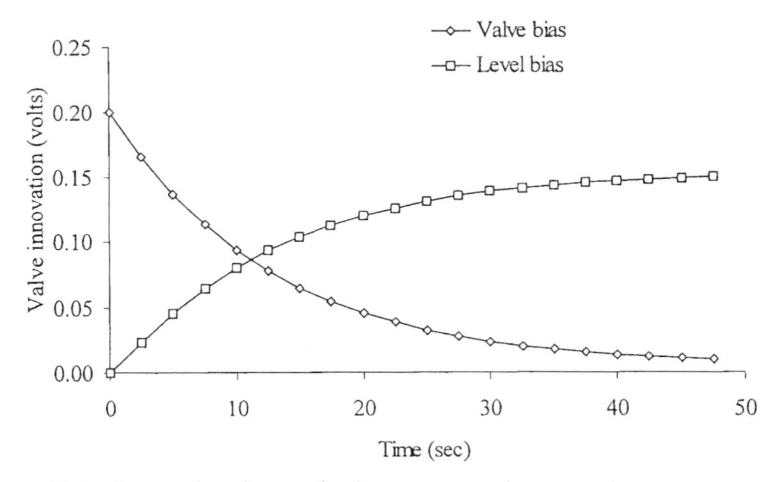

Figure 9-6. Expected evolution of valve innovation for sensor biases.

of magnitude 0.5 volts in valve position were simulated at initial time and the GLR test statistics were computed for window lengths of 10 and 20, respectively.

The GLR test statistics for bias in level and bias in valve position were found to be 29.37 and 4.89, respectively, for a window length of 10 and 40.58 and 7.21, respectively, for a window length of 20. Since the maximum test statistic occurs for bias in level hypothesis, and the test statistic also exceeds the test criterion (3.84 at 5% level of significance), a bias in level is identified. The bias magnitude was estimated as 0.589 for a window length of 10 and 0.516 when a window length of 20 is used. The ability of the GLR method to identify the bias as well as to obtain a more accurate estimate of its magnitude increases with the window length, as expected.

In the above derivation, it is implicitly assumed that the time t_0 at which a sensor bias is presumed to have occurred is known precisely. In practice, only an estimate of this time can be obtained. The procedure described in the preceding section which makes use of the global test can be used for this purpose. Alternatively, Willsky and Jones [3] used the GLR test itself to estimate the time of occurrence of the gross error by treating it as a parameter (similar to the unknown bias magnitude) and obtaining the maximum likelihood ratio over all possible values of t_0 within the time window being considered. This can result in a significant computation burden, especially for large systems unless the system matrices are independent of time.

An on-line algorithm which uses the Kalman filter for estimating the state variables at each time, the global test for detecting the time of occurrence of a gross error and the GLR method for identifying the location and estimating the magnitude of the gross error is as follows. We assume that we are currently at time $k=0$ and have initial estimates of the state variables and the covariance matrix of the estimates.

Step 1. Increment time counter *k,* and use the Kalman filter equations for current time instant *k,* to compute the state variables and the covariance matrix of the state estimates using Equations 6-13 through 6-17. Also compute the innovations ν_k.

Step 2. Apply the global test using the test statistic (Equation 9-7). If the GT rejects the null hypothesis, initialize all elements of the vector **d**, matrix **C** and matrices **T**, **G**, and **J** (which are required for computing the GLR test statistic), and the quantity $\overline{\gamma}$ (required for computing the global test statistic of Equation 9-8) to zero. Set the time index $t_0 = k$ and go to Step 3; or else return to Step 1.

Step 3. Update the matrices **T, G** and **J** using Equations 9-10 through 9-12. Update **d**, **C**, and $\bar{\gamma}$ using the following equations:

$$\mathbf{d} = \mathbf{d} + \mathbf{G}^T \mathbf{V}_k^{-1} \nu_k \quad (9\text{-}16)$$

$$\mathbf{C} = \mathbf{C} + \mathbf{G}^T \mathbf{V}_k^{-1} \mathbf{G} \quad (9\text{-}17)$$

$$\bar{\gamma} = \bar{\gamma} + \nu_k^T \mathbf{V}_k^{-1} \nu_k \quad (9\text{-}18)$$

Step 4. Increment time counter k and compute the state estimates and covariance matrix of state estimates using Kalman filter equations. If the time index $k = t_0 + N$ go to Step 5; or else return to Step 3.

Step 5. Apply the global test using $\bar{\gamma}$. If the GT rejects the null hypothesis, then it confirms a gross error did occur at time t_0. Compute the GLR test statistics $T_i = (d_i)^2/C_{ii}$, where d_i is the ith element of **d** and C_{ii} is the ith diagonal element of **C**. Identify a bias of magnitude d_{i*}/C_{i*i*} in the measurement i^* which gives the maximum value of T_i among all the measurements. Recalibrate the sensor if required. Return to Step 2.

Since the global test which is used to determine the time of occurrence of a gross error in Step 2 may commit a Type I error, it is again applied at Step 5 using all the innovations during the elapsed time window of N measurements to confirm whether a gross error did occur N time steps before. This causes a delay of N time sampling instants before a gross error is detected. It should be noted that the GT (confirmatory test) applied in Step 5 can also commit a Type I error or a Type II error. If a gross error did occur at time t_0, and the GT at Step 5 does not detect this, then we incorrectly conclude that no gross errors are present in the measurements made so far and resume the on-line monitoring procedure from Step 2. This causes a further delay of at least N time steps before the gross error is detected. Other variations of this on-line scheme are also possible.

In the GLR method described above, we have only considered the detection and identification of gross errors due to measurement biases. This approach can also be used to detect gross errors or faults due to biases in actuators, process leaks, or even complete failure of sensors and actuators [9]. Once we expand the definition of gross errors, however, to include these types of faults it is only proper to also critically examine a whole host of methods developed in the general area of fault diagnosis to evaluate their suitability. In the following section, we give a brief introduction to some of the fault diagnosis techniques and recommend to the interested reader the books by Patton et al. [1], Basseville and Nikiforov [10] and Gertler [2] for a more detailed treatment.

FAULT DIAGNOSIS TECHNIQUES

The term *faults* cover a wide range of malfunctions associated with sensors, actuators, and process equipment. They include both *soft faults,* such as biases in sensors or actuators; degradation in equipment performance, such as fouling of heat exchangers or catalyst deactivation, or partial blockages of pipes; and also *hard faults,* such as failure of sensors and actuators or unacceptable leaks from pipes and other process units. Several different techniques have been developed to detect and identify these types of faults. These methods can be grouped into different classes which are briefly described here.

Faults which can be associated with different parameters can be detected and identified by directly estimating these parameters as part of a general state and parameter estimation method. These parameters can be compared with their nominal design values to determine whether a fault exists. For example, fouling of heat exchangers can be detected and identified by estimating the overall heat transfer coefficient of heat exchangers. Similarly, deactivation of a catalyst can be detected and identified by estimating the rate constants of reactions.

A survey of such methods has been presented by Isermann [11]. Himmelblau and coworkers [12, 13] have described the application of this technique for fouling of heat exchangers and fouling of catalyst using a simulated example of a reactor with heat exchange. This technique can also be applied for detecting and identifying sensor or actuator biases by assuming these biases to be present and estimating their magnitudes. Based on the estimated magnitude, a decision can be made whether the bias is large enough to warrant corrective measures.

Bellingham and Lees [14] used this approach for detecting and estimating sensor biases for a simulated example of a level control process. In general, these techniques are more useful when the model is derived from first principles and it is easy to associate the parameters with different faults. The technique, however, can also be applied for detecting and identifying faults or changes in parameters of empirical models from their nominal values [2], but it is difficult to associate these with actual equipment-related faults. The book by Himmelblau [15] describes several techniques for fault diagnosis and applications to chemical processes.

A second class of methods is based on design of observers for fault diagnosis, where the observer can be regarded as a state estimator which has a similar form as a Kalman filter for linear systems, but with the Kalman gain matrix chosen based on other requirements rather than on minimum variance estimation considerations. The innovations obtained from such an observer can be defined by a similar equation as Equation 9-3.

For fault diagnosis using these innovations, the observer or elements of the gain matrix is designed so that the innovations become more sensitive to those faults which we wish to detect. A linear transformation of the innovations may also be used with the transformation matrix being designed to meet the requirements. This technique is equivalent to fault diagnosis techniques based on what are known as parity equations that make use of input-output models rather than a state space representation [2]. Gertler and Luo [16] have described the design of parity equations for a distillation column to make them sensitive to sensor faults and insensitive to unmeasured disturbances in the feed flow and feed compositions.

Another class of fault diagnosis methods are based on designing or structuring residuals such as innovations as parity equations so that each element of the residuals respond (differ from zero significantly) to a particular fault or set of faults but not the others. Thus, the effect of each fault on the residuals can be described by a binary vector known as the *fault signature.* When a fault occurs, a test is applied to each element of the residuals to decide whether it is significantly different from zero or not. Based on these decisions and comparison with the fault signatures, a fault may be detected and identified. Although these methods have not been tried out in chemical processes, it should be pointed out that the concept of a fault signature has been utilized even in the GLR method.

Finally, a relatively new method for sensor fault detection and identification by Dunia et al. [17] uses a principal component model. The use of multivariate *principal component analysis* (PCA) for sensor fault identification via reconstruction provides a reliable technique for fault diagnosis when there is sufficient correlation among the measurements of process variables. The principal component model captures measurement correlations and reconstructs each variable by a successive substitution and optimization. Sensor reconstruction is used to validate the sensor measurement via the PCA model. The procedure proposed by Dunia et al. [17] assumes that one sensor has failed, and that the remaining ones are used for reconstruction. A sequential procedure is used to analyze and validate all sensors.

THE STATE OF THE ART

Extensive research in the area of gross error detection and identification in dynamic chemical processes has not been carried out. Even the few techniques that have been attempted are applicable to linear systems or linearized approximations of nonlinear processes. Bagajewicz and Jiang [18] have proposed an integral dynamic measurement test for gross error detection in linear dynamic processes. This test is essentially the use of the measurement test developed for a steady-state process, which is applied to a dynamic flow process.

In their method, since the integral form of the dynamic flow balances are converted to algebraic equations by modeling the flows and levels as polynomial in time, linear data reconciliation and gross error detection tests developed for steady state processes can be applied. Albuquerque and Biegler [19] have considered the problem of gross error detection in nonlinear dynamic processes. Their method is significantly different from the traditional methods of data reconciliation and gross error detection, and is based on robust estimation of the state variables in the presence of gross errors.

In order to obtain good state estimates even in the presence of gross errors, suitable forms of the objective function for reconciliation are chosen. Although the authors demonstrate that their methods work well for selected examples, extensive testing still needs to be done. Moreover, these methods can only be applied for treating measurement biases and not other types of gross errors or faults.

In summary, it will take more years of research and development effort before industrial applications of gross error detection for dynamic processes can be taken up. Since significant developments have occurred in the field of fault diagnosis, some of which have been applied in the nuclear and aerospace engineering, it is also worthwhile examining how these techniques can be adapted and used for chemical processes also.

SUMMARY

- Gross error detection in dynamic systems involves not only the detection and identification of gross errors but also the time at which a gross error has occurred.
- The innovations used for Kalman filtering for discrete linear dynamic models are usually normally distributed with zero mean and a certain covariance matrix and they can be used to construct statistical tests.
- Global test statistic for gross error detection can be developed for linear dynamic processes. This test can either be for each time instant or for a time window of measurements.
- The time of occurrence of a gross error can be detected either by a global test, a sequential probability ratio test, or the generalized likelihood ratio test.
- The GLR test can be used to detect, identify, and estimate gross errors.
- An on-line gross error detection algorithm which uses Kalman filtering for estimating the state variables, the dynamic global test for detecting the time when a gross error occurs and the dynamic GLR test for gross error identification and estimation of magnitudes of the gross errors for linear dynamic processes is available.
- Techniques developed for fault diagnosis in dynamic systems can be adapted for gross error detection and identification.

REFERENCES

1. Patton, R. et al. *Fault Diagnosis in Dynamic Systems—Theory and Applications.* Englewood Cliffs, N. J.: Prentice-Hall, 1989.

2. Gertler, J. J. *Fault Detection and Diagnosis in Engineering Systems.* New York: Marcel Dekker, 1998.

3. Willsky, A. S., and H. L. Jones. "A Generalized Likelihood Ratio Approach to the Detection and Estimation of Jumps in Linear Dynamic Systems." *IEEE Trans. Automatic Control* AC-21 (1976): 108–112.

4. Maybeck, P. S. *Stochastic Models, Estimation and Control—Vol. 2.* New York: Academic Press, 1982.

5. Rollins D. F., and S. Devanathan. "Unbiased Estimation in Dynamic Data Reconciliation." *AIChE Journal* 39 (1993): 1330–1334.
6. Wald, A. *Sequential Analysis.* New York: Wiley, 1947. Reprint. Dover, 1973.
7. Montgomery, R. C., and J. P. Williams. "Analytic Redundancy Management for Systems with Appreciable Structural Dynamics," in *Fault Diagnosis in Dynamic Systems—Theory and Applications.* (edited by R. Patton, P. Frank, and R. Clark), pp. 361–386. Englewood Cliffs, N. J.: Prentice-Hall, 1989.
8. Mehra, R. K., and J. Peschon. "An Innovations Approach to Fault Detection and Diagnosis in Dynamic Systems." *Automatica* 7 (1971): 637–640.
9. Narasimhan, S., and R.S.H. Mah. "Generalized Likelihood Ratios for Gross Error Identification in Dynamic Processes." *AIChE Journal* 34 (1988): 1321–1331.
10. Basseville, M., and I. V. Nikiforov. *Detection of Abrupt Changes—Theory and Applications.* Englewood Cliffs, N. J.: Prentice-Hall, 1993.
11. Isermann, R. "Process Fault Diagnosis Based on Modeling and Estimation Methods: A Survey." *Automatica* 20 (1984): 387–404.
12. Watanabe, K., and D. M. Himmelblau. "Fault Diagnosis in Nonlinear Chemical Processes—Parts I and II." *AIChE Journal* 29 (1983): 243–260.
13. Park, S., and D. M. Himmelblau. "Fault Detection and Diagnosis via Parameter Estimation in Lumped Dynamic Systems." *Ind. Eng. Chem. Des. Dev.* 22 (no. 3, 1983): 482–487.
14. Bellingham, B., and F. P. Lees. "The Detection of Malfunction Using a Process Control Computer: A Kalman Filtering Technqiue for General Control Loops." *Trans. IChemE* 55 (1977): 253–265.
15. Himmelblau, D. M. *Fault Detection and Diagnosis in Chemical and Petrochemical Processes.* Amsterdam: Elsevier, 1978.
16. Gertler, J. J., and Q. Luo. "Robust Isolable Models for Failure Diagnosis." *AIChE Journal* 35 (1989): 1856–1868.
17. Dunia, R., S. J. Qiu, T. F. Edgar, and T. J. McAvoy. "Identification of Faulty Sensors Using Principal Component Analysis." *AIChE Journal* 42 (no. 10, 1996): 2797–2812.
18. Bagajewicz, M. J., and Q. Jiang. "Gross Error Modeling and Detection in Plant Linear Dynamic Reconciliation." *Computers & Chem. Engng.* 22 (no. 12, 1998): 1789–1809.
19. Albuquerque, J. S., and L. T. Biegler. "Data Reconciliation and Gross Error Detection for Dynamic Systems." *AIChE Journal* 42 (1996): 2841–2856.

10

Design of Sensor Networks

The principal objective of data reconciliation and gross error detection is to improve the accuracy and consistency of estimates of process variables. These techniques certainly reduce the error content in measurements if redundancy exists in the measurements. The extent of improvement that can be achieved depends crucially on (1) the accuracy of the sensors which is specified by the variance in the measurement errors and (2) the number of variables and their type which are measured. Different sensors may be available for measuring a variable with widely varying capabilities such as the range over which it can measure, reliability, and accuracy. The cost of the sensor will be a function of its capabilities. This information must typically be obtained from instrumentation manufacturers or suppliers [1].

If we consider all the different variables such as flow rates, temperatures, pressures and compositions of the streams in a process, these could be of the order of several thousands in number. Clearly, from the viewpoint of cost, complexity or technical feasibility it is not possible to measure each and every variable. Only a subset of these variables are usually measured. It is estimated that the cost of instrumentation (including control elements) is about 2–8% of the total fixed capital cost of a plant [2]. During the design phase of a plant, the decisions regarding which variables should be measured are generally taken when the piping and instrumentation diagrams for the process are prepared [3].

Current practices, however, indicate that these decisions are made based on previous experience with similar processes and thumb rules. We refer to the problem of selecting the variables to be measured as the *design of a sensor network*. Although this problem is an important one in design of new plants, it can also be used to retrofit the measurement structure of existing plants by identifying new variables that need to be measured for improved monitoring and control of the process.

The design of a sensor network is influenced by different considerations, such as controllability of the plant, safety, reliability, environmental regulations, and accurate estimation of all important variables. If the estimates of variables are used in control, then the accuracy of estimation also has an effect on control performance. Keeping the scope of this text in mind, in this chapter we only consider the design of sensor networks for maximizing accuracy of estimation through data reconciliation, while giving due considerations to the cost of the design.

Moreover, the treatment in this chapter is limited to linear (flow) processes only. It should be noted that the objective of maximizing estimation accuracy is only one of the important considerations and a comprehensive design should also take into account other requirements mentioned above. This problem is receiving increasing attention from different researchers in recent years and new solution strategies are being developed. It may require several years of additional effort before these solutions are implemented in practice.

ESTIMATION ACCURACY OF DATA RECONCILIATION

Before we discuss the mathematical formulation of the sensor network design problem, we first examine the estimation accuracy obtained through data reconciliation and the effect that the choice of measured variables have on it. The flow reconciliation example discussed in Example 1-1 in Chapter 1 does highlight some of these issues. We reexamine this problem in greater depth.

Example 10-1

The reconciled estimates of the stream flows for the process shown in Figure 1-2 were presented in Tables 1-1 and 1-2 for different choices of measured variables. Let us consider the results of Table 1-1 for which all flows are measured, and Case 2 of Table 1-2 for which the only the flows of streams 1 and 2 are measured. The difference between the estimated and true values of all streams (estimation errors) can be computed from these results and are shown in Table 10-1, along with the sum of squares of the estimation errors.

From these results, we can observe that the error in the flow estimates of streams 1, 3, 5, and 6 are much less for the case when all flows are measured as compared to the case when only the flows of streams 1 and 2 are measured. The estimation errors, however, for streams 2 and 4 are marginally more when all flows are measured. Although, from a purely intuitive viewpoint, we expect the estimation errors to be reduced if more measurements are available, it is clear from this example that *the estimation errors for all variables are not reduced when more measurements are made.* This is more forcefully brought out by the example presented by Mah [4] where it was shown through simulation that, as more measurements are made, a larger fraction of the reconciled estimates have smaller errors. It is thus clear that it is not appropriate to focus on any particular variable for the purpose of designing sensor networks to increase accuracy of estimation.

Table 10-1
Estimation Errors in Reconciled Flows for Process in Figure 1-2

Stream	Estimation Errors	
	All Flows Measured	Flows 1, 2 Measured
1	0.22	1.91
2	0.50	0.45
3	0.28	1.46
4	0.50	0.45
5	0.28	1.46
6	0.22	1.91
Sum Squares of Estimation Errors	0.7536	11.9644

As an overall measure of estimation accuracy, we can use the sum of squares of the estimation errors of all variables (which represents the overall inaccuracy). Table 10-1 shows that the sum of square of estimation errors is less when more measurements are available. We can therefore use this measure in order to design sensor networks. This measure, however, also depends on the measured values which can be different each time due to their random characteristics. The appropriate measure that we can use for design purposes is the expected value of the sum of squares of the estimation errors which will be independent of the actual outcome of the measurements and will depend only on the sensor network design as well as the inherent process structure.

The overall measure for estimation accuracy was first proposed by Kretsovalis and Mah [5] and is defined by

$$J = E[(\hat{\mathbf{x}} - \mathbf{x})^T (\hat{\mathbf{x}} - \mathbf{x})] \tag{10-1}$$

It should be noted that as *J decreases* the estimation accuracy *increases* and therefore we refer to it as the *measure* of the overall estimation accuracy. It is implicitly assumed in the above definition that all variables are observable. If there are unobservable variables, then the measure of overall estimation accuracy can be written as the expected sum of squares of estimation errors for observable variables only. We will ignore this modification and restrict our considerations throughout this chapter to the design of sensor networks which ensure the observability of all variables. It can be proved [5] that the overall estimation accuracy given by Equation 10-1 for a data reconciliation solution is given by

$$J = Tr(\mathbf{S}) \tag{10-2}$$

where **S** is the covariance matrix of estimation errors and the operator *Tr* is the *trace* of the matrix. It should be noted that the diagonal elements of **S** are the variances of the estimation errors and *J* is therefore the sum of the variances of estimation errors, which is equivalent to the expected sum of squares of the estimation errors. Although it is possible to derive the estimation error covariance matrix from the data reconciliation solutions for the measured and unmeasured variables given in Chapter 3, we describe later an alternative sequential update procedure which is more useful in the context of the sensor network design problem. Different approaches have been developed to solve the sensor network design problem. In the following section, we discuss these methods which consider objectives of estimation accuracy, observability and cost.

SENSOR NETWORK DESIGN

Methods Based on Matrix Algebra

Let us consider a linear flow process for which the material balances are given by Equation 3-2:

$$\mathbf{Ax} = \mathbf{0} \tag{10-3}$$

where the variables $\mathbf{x}$ represent the stream flows. In general, only some of these flow variables are measured and the relationships between the measurements and stream flows can in general be represented using

$$\mathbf{y} = \mathbf{Hx} + \mathbf{v} \tag{10-4}$$

where each row of matrix $\mathbf{H}$ is a unit vector with unit in the column corresponding to the flow variable which is measured, the number of rows being equal to the number of measurements. We have chosen Equations 10-3 and 10-4 to represent a partially measured linear process rather than the equivalent alternative model Equations 3-1 and 3-11, because it is more convenient for the purposes of designing sensor networks.

Minimum Observable Sensor Networks

We are interested in sensor networks which ensure the observability of all variables. We, therefore, first address the question of the minimum number of measurements to be made in order to ensure that every flow variable is observable. We will for convenience refer to such a design as a *minimum observable sensor network.* If there are n stream flows to be estimated and we have m flow constraints, then it is evident that at least n–m flows must be specified. In other words, the minimum number of measurements is n–m.

For the flow process considered in Example 1-1, the minimum of measurements to be made in order to ensure that all stream flows are observable is 2, since there are 6 streams and 4 flow balances. Case 2 of Example 1-2 is a specific instance of a minimum observable sensor network for this process. An additional point to be noted is that in a minimum observable sensor network none of the measured variables is redundant and the reconciled values of these variables are exactly equal to their respective measured values.

Not every combination of n–m measurements will give rise to an observable system, however. For example, in Case 3 of Example 1-2, although two measurements are made, the flows of streams 2 to 5 are unobservable. The condition that a sensor network must satisfy in order to ensure observability of all variables in a linear process is discussed as follows:

Let us separate the flow variables into a set of *n–m independent* variables $\mathbf{x}_I$ and a set of *m* dependent variables $\mathbf{x}_D$, and recast Equation 10-3 as

$$\mathbf{A}_I\mathbf{x}_I + \mathbf{A}_D\mathbf{x}_D = \mathbf{0} \tag{10-5}$$

The dependent variables are chosen in such a way that the matrix $\mathbf{A}_D$ is non-singular. If we measure only the independent variables, then we can use Equation 10-5 to compute unique estimates of the dependent variables as

$$\hat{\mathbf{x}}_D = \mathbf{F}\hat{\mathbf{x}}_I \tag{10-6}$$

where

$$\mathbf{F} = -\mathbf{A}_D^{-1}\mathbf{A}_I \tag{10-7}$$

It is thus clear that this sensor network is a minimum observable sensor network. Therefore, the condition to be satisfied by a minimum sensor network in order to give rise to an observable system is that the sub-matrix corresponding to the unmeasured variables should be nonsingular. Note that this implies that the columns of the constraint matrix corresponding to unmeasured variables are linearly independent (which is the observability condition in Exercise 3-3).

If more measurements are made than the minimum required to ensure observability of all variables, then we obtain a *redundant sensor network design.* Even if a redundant sensor network is designed, it does not automatically imply that all flows are observable. There could be subsets of variables which are unobservable while the rest are redundant. A redundant sensor network gives rise to an observable process if and only if we can choose *n-m* independent variables from among the set of *measured* variables such that the constraint submatrix corresponding to the remaining variables is nonsingular. In this case, the dependent set contains one or more measured variables. We refer to such a design as a *redundant observable sensor network.* We can always obtain a redundant observable sensor network starting from a minimum observable sensor network by choosing to additionally measure one or more of the unmeasured variables in the minimum sensor network.

Estimation Accuracy of Minimum Observable Sensor Networks

We will now consider a minimum observable sensor network design and obtain the overall estimation accuracy for reconciled estimates. Based on the discussion above, we can choose the measured variables as the independent variables. For any observable sensor network (non-redundant or otherwise), the estimates obtained using data reconciliation must satisfy constraint Equation 10-3. Thus, Equation 10-6 can also be used to relate the reconciled estimates for an appropriate choice of the independent and dependent variables.

Since there is no redundancy in a minimum observable sensor network, the estimates of the independent (measured) variables are equal to their respective measurements. This implies that the covariance matrix of estimation errors in the independent variables is equal to $\mathbf{Q}_I$, which is the covariance matrix of measurement errors of the independent variables. Let us denote the covariance matrix of estimation errors corresponding to the independent and dependent variables by $\mathbf{S}_I$ and $\mathbf{S}_D$, respectively. Then we obtain from the preceding arguments that

$$\mathbf{S}_I = \mathbf{Q}_I \tag{10-8}$$

Using Equations 10-8 and 10-6 we can show that

$$\mathbf{S}_D = \mathbf{F}\mathbf{S}_I\mathbf{F}^T \tag{10-9}$$

Combining Equations 10-2, 10-8, and 10-9, the measure for overall estimation accuracy for a minimum observable sensor network can be expressed as

$$J = Tr(\mathbf{S}_I) + Tr(\mathbf{S}_D) = Tr(\mathbf{Q}_I) + Tr(\mathbf{F}\mathbf{Q}_I\mathbf{F}^T) \tag{10-10}$$

Maximum Estimation Accuracy Sensor Network Designs

A **minimum observable sensor network design** that minimizes J defined by Equation 10-10 is desired. In order to solve this problem, a mixed integer optimization problem can be used, which is described later in this chapter. Here we will use a naïve approach and examine every *feasible* combination to determine the optimal solution. We can select every combination of n-m independent variables (such that the submatrix corresponding to the dependent variables is nonsingular). For each combination, the independent variables can be chosen as the measured variables and the measure J for each sensor network design can be computed using Equation 10-10. The combination that gives the least J is the optimal sensor network design that we seek.

Example 10-2

We will illustrate the minimum observable sensor network design that maximizes estimation accuracy for the ammonia process shown in Figure 10-1. We will limit our consideration only to the overall mass flows of this process. For simplicity, let us consider the case when the flow sensors used for measuring any stream have an error with variance equal to 1. Since there are 8 streams and 5 process units, we require a minimum of 3 sensors to observe all variables.

The different feasible combinations of sensor locations along with the corresponding measures of estimation accuracy are shown in Table 10-2. We can observe that there are 6 optimal sensor network designs corresponding to sensor locations (1, 2, 6), (2, 5, 7), (1, 3, 6), (3, 5, 7), (1, 4, 6), and (4, 5, 7) with a minimum expected sum square of estimation errors equal to 11 units. It can also be observed that, although there are

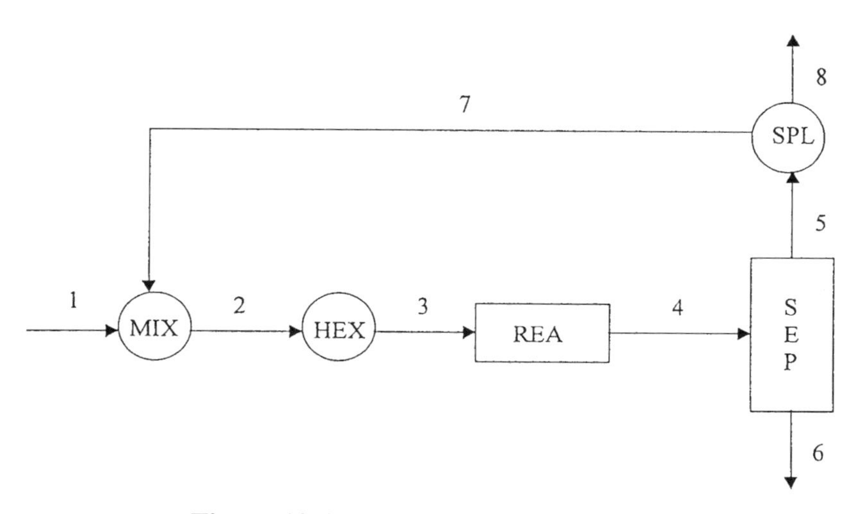

Figure 10-1. Simplified ammonia process.

56 combinations of choosing 3 sensors locations out of 8 sensors locations, only 32 of these combinations give rise to observable sensor network designs.

Table 10-2
Minimum Observable Sensor Network Designs for Ammonia Process

No.	Measured Variables	Overall Expected Estimation Error
1	1, 2, 5	12
2	1, 2, 6	11
3	1, 2, 8	12
4	2, 5, 7	11
5	2, 5, 8	12
6	2, 6, 7	12
7	2, 6, 8	12
8	2, 7, 8	12
9	1, 3, 5	12
10	1, 3, 6	11
11	1, 3, 8	12
12	3, 5, 7	11
13	3, 5, 8	12
14	3, 6, 7	12
15	3, 6, 8	12
16	3, 7, 8	12
17	1, 4, 5	12
18	1, 4, 6	11
19	1, 4, 8	12
20	4, 5, 7	11
21	4, 5, 8	12
22	4, 6, 7	12
23	4, 6, 8	12
24	4, 7, 8	12
25	1, 5, 6	14
26	1, 5, 7	14
27	1, 5, 8	16
28	1, 6, 7	14
29	1, 7, 8	13
30	5, 6, 7	14
31	5, 6, 8	13
32	6, 7, 8	16

Redundant observable sensor networks. The measure of estimation accuracy for redundant observable sensor networks, can be computed using simple update formulae developed by Kretsovalis and Mah [5]. Let us begin with a minimum observable sensor network corresponding to a set of *n-m* measured independent variables, $\mathbf{x}_I$ and the remaining unmeasured variables $\mathbf{x}_D$. The measure of estimation accuracy for this sensor network is given by Equation 10-10. Let us consider the addition of a new sensor to measure one of the variables in the set $\mathbf{x}_D$. Let q be the variance in the error of this new measurement. As in Eqazuation 10-4, the new measurement y can be related to the variables $\mathbf{x}$ by

$$y = \mathbf{h}^T\mathbf{x} + v = \bar{\mathbf{h}}^T\mathbf{x}_I + v \tag{10-11}$$

where

$$\mathbf{h}^T = \mathbf{h}_I^T + \mathbf{h}_D^T\mathbf{F} \tag{10-12}$$

and $\mathbf{h}^T$ is a unit row vector with unity in the column position corresponding to the new øvariable being measured. The expected estimate error covariance matrices of the independent and dependent variables after the addition of this new measurement, $\tilde{\mathbf{S}}_I$, and $\tilde{\mathbf{S}}_D$ respectively are given by [5]

$$\tilde{\mathbf{S}}_I = \mathbf{S}_I - k_1\mathbf{S}_I\bar{\mathbf{h}}\bar{\mathbf{h}}^T\mathbf{S}_I \tag{10-13}$$

$$\tilde{\mathbf{S}}_D = \mathbf{F}\tilde{\mathbf{S}}_I\mathbf{F}^T \tag{10-14}$$

where

$$k_1 = \frac{1}{q + \bar{\mathbf{h}}^T\mathbf{S}_I\bar{\mathbf{h}}} \tag{10-15}$$

The change in the measure of estimation accuracy due to the addition of this new measurement can be shown to be

$$\Delta J = -k_1\bar{\mathbf{h}}^T\mathbf{S}_I\mathbf{C}^T\mathbf{C}\mathbf{S}_I\bar{\mathbf{h}} \tag{10-16}$$

where

$$\mathbf{C} = \begin{bmatrix} \mathbf{I}_{n-m} \\ \mathbf{F} \end{bmatrix} \tag{10-17}$$

Equations 10-12, 10-15, and 10-16 can be directly used to compute the change in the measure of estimation accuracy due to the addition of a new measurement using the covariance matrix of estimate errors in the preceding sensor network design solution. Thus, starting from a minimum sensor network design solution the measure of estimation accuracy for a redundant sensor network design can be obtained by successively adding the required measurements, and using the update equations after each addition. Similar equations can also be derived for deletion of a measurement from a redundant sensor network design solution. In this case, the change in the measure of estimation accuracy is given by

$$\Delta J = k_2 \bar{\mathbf{h}}^T \mathbf{S}_I \mathbf{C}^T \mathbf{C} \mathbf{S}_I \bar{\mathbf{h}} \tag{10-18}$$

where

$$k_2 = \frac{1}{q - \bar{\mathbf{h}}^T \mathbf{S}_I \bar{\mathbf{h}}} \tag{10-19}$$

and the updated covariance matrix of estimate errors in the independent variables is given by

$$\tilde{\mathbf{S}}_I = \mathbf{S}_I + k_2 \mathbf{S}_I \bar{\mathbf{h}} \bar{\mathbf{h}}^T \mathbf{S}_I \tag{10-20}$$

It should be noted that the set of independent variables and dependent variables do not change as new measurements are added or deleted, so that after a series of additions or deletions each of these sets can contain a mixture of measured and unmeasured variables. Care must be taken, however, when a measurement is deleted to ensure that an unobservable design is not obtained. In fact, if a measurement is deleted which can lead to an unobservable process, then the denominator in Equation 10-19 will become zero and this can be used as an indicator to avoid such choices.

Example 10-3

We will consider the ammonia process example and compute the decrease in the expected error in estimates for the addition of a single measurement to a minimum observable sensor network design. The variances of all sensor errors are taken as unity as before. For this purpose we will start with the optimal minimum observable sensor design obtained in Example 10-2 in which the variables 2, 5, 7 are measured.

Choosing these measured variables as the independent variables, the covariance matrix of estimation errors in the independent variables is the identity matrix (of dimension 3). The matrix **F** is given by

$$\mathbf{F} = \begin{array}{c} \begin{matrix} x_2 & x_5 & x_7 \end{matrix} \\ \begin{bmatrix} -1 & 0 & 1 \\ -1 & 0 & 0 \\ -1 & 0 & 0 \\ -1 & 1 & 0 \\ 0 & -1 & 1 \end{bmatrix} \begin{matrix} x_1 \\ x_3 \\ x_4 \\ x_6 \\ x_8 \end{matrix} \end{array}$$

If we choose, in addition, to measure the flow of stream 1, then the vector which relates this measurement to the independent variables is given by

$$\bar{\mathbf{h}}^T = \begin{bmatrix} -1 & 0 & 1 \end{bmatrix}$$

The value of k_1 from Equation 10-15 is equal to $1/(1+2) = 1/3$ and hence the decrease in estimation error can be calculated from Equation 10-16 as –3.3333. The updated covariance matrix of estimate errors of independent variables is given by

$$\mathbf{S}_I = \begin{bmatrix} 0.6667 & 0 & 0.3333 \\ 0 & 1 & 0 \\ 0.3333 & 0 & 0.6667 \end{bmatrix}$$

In order to design an optimal redundant observable sensor network for a specified number of sensors, say r ($r > n-m$) we can start with any minimum observable sensor network design and add $r-n+m$ additional sensors, one at a time and update the measure of estimation accuracy using Equations 10-12, 10-14, and 10-15. We can then *relocate* the sensors by, in turn, adding a new measurement and deleting an existing measurement to get a new redundant observable sensor network design consisting of r sensors. Equations 10-17 through 10-19 can be used for updating the measure of estimation accuracy when a measurement is deleted.

In this manner, all feasible combinations of r sensor locations can be examined in order to find the design which gives the maximum expected estimation accuracy. This will result, however, in an exponential number of solutions to be examined for a general problem. Kretsovalis and Mah [5] outlined two sub-optimal design procedures for a redundant sensor network design for a specified number of sensors, as described on the following page.

Algorithm 1

Step 1. Determine the optimal minimum observable sensor network.

Step 2. Add a new sensor in turn to each of the remaining unmeasured variables and compute the reduction in estimation error using Equation 10-16.

Step 3. Based on the results of Step 2, select the best $r–n+m$ sensor locations (the locations that give maximum reduction in estimation error) to obtain the redundant sensor network design.

Algorithm 2

Step 1. Same as in Algorithm 1.

Step 2. Same as in Algorithm 1.

Step 3. Determine the sensor placement that gives the maximum reduction in expected estimation error from the results of Step 2 and add it to measured set of variables. Stop if the number of measurements made so far is r; or else return to Step 2.

Both the above algorithms do not necessarily give the sensor network design that maximizes estimation accuracy, but reduces the computational burden significantly.

Example 10-4

We apply the above two algorithms for designing redundant observable sensor networks using six measurements for the ammonia process. From Example 10-2, the optimal minimum observable sensor network design corresponding to measured variables 2, 5, 7 is chosen. We have to select three additional variables to be measured with the objective of reducing the expected estimation error as much as possible. Table 10-3 shows the expected decrease in estimation error achieved by adding one, two, and three additional sensors for different combinations of the variables selected to be measured. The maximum estimate error reduction is achieved by choosing to measure additionally the variables 1, 6, 8.

If we apply Algorithm 1 above we would select the variables 1, 6, 8 to be measured since these give the maximum estimate error reduction for addition of a single sensor as observed from column 1 of Table 10-3. On the other hand, if we apply Algorithm 2, then we would first select variable 1 (or 6) to be measured since this gives the maximum estimate error reduction (column 1). In the next iteration, we select variable 6 (or 1) to be measured since this gives maximum estimate error reduction (column 2) among all remaining variables, and, finally, variable 8 is chosen to be measured due to the same reason. In this example, both algorithms give the optimum sensor network design, although in general this may not be the case.

Table 10-3
Expected Estimation Error of Redundant Sensor Network Design for Ammonia Process

Change in Expected Estimation Error (Measurements Added)		
–3.33 (1)	–4.60 (1 3)	–5.14 (1 3 4)
–2.50 (3)	–4.60 (1 4)	–6.67 (1 3 6)
–2.50 (4)	–6.00 (1 6)	–6.77 (1 3 8)
–3.33 (6)	–5.75 (1 8)	–6.67 (1 4 6)
–2.67 (8)	–3.33 (3 4)	–6.77 (1 4 8)
	–4.60 (3 6)	–7.00 (1 6 8)
	–5.17 (3 8)	–5.14 (3 4 6)
	–4.60 (4 6)	–6.00 (3 4 8)
	–5.17 (4 8)	–6.77 (3 6 8)
	–5.75 (6 8)	–6.77 (4 6 8)

Minimum Cost Sensor Network Designs

Instead of maximizing estimation accuracy, a minimum cost sensor network may be designed that ensures observability of all variables. This objective function was considered by Madron and Veverka [6] for sensor network design. Although several other issues were considered in their work, we limit our consideration to the design of minimum observable sensor networks at minimum total cost. The design algorithm proposed by Madron and Ververka [6] essentially attempts to obtain a set of dependent variables such that the measured independent variables will have the least total cost.

The columns of the constraint matrix are first arranged in decreasing order of the cost of the sensor for measuring the corresponding variables. A Gaussian elimination procedure is applied, with the pivot element being chosen from the next available column if possible, and reordering of the rows and columns is done if required. This procedure stops once the first *m* columns form an identity matrix. The least cost minimum observable sensor network design is obtained by measuring the variables corresponding to the remaining *n-m* columns of the constraint matrix. We will illustrate this procedure by means of the following example.

Example 10-5

We consider the ammonia process with the sensor cost data for measuring different variables given by Table 10-4.

Table 10-4
Flow Sensor Costs for Ammonia Process

Stream	Sensor Cost
1	2.5
2	4.0
3	3.5
4	3.0
5	1.0
6	2.0
7	2.0
8	1.5

The constraint matrix for this process is given by

$$\mathbf{A} = \begin{array}{c} \begin{matrix} x_2 & x_3 & x_4 & x_1 & x_6 & x_7 & x_8 & x_5 \end{matrix} \\ \begin{bmatrix} -1 & 0 & 0 & 1 & 0 & 1 & 0 & 0 \\ 1 & -1 & 0 & 0 & 0 & 0 & 0 & 0 \\ 0 & 1 & -1 & 0 & 0 & 0 & 0 & 0 \\ 0 & 0 & 1 & 0 & -1 & 0 & 0 & -1 \\ 0 & 0 & 0 & 0 & 0 & -1 & -1 & 1 \end{bmatrix} \end{array} \begin{matrix} 1 \\ 2 \\ 3 \\ 4 \\ 5 \end{matrix}$$

where the columns are arranged in order of decreasing sensor costs for measuring the corresponding stream flows and the rows are the flow balances for nodes 1 to 5.

After applying Gaussian elimination to obtain an identity matrix in the first *m* columns, we get the following modified matrix

$$\mathbf{A} = \begin{array}{c} \begin{array}{cccccccc} x_2 & x_3 & x_4 & x_1 & x_7 & x_6 & x_8 & x_5 \end{array} \\ \left[\begin{array}{cccccccc} 1 & 0 & 0 & 0 & 0 & -1 & 0 & -1 \\ 0 & 1 & 0 & 0 & 0 & -1 & 0 & -1 \\ 0 & 0 & 1 & 0 & 0 & -1 & 0 & -1 \\ 0 & 0 & 0 & 1 & 0 & -1 & -1 & 0 \\ 0 & 0 & 0 & 0 & 1 & 0 & 1 & -1 \end{array}\right] \begin{array}{c} 1 \\ 2 \\ 3 \\ 4 \\ 5 \end{array} \end{array}$$

The order in which the pivots were selected for Gaussian elimination are (1,1), (2,2), (3,3), (4,4), and (5,6), where the elements within brackets indicate the row and column index of the pivot element. It should be noted that for selecting the pivot elements the columns had to be rearranged since a nonzero pivot element was not available in the next column. The least cost minimum observable sensor network design is obtained by measuring the variables 6, 8, and 5 corresponding to the last three columns of modified matrix A.

Madron and Veverka [6] also considered constraints on the sensor location problem such as specifications of which variables are unmeasureable and which variables were required to be estimated. They also considered the problem of locating additional sensors in a given partially measured process in order to obtain an observable sensor network design at minimum additional cost. In order to solve these problems, the columns of the constraint matrix **A** have to be ordered appropriately before applying Gaussian elimination. The details of the procedure may be obtained from the publication by Madron and Veverka [6].

Methods Based on Graph Theory

Sensor networks for linear flow processes can be designed elegantly using graph-theoretic techniques. Unlike other methods, powerful insights are obtained concerning the structure of the sensor network which make it possible to develop efficient algorithms for solving the design problem. We will again consider the design of sensor networks for maximizing estimation accuracy or for minimizing the total cost. The additional graph-theoretic concepts required for understanding the method discussed in this section can be found in Appendix B.

Maximum Estimation Accuracy Sensor Network Design

Minimum observable sensor networks. In the preceding section we stated that a minimum observable sensor network can be designed by choosing $n-m$ independent variables to be measured such that the constraint submatrix corresponding to dependent variables is nonsingular. In other words, our choice of independent variables should make it possible to express each of the dependent variables as a linear combination of independent variables only.

In Chapter 3, we showed that all unmeasured variables are observable, if no cycle containing only unmeasured variables exists in the process graph. We also showed that in order to ensure observability of all unmeasured variables using a minimum number of measurements, the unmeasured variables should form a spanning tree of the process graph. In other words, a minimum observable sensor network can be designed by simply constructing any spanning tree of the process graph and choosing the flows of chords of the spanning tree as the measured variables. In this case, the chord stream flows are the independent variables and the branch stream flows are the dependent variables. Note that this is similar to the choice of independent and dependent variable choice made in Simpson's method for solving bilinear data reconciliation problems efficiently that was discussed in Chapter 4.

The relationship between dependent and independent variables can also be obtained easily using the fundamental cutsets of the spanning tree. As described in Appendix B, a fundamental cutset, with respect to a branch of the spanning tree, contains one or more chords and the stream flow corresponding to the branch can be written in terms of these chord streams flows as

$$x_i = \sum_{j \in K_i^f} p_{ij} x_j \quad i = 1 \ldots \text{branches} \tag{10-21}$$

where K_i^f is the fundamental cutset with respect to branch i. The elements p_{ij} are 0 if chord j is not an element of K_i^f; otherwise they are +1 or −1 depending on whether chord j has the same or opposite orientation as branch i. If the variance in the measurement error of chord flow j is σ_j^2, then from Equation 10-21, the expected variance of the estimate error of branch flow i can be obtained as

$$\sigma_i^2 = \sum_{j \in K_i^f} \sigma_j^2 \tag{10-22}$$

For a minimum observable sensor network, the estimate of the measured stream is given by the measured values themselves. Thus, the expected variance in the estimate of a measured variable is equal to its measurement error variance. The overall expected estimation error (measure of estimation accuracy) is the sum of all the expected variances in the estimate of all variables. Using Equation 10-22, we get the overall measure of estimation accuracy as

$$J = \sum_{j \in chords} (k_j + 1)\sigma_j^2 \tag{10-23}$$

where k_j is the number of fundamental cutsets of the spanning tree in which chord *j* occurs. Equation 10-23 is exactly equivalent to Equation 10-10 except that it uses spanning tree and fundamental cutset concepts instead of their matrix equivalents.

Example 10-6

The process graph of the ammonia process considered in Example 10-2 is depicted in Figure 10-2. A spanning tree of the process graph is shown in Figure 10-3 which consists of branches 2, 3, 4, 5, and 8 and chords 1, 6, and 7. Corresponding to this spanning tree, in the minimum observable sensor network design the flows of streams 1, 6, and 7 are measured. The fundamental cutsets of this spanning tree are [$\underline{2}$, 1, 7], [$\underline{3}$, 1, 7], [$\underline{4}$, 1, 7], [$\underline{5}$, 1, 6, 7], and [$\underline{8}$, 1, 6] where the branch in each fundamental cutset are denoted by an underscore. Chord 1 occurs in five fundamental cutsets, chord 6 in two, and chord 7 in four. If we assume all measurement error variances as unity, then from Equation 10-23 we get the overall expected estimation error for this sensor design as 14. This value may be compared with the solution given in Table 10-1.

A process graph can contain several spanning trees. The number of spanning trees which is equal to the number of feasible minimum observable sensor network designs can be as large as n^{n-2}, where *n* is the number of nodes in the process graph [7]. There are several algorithms for constructing a spanning tree of a process graph and for finding the fundamental cutsets of the spanning tree. Some of these algorithms along with computer programs in FORTRAN language have been described in Deo [7]. It should be kept in mind that, for the purposes of constructing a spanning tree the direction of the streams are ignored, that is, the process

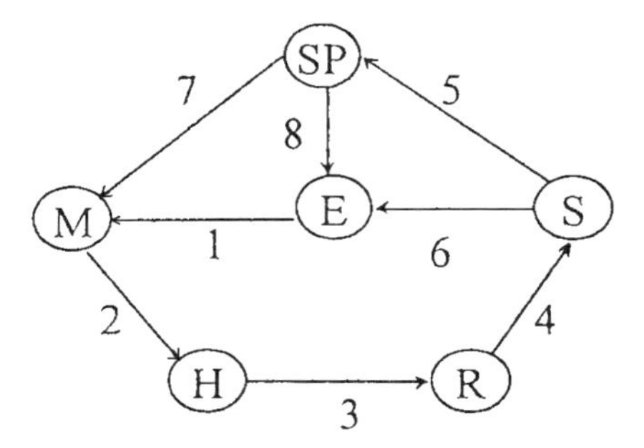

Figure 10-2. Process graph of simplified ammonia plant.

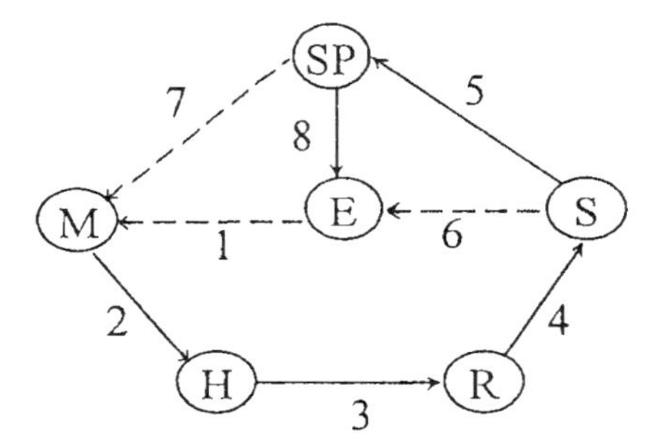

Figure 10-3. Minimum observable sensor network.

graph is treated as an undirected graph. The directions of the streams are used only to obtain the coefficients, p_{ij} in Equation 10-21 if required.

The problem of designing a minimum observable sensor network that maximizes estimation accuracy (or equivalently minimizes J) can be restated as the problem of constructing a spanning tree of the process graph which gives the least value of J as defined by Equation 10-23. Starting with a spanning tree, we can generate a new spanning tree by means of a chord-branch interchange described in Appendix B as an elementary tree transformation.

In this technique, if we add a chord to the spanning tree, then we should delete a branch from the fundamental cycle formed by the chord, in order to obtain a new spanning tree. This elementary tree transformation implies that the sensor measuring the chord stream flow is removed and instead a sensor is used to measure the branch flow deleted from the initial spanning tree solution.

We can start with an arbitrary spanning tree and use the elementary tree transformation technique to successively obtain new spanning trees which gives improved overall estimation accuracy. The only issue to be resolved is to select the chord and branch to be interchanged to improve estimation accuracy. An algorithm for this purpose is outlined below.

Algorithm 3

Step 1. Construct an arbitrary spanning tree of the process graph.

Step 2. Determine all fundamental cutsets of the current spanning tree solution and compute its overall estimation error using Equation 10-23.

Step 3. Find the number of occurrences of each chord in the fundamental cutsets and compute the contribution $(k_i+1)\sigma_i^2$ of each chord *i* to the overall estimation inaccuracy, and rank the chords in decreasing order of their contribution.

Step 4. Select the next ranked chord, say c_i, from the ordered set of chords and find the fundamental cycle formed by chord c_i. Stop if there are no more chords to be examined; else rank the branches in the fundamental cycle in increasing order of their measurement error variances.

Step 5. Select the next ranked branch, say b_j, from the fundamental cycle and interchange chord c_i and branch b_j to obtain a new spanning tree. If there are no more branches to be examined, return to Step 4.

Step 6. Obtain the fundamental cutsets with respect to the new spanning tree and compute the overall estimation error using Equation 10-23 corresponding to this new solution. If the overall estimation error of the new solution is less than that of the old spanning tree, replace old solution with new spanning tree and return to Step 2; or else restore old spanning tree solution and return to Step 5.

In the above algorithm, at each stage an attempt is made to obtain a new spanning tree solution having better estimation accuracy, by a chord-branch interchange in the current spanning tree. If this attempt is successful, then the current solution is replaced by the new one and the procedure repeated. If after systematically examining all possible chord-branch interchanges an improved solution is not obtained, then the algorithm stops.

The procedure adopted in the above algorithm is known as a *local neighborhood search* technique since only the neighboring spanning tree solutions which differ from the current one in respect of only one branch are examined for obtaining a better solution at every iteration. The algorithm, therefore, gives only a local optimum solution and not the global optimum (sensor network design with least estimation error).

Example 10-7

The above algorithm is applied to the ammonia process using the initial spanning tree with branch set [2, 3, 4, 5, 8] considered in Example 10-6. From the fundamental cutsets of this spanning tree obtained in Example 10-6, the overall estimation error is computed as 14. Moreover, the chord set ranked in order of their individual contributions is [1, 7, 6]. We select chord 1 and find the branch set which forms a fundamental cycle with this chord as [2, 3, 4, 5, 8]. Since all measurement error variances are equal, we can choose any branch for the interchange.

If we arbitrarily select branch 2 and interchange with chord 1, we get a new spanning tree with branch set [1, 3, 4, 5, 8]. The overall estimation error of this solution is 12 which is less than the current solution. Therefore, we accept this new solution. If we repeat this procedure with the new solution we find that none of the chord-branch interchanges results in a better solution. Thus, the sensor network design obtained by this algorithm corresponds to the measurements of streams 2, 6, and 7 with estimation error of 12. This solution is worse than the global optimum design which has an estimation error of 11 as observed from Table 10-1.

Algorithms for redundant sensor network designs for maximizing estimation accuracy using graph theoretic techniques are yet to be developed.

Minimum Cost Sensor Network Design

The design of a minimum observable sensor network which has the least cost among all minimum observable sensor networks can be easily accomplished using graph-theoretic techniques. From the discussion in the preceding section, we note that every spanning tree corresponds to a minimum observable sensor network. If we assign a weight to each stream which is equal to the cost of the sensor required to measure the flow of the stream, then the problem considered here is to determine the maximum weight spanning tree, where the weight of a spanning tree is the sum of the weights of its branch streams. The problem of determining the maximum (or minimum) weight spanning tree is one of the classical problems of graph theory that has been well studied. Several algorithms are available for determining the maximum weight spanning tree in a straight-forward manner [7], and we choose to describe Kruskal's algorithm below [8]:

Step 1. Sort the streams (edges of the process graph) in decreasing order of their weights. Initialize the set of edges in the tree, T to be a null set.

Step 2. Pick the next edge from the sorted list.

Step 3. Check if the edge picked forms a cycle with edges of the partial tree constructed so far. If so, discard this edge or else add edge to set T. If the number of edges in T is less than n (number of process units), return to step 2, or else Stop.

The method of checking if an edge forms a cycle with other edges in a set, and the other operations to be carried out when an edge is added to T, are explained in Deo [7]. We illustrate this algorithm in the following example.

Example 10-8

We repeat Example 10-5 to find the least cost minimum observable sensor network of the ammonia process, but this time using Kruskal's algorithm. The order of the streams in decreasing order of weights (sensor costs) is [2, 3, 4, 1, 6, 7, 8, 5]. We pick edge 2 and add it to the tree being constructed. Next, we pick edge 3 from the sorted list and add it to the tree (since it does not form a cycle with edge 2). We continue to pick and add edges 4, and 1 because they do not form cycles with other edges added so far to the tree.

When we next pick edge 6, we find that it cannot be added to tree since it forms a cycle with edges 2, 3, 4, and 1 added to tree so far (can be visually verified from the ammonia process graph of Figure 10-2). So we discard it and pick the next edge in the sorted list and add it to tree since it does not form a cycle with other edges. We stop because we now have added 5 edges to the tree which is equal to the number of process units. The resulting spanning tree is shown in Figure 10-4. Corresponding to the spanning tree the sensor network design measures chord streams 5, 6, and 8. Comparing with the solution obtained in Example

10-5, we can observe that this is the minimum cost sensor network design among all minimum observable networks. The order of choosing the edges of the spanning tree is identical to the order of picking the pivots from the columns of the constraint matrix in the linear algebraic method used in Example 10-5.

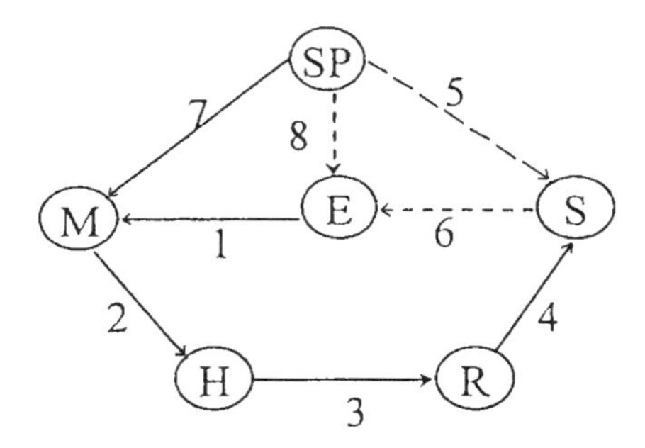

Figure 10-4. Minimum cost observable sensor network.

Methods Based on Optimization Techniques

The problem of sensor network design can also be formulated as a mathematical optimization problem and solved using appropriate optimization techniques. In the preceding sections, the design of sensor networks with the objective of either minimizing cots or maximizing estimation accuracy was considered. Optimization techniques offer the possibility of simultaneously considering different objectives and also imposing other constraints.

Furthermore, it can be extended to consider more general processes involving flows, temperatures, pressures and concentration measurements. Bagajewicz [9] was the first to formulate sensor network design optimization problem. We describe here only the formulation of the problem and refer the reader to standard texts on optimization for the details of the optimization technique used to solve the problem.

In sensor network design, the important decision to be made with regard to each stream flow variable is whether to measure it or not. In order to mathematically formulate these decisions, we can conveniently make use of binary (0-1) integer decision variables q_i, one for each stream *i* which have the following interpretation.

$$q_i = \begin{cases} 0 & \text{if } x_i \text{ is not measured} \\ 1 & \text{if } x_i \text{ is measured} \end{cases} \tag{10-24}$$

Let c_i be the cost of sensor for measuring flow of stream *i,* and let σ_i^* be the maximum allowable standard deviation of the error in the estimate of variable *i.* A minimum cost sensor network design which satisfies the constraints on estimation error can be formulated as

$$\underset{q_i}{\text{Min}} \sum_{i=1}^{n} c_i q_i \tag{10-25}$$

subject to

$$\hat{\sigma}_i \leq \sigma_i^* \qquad i = 1 \ldots n \tag{10-26}$$

$$q_i = 0 \quad \text{or} \quad 1 \quad i = 1 \ldots n \tag{10-27}$$

where $\hat{\sigma}_i$ is the standard deviation of the error in the estimate of flow of stream *i* and is the square root of the diagonal element of covariance matrix of estimation errors, **S,** which can be computed for any choice of sensor locations (defined by the values chosen for the binary variables q_i) using Equation 3-10. The above problem is a mixed integer optimization problem and can be solved using techniques such as branch and bound. Alternatively, commercial optimization packages such as GAMS, GINO or MINOS which have a suite of optimization techniques can be effectively used for solving the above problem.

In the above optimization problem formulation, minimization of the sensor network cost was used as the objective (Equation 10-25), subject to a minimum accuracy specification for the estimates. Alternatively, we can choose to maximize the overall estimation accuracy subject to a maximum limit on the cost of the sensor network. Other constraints, such as minimum desired reliability, can also be included in the problem formulation.

DEVELOPMENTS IN SENSOR NETWORK DESIGN

The earliest statement of the sensor network design problem for maximizing estimation accuracy through data reconciliation was given by Vaclavek [10] for linear (flow) processes. Later, Vaclavek and Loucka

[11] proposed algorithms for sensor network design for ensuring observability of important variables in linear as well as multicomponent (bilinear) processes. Almost two decades later, Kretsovalis and Mah [5] proposed a systematic strategy for solving this problem, where a measure of estimation accuracy was formally defined and algorithms proposed for design of redundant sensor networks for maximizing estimation accuracy.

Ali and Narasimhan [12, 13, 14] addressed the problem of sensor network design for maximizing reliability and developed graph-theoretic algorithms for this purpose. Matrix methods for sensor network design for maximizing reliability were independently developed by Turbatte et al. [15]. Observable sensor network designs for linear and bilinear processes using matrix methods were also addressed by their group [16]. More recently, the issue of sensor network design for improving diagnosability and isolability of faults have been tackled by Maquin et al. [17] and Rao et al. [18].

Independently, Madron and Veverka [6] tackled the problem of minimum cost observable sensor network design using matrix methods. Bagajewicz [9] formulated the sensor network design as an optimization problem. The use of generic optimization algorithms for sensor network design considering different objectives such as cost, estimation accuracy, and reliability was reported by Sen et al. [19]. Recently, Bagajewicz and Sanchez [20] presented a methodology for designing or upgrading a sensor netork in a process plant with the goal of achieving a certain degree of observability and redundancy for a specific set of variables. Although significant progress has been made in the design of sensor networks, a comprehensive strategy simultaneously considering different objectives still has to be developed.

SUMMARY

- The location and accuracy of sensors determine the estimation accuracy of data reconciliation and performance of gross error detection methods.
- The unmeasured flows in a minimum observable sensor network design for a flow process forms a spanning tree structure.
- A minimum cost minimum observable sensor network design is equivalent to the minimum weight spanning tree.
- The general sensor network design can be formulated and solved as a mixed integer nonlinear optimization problem.

REFERENCES

1. Liptak, B. G. *Instrument Engineers' Handbook—Process Measurement and Analysis,* 3rd ed. Oxford: Butterworth-Heinemann, 1995.
2. Peters, M. S., and K. D. Timmerhaus. *Plant Design and Economics for Chemical Engineers.* New York: McGraw-Hill, 1980.
3. Coulson, J. M., J. F. Richardson, and R. K. Sinnott. *Chemical Engineering—Vol. 6. Design.* Oxford: Pergamon, 1983.
4. Mah, R.S.H. *Chemical Process Structures and Information Flows.* Boston: Butterworths, 1990.
5. Kretsovalis, A., and R.S.H. Mah. "Effect of Redundancy on Estimation Accuracy in Process Data Reconciliation." *Chem. Eng. Sci.* 42 (1987): 2115–2121.
6. Madron, F., and V. Veverka. "Optimal Selection of Measuring Points in Complex Plants by Linear Models." *AIChE Journal* 38 (1992): 227–236.
7. Deo, N. *Graph Theory with Applications to Engineering and Computer Science.* Englewood Cliffs, N.J.: Prentice-Hall, 1974.
8. Kruskal, J. B., Jr. "On the Shortest Spanning Subtree of a Graph and the Travelling Salesman Problem." *Proc. Am. Math. Soc.* 7 (1956): 48–50.

9. Bagajewicz, M. "Design and Retrofit of Sensor Networks for Linear Processes." *AIChE Journal* 41 (1997): 2300–2306.
10. Vaclavek, V. "Studies on System Engineering—III. Optimal Choice of the Balance Measurements in Complicated Chemical Engineering Systems." *Chem. Eng. Sci.* 24 (1969): 947–955.
11. Vaclavek, V., and M. Loucka. "Selection of Measurements Necessary to Achieve Multicomponent Mass Balances in Chemical Plants." *Chem. Eng. Sci.* 31 (1976): 1199–1205.
12. Ali, Y., and S. Narasimhan. "Sensor Network Design for Maximizing Reliability of Linear Processes." *AIChE Journal* 39 (1993): 820–828.
13. Ali, Y., and S. Narasimhan. "Redundant Sensor Network Design for Linear Processes." *AIChE Journal* 41 (1995): 2237–2306.
14. Ali, Y., and S. Narasimhan. "Sensor Network Design for Maximizing Reliability of Bilinear Processes." *AIChE Journal* 42 (1996): 2563–2575.
15. Turbatte, H. C., D. Maquin, B. Cordier, and C. T. Huynh. "Analytical Redundancy and Reliability of Measurement System," presented at IFAC/IMACS Symposium SafeProcess '91, Baden-Baden, Germany, 1991, 49–54.
16. Ragot, J., D. Maquin, and G. Bloch. "Sensor Positioning for Processes Described by Bilinear Equations." *Diagnostic et surete de fonctionment* 2 (1992): 115–132.
17. Maquin, D., M. Luong, and J. Ragot. "Fault Detection and Isolation and Sensor Network Design." *RAIRO—APII—JESA* 31 (1997): 393–406.
18. Rao, R., M. Bhushan, and R. Rengaswamy. "Locating Sensors in Complex Chemical Plants Based on Fault Diagnostic Observability Criteria." *AIChE Journal* 45 (1999): 310–322.
19. Sen, S., S. Narasimhan, and K. Deb. "Sensor Network Design of Linear Processes Using Genetic Algorithms." *Computers Chem. Engng.* 22 (1998): 385–390.
20. Bagajewicz, M. J., and M. C. Sanchez. "Design and Upgrade of Nonredundant and Redundant Linear Sensor Networks." *AIChE Journal* 45 (1999): 1927–1938.

11

Industrial Applications of Data Reconciliation and Gross Error Detection Technologies

Data reconciliation technology is widely applied nowadays in various chemical, petrochemical, and other material processing industries. It is applied offline or in connection with on-line applications, such as process optimization and advanced process control.

This chapter presents a review of major industrial applications of data reconciliation and gross error detection reported in the literature. Based on this published information, we only describe the broad features of the applications without going into the details about the particular solution technique or the software used. However, we describe in greater detail two applications (with which the authors were personally associated) in order to highlight some practical problems and their resolution. Although there are many other industrial implementations and software for data reconciliation applications, they are either proprietary (and no detailed information is publicly disclosed) or the published source of information is not easily accessible.

The analysis in this chapter is organized according to the major industrial types of applications for data reconciliation technology. From the multitude of industrial data reconciliation applications, we can distinguish three major types of applications:

1. Process unit balance reconciliation and gross error detection
2. Parameter estimation and data reconciliation
3. Plant-wide material and utilities reconciliation

PROCESS UNIT BALANCE RECONCILIATION AND GROSS ERROR DETECTION

Process unit reconciliation (material and energy balancing for process units) was the first type of application of data reconciliation and gross error detection. The interest in applying data reconciliation methods to industrial data started in the late 1980s, when plant management realized the benefits of using a data reconciliation system and commercial software for data reconciliation and gross error detection became available. Previously, manual operations for material data reconciliation used to be performed by plant engineers for accounting purposes. Raw, unconditioned data was used for process modeling, simulation, and optimization. Some data filtering was used mainly for process control applications. The performance of all these systems using process data was poor, since erroneous data was allowed as inputs.

Currently, most integrated systems for process simulation, optimization and control include a data reconciliation system, which precedes all applications that make use of process data. The use of reconciled data, which are free of gross errors and which satisfy all plant material and energy balances, enhances dramatically the accuracy of process models used for optimization and increases the quality of control operations.

Many implementations of data reconciliation software exist in refineries and petrochemical plants. The simplest industrial applications are for reconciliation of data around single units, especially for distillation or separation columns such as aromatics recovery unit, naphtha cracker downstream units, and crude units (atmospheric and vacuum distillation) [1]. In these off-line applications, the flows and compositions of feed and product streams are reconciled using overall material balances around the column.

In the case of crude units, the petroleum streams are modeled using pseudo-components and material balances for each pseudo-component are written by making use of distillation curve data for each stream obtained from the laboratory. These applications are generally useful before any off-line simulation of the process unit is carried out. The use of reconciled data in the simulation model leads to an improved prediction and optimization of the steady-state performance of the process unit. Application of data reconciliation to reactors such as a catalytic reformer [1] and a fluid catalytic cracker unit [2] have also been reported.

For on-line optimization of processes, it is more appropriate to perform data reconciliation for a set of interconnected process units constituting a subsystem. Data reconciliation of a subsystem comprising the crude distillation tower along with the crude preheat train of exchangers in a refinery has been applied [3]. A similar application was developed for a crude and vacuum system consisting of a crude preheat train, crude heater, atmospheric tower strippers, and dehexenizer [4].

Overall, material and energy balances were used to reconcile the measured flows and temperatures in this subsystem. Other applications include overall material and heat balance refinery reconciliation [5], reconciliation of a multiunit complex of two hydrotreaters and two reformers [3, 6], a hydrogen and sulphur plant [7], a NGL Recovery/ Nitrogen Rejection Unit (NGL/NRU) [8], an ethylene plant [9], and the demethanizer sector of an ethylene plant [10].

In the chemical production, the following examples have been reported: an ethylene dichloride plant [11], catalytic processes for ammonia, methanol and synthesis gas [12], a chemical extraction plant [13], a vinyl acetate plant and a ketene plant [14], and various units of an ammonia plant [15]. Other industrial data reconciliation applications include: mineral processing [16], the screening and cleaning subsystem of a chemical pulping process [17], and a beverage alcohol distillation plant [18].

The above examples are just a few of the actual number of data reconciliation technology implementations. Many of the above examples are *on-line implementations.* Process data is usually provided by a DCS, via a data historian. Data prescreening or filtering is necessary to remove the large outliers.

Most on-line implementations are for *steady-state operations.* Average data (usually for one hour period) are used for steady-state reconciliation. A steady-state detection for process data is usually a part of the steady-state on-line data reconciliation. One method to detect steady-state changes in process plants is using the Hotelling T^2 test or the theory of evidence [19, 20]. The implementation for the FCCU example above [2] is the only one in this list which was tailored for dynamic processes, in connection with an advanced process control system. Its implementation, however, is similar to a steady-state application because the material and energy accumulation terms are approximations based on the observed changes in the levels of the vessels. A pseudo steady-state model is obtained, which works well if the process dynamics are relatively slow.

Various implementation guidelines for data reconciliation software have been summarized by Charpentier et al. [4] as follows:

- Instrument standard deviations are very important in data reconciliation, and a systematic estimation procedure for the standard deviations should be employed.
- Redundancy is another important factor in data reconciliation solution and capability of detecting gross errors. Additional instrumentation may often be necessary to achieve a satisfactory level of redundancy. An optimal sensor location design software is ideal for data reconciliation applications.
- Gross error detection should be always followed by instrument checking and correction. Uncorrected instrument problems deteriorate the quality of the data reconciliation solution.
- Data reconciliation/validation is a complex problem that might require more than one solution technique. Since important flows and temperatures in the plant may be nonredundant, data filtering and validation can be used to provide a quality solution overall.
- A satisfactory data reconciliation system should have enough flexibility to handle process configuration changes, variable standard deviations, missing measurements and to accept various kinds of equations, including inequality process constraints.

More guidelines and challenges still facing data reconciliation technology have been pointed out by Bagajewicz and Mullick [3]:

- For refinery applications with a large variety of stream compositions, proper assay characterization is the key to a successful data reconciliation. With inaccurate compositions, results may not satisfy material balances and good measurements may be identified as gross errors.
- For more accurate data reconciliation, material and energy balance reconciliation is necessary. Heat balances enhance the redundancy in the flow measurements and an improved accuracy in the reconciled flow rates is obtained.
- The steady-state assumption and the data averaging might not be satisfactory for some processes and material balances cannot be accurately closed. Dynamic data reconciliation software needs to be developed for such processes and especially for advanced process control applications.
- Rigorous models do not necessarily increase the accuracy in the data reconciliation solution, but they enable merging data reconciliation and the associated parameter estimation problem in one run.
- Increased accuracy in gross error detection is still a current need, since none of the existent methods and strategies provide effective gross error detection for all types of errors and error locations simultaneously.

Typical software for process units material and energy data reconciliation and gross error detection are: DATACON™ (Simulation Sciences Inc.) [3, 5, 6], DATREC (Elf Central Research) [4], RECON (part of RECONSET of ChemPlant Technology s.r.o., Czech Republic) [21], VALI (Belsim s.a.) [14], and RAGE (Engineers India Limited) [1, 16]. Many NLP-based optimization packages designed for on-line applications have data reconciliation capabilities. They mostly use rigorous models, making the gross error detection more challenging. For that reason, only few have some sort of gross error detection. ROMeo™, a new product of Simulation Sciences Inc. designed for closed loop on-line optimization has data reconciliation and gross error detection capability [22].

PARAMETER ESTIMATION AND DATA RECONCILIATION

A problem associated with data reconciliation is the estimation of various model parameters. Data reconciliation and gross error detection algorithms make use of plant models, which might have totally unknown parameters, or parameters that are changing during the plant operation. Most of these parameters—such as heat transfer coefficients, fouling factors, distillation column tray efficiencies, compressor efficiencies, etc.—are fixed values for the process optimization; therefore, a high accuracy in their estimated values is required.

One approach to parameter estimation problem is to solve it *simultaneously* with the reconciliation problem. The model parameters can be treated as regular unmeasured variables, or as tuning parameters that are adjusted in NLP-type algorithms to match the plant measurements. The major problem with this approach is that in the presence of gross errors, the parameters may be adjusted to wrong values or some measurements can wrongly be declared in gross error because of errors in model parameters. To obtain an accurate solution for both measured variables and model parameters, an iterative process is usually required, which is time consuming, especially with rigorous models.

An alternative approach is to *separate and sequentially solve* the two problems. First, data reconciliation and gross error detection is performed using only overall material and energy balances. The model parameters are then estimated using the reconciled values. This is similar to projecting out the unmeasured model parameters from the reconciliation problem along with their associated model equations. The parameter estimates obtained using the sequential approach are identical to those obtained using the simultaneous approach *if there are no* a priori *estimates of the parameters available.* Moreover, the parameter estimates obtained using the sequential approach may not always satisfy bounds on parameter values. An iterative procedure may be used to eliminate such problems. This approach was applied to parameter estimation problems in connection with advanced process control applications. The computational time is a serious constraint for such applications, and usually only one iteration is applied [2].

PLANT-WIDE MATERIAL AND UTILITIES RECONCILIATION

Plant-wide reconciliation is a very important tool for material and utilities accounting, yield accounting, or monitoring of energy consumed by the process. Many refineries are already saving a significant amount of money by using a production accounting and reconciliation system. The usual term for a plant-wide reconciliation system is *production accounting*; therefore, we will adopt this term for the description in this section. *Yield accounting* is another frequently used term [23, 24].

A production accounting system interacts with various groups in the plant. The operations department provides the input information and collects the reconciled results. The instrumentation group obtains instrument status and performs instrument recalibration and correction if necessary. Process engineers, accounting and financial personnel, and planning and scheduling management retrieve periodic reports for their various needs. Daily, weekly, or monthly reports are standard requirements for a production accounting system.

Various types of data are required for plant reconciliation as indicated in Table 11-1. For better data quality and timely information, a production accounting system is usually integrated with the plant historian and the entire process information/management system. Some data is retrieved automatically from a historian, but other data is entered manually. Human error is a factor affecting the data accuracy (and the reconciled results). For that reason, some sort of **data and model validation** is very important.

Veverka and Madron [25] describe an empirical procedure for detecting topology errors, such as a missing stream or a wrong stream orientation. They used the balance residual (or *deficit*), defined as:

$$r = \text{inputs} - \text{outputs} - \text{accumulation} \tag{11-1}$$

Table 11-1
Data Types for a Production Accounting System

Data Type	Description
Plant Topology	Process units, tanks, and their connecting streams
Process Data	Process data (e.g., compensated flows) from the process and utility units
Tanks Inventories	Inventory from each tank
Movements (from unit to unit/tank) and Transfers (from tank to tank)	Movement data—start/stop time, source node, destination node, quantity transferred
Blends	Blend data—start/stop time, source node, destination node, tank volumes and blending quantities
Receipts	Receipt data—start/stop time, source node, destination node, quantity received
Shipments	Receipt data—start/stop time, source node, destination node, quantities measured
Meters/Sensors Accuracy Factors	Instrument accuracy factors (tolerance, reliability, etc.) for each of the measurement devices
Laboratory	Lab test results (density, %H_2O, compositions, etc.)
Additional Data Entry	Unspecified or adjusted data to be used by the model prior to its balance calculation. Includes missing values, specified values, adjustments, etc.

The balance deficit for each node is compared with a critical value, say r_{crit}. If for a particular node $r > r_{crit}$, then the balance around that node is declared inconsistent. The major problem is how to set the r_{crit} value for each node. A good portion of the node imbalance can be attributed to errors in data, and is therefore admissible. The remaining of the difference is considered modeling error. The value r_{crit} can be obtained from the statistical analysis of the balance residual **r,** which is a random variable (similar to the nodal test for gross error detection described in Chapter 7).

Serious imbalance problems occur due to frequent (daily) changes in some movements [26]. The plant topology for many refinery processes is rather dynamic. There are many "temporary flows" that one day have a nonzero value, and in another day becomes zero (closed valve) or the flow is redirected to a different tank or unit. Mistakes in the reconciliation topology input can very easily be produced. Kelly [26] proposed a more complex strategy for finding the wrong material constraints followed by detection of gross errors in measured data. His strategy is based on a previous algorithm for deleting different combinations of measurements in order to assess the reduction in the objective function, developed by Crowe [27]. Since the deletion process gives rise to a large combinatorial problem, Kelly designed an algorithm to narrow down the number of possible combinations to delete.

A good method for gross error detection is crucial for a production accounting system, which is exposed to many sources of errors. In the presence of significant topology and measurement errors, the reconciled result may become meaningless. The gross error detection task, however, is very challenging for this type of problem because it is very difficult to distinguish between the true measurement errors and topology errors. Leaks and losses and existence of unmeasured flows create an even higher level of complexity. A lot of research effort in data reconciliation is done to resolve these issues and provide more accurate production accounting algorithms.

Some unmeasured flows are observable and can be estimated based on their relationship with measured flows. But the precision of such estimation is often unsatisfactory, due to propagation of errors [25]. An alternative way to get an estimate for unmeasured flows and other variables involved in the plant reconciliation is by using appropriate chemical engineering calculations, which is best accomplished with a process simulator [4, 25]. Process unit data reconciliation performed before plant-wide reconciliation is another possible approach [28].

A more complicated problem for plant reconciliation problems is the estimation of material and energy losses [26, 29]. There are many sources for material and energy losses in a refinery or chemical plant such as flares, fugitive emissions (from volatile organic compounds), leaking valves, fittings, pumps, or heat exchangers, and tank losses (by evaporation or liquid leaks). In addition to the real losses mentioned above, there are apparent losses caused by measurement errors, lab density errors, line fills, or timing errors due to unsynchronized readings in tank gauges and meters.

Many loss models and loss estimation formulas are available. Governmental agencies in the United States such as the American Petroleum Institute (API) and the U.S. Environmental Protection Agency (EPA), provide publications with procedures for predicting fugitive emissions, tank losses, and various leaks. For plant-wide data reconciliation, it is important to clarify how to use these estimates. There are three major ways of including the loss estimates in the data reconciliation model:

1. Treating the losses as unmeasured flows, and reestimating them based on the measured data. This approach does not require a "good" estimate of the loss flows, but requires observability of all loss flows, which is very unlikely to be obtained in a real plant.
2. Modeling leaks separately as explained in Chapter 7. A GLR test procedure can be used to detect leaks and estimate the order of magnitudes of the leaks (or losses). The method might have practical limitations if too many leaks are included in the model (it becomes a large combinatorial problem; also there might not be enough redundancy to accurately estimate the magnitudes of all leaks and losses).
3. Treating the losses and leaks as "pseudo-measured" flows. The estimated loss or leak value is used as a "measured" value, and a relatively higher standard deviation than that for the real measured flows is given to the loss flow. The values of the flow rates for the loss flows are reconciled together with the other measured flows.

Typical software for plant-wide reconciliation and yield (production) accounting is [21]: OpenYield (Simulation Sciences Inc.) and SIGMAfine (KBC Advanced Technologies).

CASE STUDIES

Reconciliation of Refinery Crude Preheat Train Data [16]

A crude preheat train is an important subsystem of a refinery used to minimize the external energy consumption required for heating crude oil. It consists of a network of heat exchangers which are used to preheat crude oil before the crude is sent to a furnace for further heating prior to distillation. The hot streams used for preheating the crude are the distillate streams from the downstream columns. Figure 11-1 shows the crude preheat train of a refinery consisting of 21 exchangers in which the crude is heated by 11 distillate streams.

The flow of the crude before the splitter, as well as the two split flows of the crude, are measured. The inlet flows of all distillate streams are measured as are also the inlet and outlet temperatures of both hot and cold stream to every exchanger. Table 11-2 shows a typical set of measured flows and Table 11-3 shows the measured temperatures. The motivation for reconciling these measurements arises from the need to optimize the crude split flows every few hours. In Chapter 1, we have already described this application and here we will focus only on the reconciliation problem.

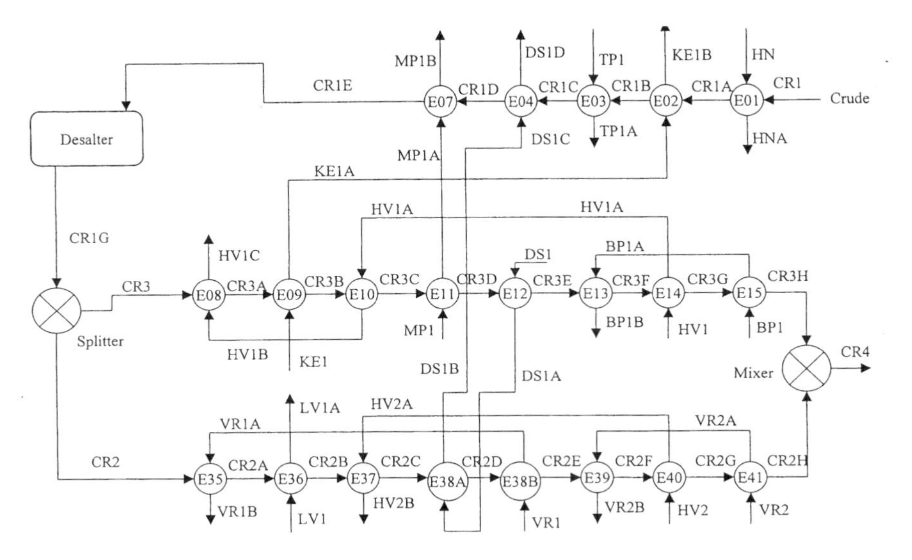

Figure 11-1. Crude preheat train of a refinery.

Table 11-2
Measured and Reconciled Flows of Crude Preheat Train

Stream	Measured Flow (tons/hr)	Reconciled Flows (tons/hr) Before GED	After GED
CR1	370.068	399.2352	409.5705
CR2	151.602	153.1427	151.8511
CR3	255.803	246.0925	257.7194
HN	8.388	8.3865	8.3885
KE1	58.494	58.4888	58.6845
TP1	268.555	267.0283	267.7579
DS1	81.781	80.5460	81.5042
MP1	455.723	454.8024	455.0960
HV1	54.480	54.5657	54.5277
BP1	209.950	209.2158	209.4634
VR1	109.801	106.7616	109.9967
HV2	106.764	106.6069	106.7458
VR2	171.340	170.3060	171.2554

Table 11-3
Measured and Reconciled Temperatures of Crude Preheat Train

Stream	Measured Temperature (C)	Reconciled Temperature (C) Before GED	After GED
CR1	43.075	42.9546	43.1213
CR1A	45.417	45.6173	45.7183
CR1B	64.275	63.2519	63.1073
CR1C	90.442	88.9492	90.4678
CR1D	117.750	122.2442	118.4672
CR1E	128.625	130.4688	129.6710
CR1G	127.608	126.3370	128.3280
CR2	127.608	126.3371	128.3280
CR2A	177.358	184.9598	175.6647
CR2B	190.033	192.5826	190.5285
CR2C	217.608	223.7539	217.7045
CR2D	216.992	211.6510	216.3726
CR2E	230.875	235.3471	232.7172
CR2F	255.192	252.8019	252.1801
CR2G	276.150	281.1727	279.2358
CR2H	291.292	292.6211	290.8145
CR3	127.608	126.3371	128.3280
CR3A	138.225	136.1523	137.1301
CR3B	144.142	143.0534	143.4853
CR3C	152.933	151.8781	152.1530
CR3D	176.192	176.5259	176.7185
CR3E	201.692	200.7378	200.4972
CR3F	217.458	217.6494	217.5341
CR3G	217.458	217.6557	217.6947
CR3H	258.125	262.2992	261.7873
CR4	274.8585	274.1505	272.7489
HN	155.683	155.6410	155.6987
HNA	45.417	45.4199	45.4159
KE1	194.983	195.5111	195.3419
KE1A	168.758	168.5465	169.3793
KE1B	61.750	61.7315	61.6532
TP1	144.142	140.4201	142.3496
TP1A	106.025	107.9266	106.9361
DS1	295.225	294.0763	294.8756
DS1A	224.333	224.9355	224.6271
DS1B	254.908	248.2045	227.1742
DS1C	94.325	95.0369	94.5100
MP1	195.533	192.1993	193.5854
MP1A	181.083	179.3519	180.2033
MP1B	168.833	172.7303	170.9864

(table continued on next page)

Table 11-3 *(continued)*
Measured and Reconciled Temperatures of Crude Preheat Train

Stream	Measured Temperature (C)	Reconciled Temperature (C) Before GED	After GED
HV1	302.300	301.0654	301.4735
HV1A	299.800	301.0387	300.7566
HV1B	263.800	266.6501	265.2976
HV1C	229.492	227.4435	228.3426
BP1	309.458	304.7259	306.4544
BP1A	253.525	253.0694	253.1749
BP1B	230.667	233.4092	232.4505
VR1	283.885	277.3424	283.3504
VR1A	258.167	242.7938	260.7749
VR1B	152.550	158.8106	198.7786
LV1	212.142	206.1244	211.7749
LV1A	190.267	195.0234	190.5526
HV2	302.300	295.0904	300.2211
HV2A	273.617	268.4812	275.4542
HV2B	233.075	240.6166	251.9629
VR2	345.925	345.4640	347.0395
VR2A	345.925	345.4640	347.0395
VR2B	345.925	345.4640	347.0395

The problem in this case is to reconcile all the flows and temperatures so as to satisfy material and energy balances of each process unit of this subsystem. In addition, it is required to estimate the overall heat transfer coefficient of each exchanger given the area and number of tube and shell passes. It is assumed that all the streams are single phase fluids and the specific heat capacity of each stream is given by

$$C_{pi} = a_i + b_i T_i \tag{11-2}$$

where a_i and b_i are constants and T_i is the temperature of the stream in degrees C. The constants a_i and b_i for the different streams are given in Table 11-4. The area and the number tube passes for each exchanger are given in Table 11-5, while all exchangers have a single shell pass.

Table 11-4
Constants for Specific Heat Capacity Correlation

Stream	a	b x 100
CRUDE	0.4442	0.1011
HN	0.4581	0.1036
KE1	0.4455	0.1011
TP1	0.4819	0.1081
DS1	0.4263	0.0975
MP1	0.4455	0.1011
HV1	0.4143	0.0959
BP1	0.4263	0.0975
VR1	0.4092	0.0962
LV1	0.4285	0.0986
HV2	0.4143	0.0959
VR2	0.4062	0.0957

Table 11-5
Heat Exchanger Areas and Number of Tube Passes for a Crude Preheat Train

Exchanger	Area (m^2)	Tube Passes
E01	81.5	1
E02	302	1
E03	374	1
E04	666	2
E07	148	1
E08	148	1
E09	28	1
E10	245	1
E11	464	2
E12	490	2
E13	320	1
E14	269	1
E15	648	2
E35	954	2
E36	206	1
E37	97	1
E38A	181	1
E38B	181	1
E39	199	1
E40	556	2
E41	310	1

Using a standard deviation of 1% of the measured values for all stream flows and temperatures, the reconciled estimates are obtained assuming that no gross errors are present in the data. In the third column of Tables 11-2 and 11-3, the reconciled flows and temperatures, respectively, are shown. (All the results of this case study were obtained using the software package RAGE.) The constraints that are used for each exchanger are the flow balances for the hot and cold streams, as well as the enthalpy balance.

For the mixer, flow and enthalpy balances are imposed while, for the splitter, a flow balance and equality of temperatures across the splitter are imposed. No other feasibility constraints or bounds on the variables are imposed. In order to remove gross errors from the data, the GLR test along with serial compensation strategy is applied for multiple gross error detection after linearization of the constraints around the reconciled estimates. The final reconciled estimates after all gross errors are identified and compensated are also shown in the last column of Tables 11-2 and 11-3.

We focus on some interesting problems/features of the measured and reconciled data. If we consider the measured temperatures of streams incident on exchanger E38A (the streams CR2C, CR2D, DS1A, and DS1B), we note that the crude stream is getting cooled from 217.608 to 216.992 degrees C, while the intended "hot" distillate stream is getting heated from 224.333 to 254.908 degrees C. Although it may be possible for the roles of hot and cold streams to be reversed depending on the prevailing flows and temperatures, what is unacceptable here is that heat is being transferred from the lower temperature crude to the higher temperature distillate stream which is thermodynamically infeasible. It can be verified that reconciliation before or after *gross error detection* (GED) does not correct this problem and the estimates for exchanger E38A still violate thermodynamic feasibility. If we use these estimates to obtain an estimate of the overall heat transfer coefficient for this exchanger, then we obtain a negative value for it which is absurd.

In order to obtain thermodynamically feasible estimates, several possibilities were examined. One general approach is to include feasibility constraints at the hot and cold end of each exchanger of the form

$$T_{hi} - T_{ci} \geq 0 \qquad (11\text{-}3)$$

$$T_{ho} - T_{co} \geq 0 \qquad (11\text{-}4)$$

where the subscript i is the inlet and subscript o is the outlet end of the exchanger. This would, however, increase the number of constraints significantly. Moreover, this presupposes knowledge of the cold and the hot streams for each exchanger and does not allow any role reversal. A simpler technique is to include the relation between overall heat transfer coefficient (U) and heat load for every exchanger and impose bounds on the overall heat transfer coefficient. If we impose a nonnegativity restriction on U, then we can ensure that thermodynamic feasibility is maintained regardless of which of the streams plays the role of the hot stream and which plays the role of the cold stream.

Using this approach, the reconciliation problem was solved again. (Note that, as explained in Chapter 5, in order to solve this problem a constrained nonlinear optimization program has to be used and the unmeasured heat transfer coefficient parameters cannot be eliminated using a projection matrix.) The reconciled temperature estimates of the four streams incident on exchanger E38A alone before GED and after GED are shown in the second column of Tables 11-6 and 11-7, respectively. For comparison, we also reconcile the problem by deleting each of the four suspect temperature measurements in turn and also after deleting all the four temperature measurements which violate thermodynamic feasibility. The reconciled temperature estimates for these four streams before and after GED are also shown in Tables 11-6 and 11-7.

Table 11-6
Reconciled Temperatures Before GED Around Exchanger E38A for Different Cases

Stream	Reconciled Temperatures (C)					
	Bounds on U	T of DSIA unmeasured	T of DS1B unmeasured	T of CR2C unmeasured	T of CR2D unmeasured	All Four T's unmeasured
DS1A	235.6707	227.4071	222.3988	222.7002	223.2842	198.7978
DS1B	235.6655	248.6476	232.2426	250.8886	250.1673	231.7550
CR2C	217.2356	223.1970	219.9628	229.2839	221.1179	227.4480
CR2D	217.2356	212.1257	214.8731	214.7073	207.0257	210.7080

Table 11-7
Reconciled Temperatures After GED Around Exchanger E38A for Different Cases

Stream	Reconciled Temperatures (C)					
	Bounds on U	T of DSIA unmeasured	T of DS1B unmeasured	T of CR2C unmeasured	T of CR2D unmeasured	All Four T's unmeasured
DS1A	225.7176	221.1880	224.4315	224.4547	223.9519	215.7303
DS1B	225.7175	255.4171	217.9601	255.2667	255.5525	223.5384
CR2C	216.8586	235.1001	217.0624	236.3756	233.8225	231.2648
CR2D	216.8586	217.0614	220.4571	220.0339	217.1436	227.2575

From the reconciled estimates, it can be observed that by imposing non-negativity bounds on U, it is possible to obtain feasible estimates before and after GED. In fact, the results show that the heat transfer coefficient for exchanger E38A is at its lower bound of zero, which implies that this exchanger is being bypassed completely by one of the streams, resulting in the temperatures of both hot and cold streams being unchanged across this exchanger.

The only other case when feasible temperature estimates were obtained was after deleting the temperature of stream DS1B and application of GED (refer to column 4 of Table 11-7). Even in this case, the stream temperatures change marginally across exchanger E38B, indicating that this exchanger is largely being bypassed. (This was also later confirmed after inspecting the manual valve positions on the crude bypass line for this exchanger.) The results clearly demonstrate that imposition of bound constraints on the parameters can be used as a generic method to obtain feasible estimates.

This case study also brought out other issues that needed to be addressed in practice:

- It was assumed that all the streams are in a single phase. A more rigorous method would require the state of the stream to be determined and an appropriate correlation to be used for determining the stream enthalpy.
- Some fraction of the crude flows was being bypassed in a few other exchangers also, but sufficient measurements (redundancy) were not available for treating the bypass fractions as unknown parameters and estimating them as part of the reconciliation problem.
- Heat losses from exchangers were not accounted for in the enthalpy balances of exchangers. In order to use the reconciled data for better

optimization, it may be necessary to include a heat loss term in the enthalpy balances. However, enough redundancy does not exist for treating the loss terms as unknowns. One possibility is to assume a specified fraction of the heat load of exchangers to be lost based on past experience or based on recommended loss estimation methods.

- As pointed out in Chapter 1, the reconciliation of the crude preheat train data was performed every four hours using averaged measured data for the preceding two hours. Since the heat transfer coefficients of exchangers cannot be expected to change dramatically from one time period to the next, it is possible to use their estimates derived in one time period as "measurements" for the next time period with a larger standard deviation. Due to this extra redundancy, better estimates can be obtained. Moreover, the heat transfer coefficient estimates change smoothly from one time period to the next and will not fluctuate wildly. A trend of the heat transfer coefficients can be used to decide when cleaning/maintenance procedures have to be initiated for the exchangers.

Reconciliation of Ammonia Plant Data [30]

Ammonia is a chemical product with many industrial applications such as refrigerants and fertilizers. Figure 11-2 shows a simplified process flowsheet diagram for the synthesis section of an ammonia process [30]. Ammonia is produced by an exothermic reaction of nitrogen and hydrogen:

$$N_2 + 3H_2 \rightarrow 2NH_3$$

The feed stream S1 to the synthesis section already contains ammonia from upstream processes. To separate it, stream S1 is cooled and sent to flash drum F1, where the ammonia-rich liquid S3 is separated from the remaining vapor S2. Before entering the reactor section, the vapor stream is preheated by a product stream. The reactor section consists of two reaction stages and two internal heat exchangers. Stream S4 is split into three streams (S5, S6, and S7) and the split fractions are used to control the reactor feed temperatures.

Stream S7 is used to quench the hot product stream from the first-stage reactor and stream S5 is used to recover some of the heat from the product of the second stage reactor (S13). The three streams are then recombined and fed to the first-stage reactor. Most of the cooled reactor product (S15) is recycled to another section of the plant (stream S16), while the remainder (S17) is further cooled with refrigerant (stream S22) to condense most of the ammonia (stream S20). The two condensed streams (S3 and S20) are combined and further purified downstream.

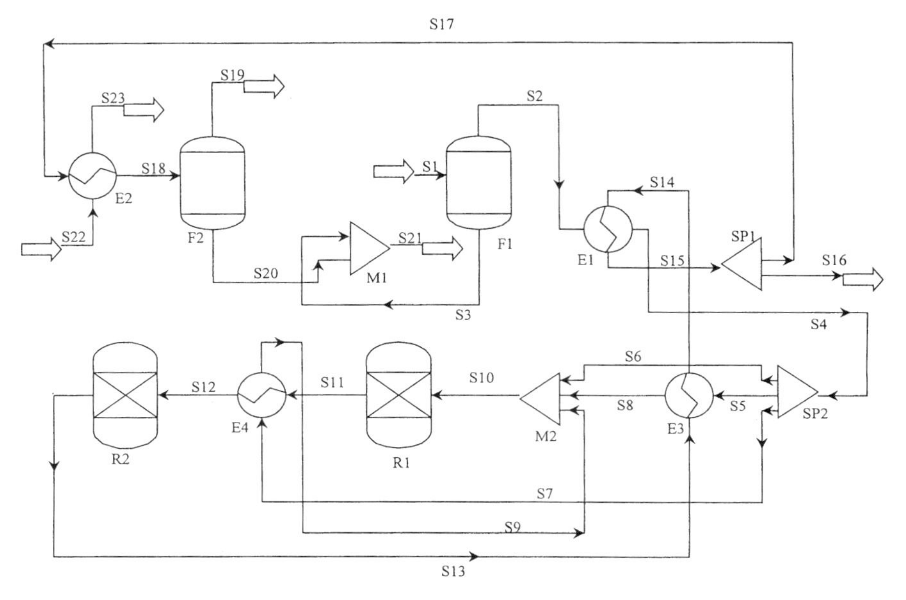

Figure 11-2. An ammonia synthesis industrial process.

The ammonia synthesis plant contains instrumentation for measuring flow rates, temperatures and various stream compositions (mole fractions). The measured values, their associated standard deviations and the reconciled values are reported in Tables 11-8 and 11-9. Tables 11-8, 11-9, and 11-10 show the reconciliation results for a case where no gross errors were found (Case A). Table 11-8 reports all stream calculation results, for both measured and unmeasured data. Other calculated values such as reaction extents, heat exchanger duties, UA values, and flash data, are also reported at the bottom of Table 11-8.

(text continued on page 354)

Table 11-8
Stream and Unit Reconciliation Solution for Ammonia Example
Case A. No Gross Errors Present in Measurements

STRM	VBL	STAT	UOM	TAG NAME	STANDARD DEVIATION	MT-STAT	MEASURED VALUE	CALC VALUE
S1								
	RATE	U	M3/HR					256.367
	TEMP	M	C	T01	1.50	1.50	–23.00	–22.46
	PRES	F	ATM					150.000
	X1	U	MOL%					32.7478
	X2	U	MOL%					11.8345
	X3	U	MOL%					7.1737
	X4	U	MOL%					2.3594
	X5	U	MOL%					45.8846
S10								
	RATE	M	M3/HR	F10	3075.000	.30	1.03E+05	1.02E+05
	TEMP	M	C	T10	5.00	.69	380.00	378.16
	PRES	F	ATM					150.000
	X1	U	MOL%					58.9875
	X2	U	MOL%					21.3170
	X3	U	MOL%					12.9218
	X4	U	MOL%					4.2499
	X5	U	MOL%					2.5238
S11								
	RATE	M	M3/HR	F11	2880.000	.07	9.60E+04	9.62E+04
	TEMP	M	C	T11	5.00	.69	465.00	466.86
	PRES	F	ATM					150.000
	X1	U	MOL%					53.8199
	X2	U	MOL%					19.6885
	X3	U	MOL%					13.6554
	X4	U	MOL%					4.4912
	X5	U	MOL%					8.3450
S12								
	RATE	U	M3/HR					9.62E+04
	TEMP	M	C	T12	5.00	.67	390.00	388.20
	PRES	F	ATM					150.000
	X1	U	MOL%					53.8199
	X2	U	MOL%					19.6885
	X3	U	MOL%					13.6554
	X4	U	MOL%					4.4912
	X5	U	MOL%					8.3450

(table continued on next page)

Table 11-8 *(continued)*
Stream and Unit Reconciliation Solution for Ammonia Example
Case A. No Gross Errors Present in Measurements

STRM	VBL	STAT	UOM	TAG NAME	STANDARD DEVIATION	MT-STAT	MEASURED VALUE	CALC VALUE
S13								
	RATE	M	M3/HR	F13	2775.000	.16	9.25E+04	9.21E+04
	TEMP	M	C	T13	5.00	1.07	450.00	453.63
	PRES	F	ATM					150.000
	X1	U	MOL%					49.5273
	X2	U	MOL%					18.3357
	X3	U	MOL%					14.2649
	X4	U	MOL%					4.6916
	X5	U	MOL%					13.1805
S14								
	RATE	M	M3/HR	F14	3000.000	.15	9.25E+04	9.21E+04
	TEMP	M	C	T14	5.00	.02	290.00	290.07
	PRES	F	ATM					150.000
	X1	M	MOL%	H2-14	1.0000	.67	49.0000	49.5273
	X2	M	MOL%	N2-14	1.0000	.39	18.0000	18.3357
	X3	M	MOL%	C1-14	1.0000	.86	15.0000	14.2649
	X4	M	MOL%	AR-14	1.0000	.36	5.0000	4.6916
	X5	M	MOL%	NH3-14	1.0000	.26	13.0000	13.1805
S15								
	RATE	U	M3/HR					9.21E+04
	TEMP	U	C					127.69
	PRES	F	ATM					150.000
	X1	U	MOL%					49.5273
	X2	U	MOL%					18.3357
	X3	U	MOL%					14.2649
	X4	U	MOL%					4.6916
	X5	U	MOL%					13.1805
S16								
	RATE	M	M3/HR	F16	3000.000	.71	9.24E+04	9.04E+04
	TEMP	U	C					127.69
	PRES	F	ATM					150.000
	X1	U	MOL%					49.5273
	X2	U	MOL%					18.3357
	X3	U	MOL%					14.2649
	X4	U	MOL%					4.6916
	X5	U	MOL%					13.1805

Table 11-8 *(continued)*
Stream and Unit Reconciliation Solution for Ammonia Example Case A. No Gross Errors Present in Measurements

STRM	VBL	STAT	UOM	TAG NAME	STANDARD DEVIATION	MT-STAT	MEASURED VALUE	CALC VALUE
S17								
	RATE	M	M3/HR	F17	48.000	1.84	1600.000	1665.585
	TEMP	U	C					127.69
	PRES	F	ATM					150.000
	X1	U	MOL%					49.5273
	X2	U	MOL%					18.3357
	X3	U	MOL%					14.2649
	X4	U	MOL%					4.6916
	X5	U	MOL%					13.1805
S18								
	RATE	U	M3/HR					1665.585
	TEMP	M	C	T18	1.00	1.50	–15.00	–15.24
	PRES	F	ATM					150.000
	X1	U	MOL%					49.5273
	X2	U	MOL%					18.3357
	X	U	MOL%					14.2649
	X4	U	MOL%					4.6916
	X5	U	MOL%					13.1805
S19								
	RATE	M	M3/HR	F19	45.000	.84	1500.000	1472.477
	TEMP	U	C					–8.15
	PRES	F	ATM					150.000
	X1	M	MOL%	H2-19	1.0000	1.34	57.0000	56.0226
	X2	M	MOL%	N2-19	1.0000	.32	21.0000	20.7403
	X3	M	MOL%	C1-19	1.0000	1.40	15.0000	16.1357
	X4	M	MOL%	AR-19	1.0000	.38	5.0000	5.3069
	X5	M	MOL%	NH3-19	1.0000	.39	2.0000	1.7945
S2								
	RATE	U	M3/HR					1.02E+05
	TEMP	U	C					–21.82
	PRES	M	ATM	P02	.500	1.50	150.000	149.998
	X1	U	MOL%					58.9875
	X2	U	MOL%					21.3170
	X3	U	MOL%					12.9218
	X4	U	MOL%					4.2499
	X5	U	MOL%					2.5238

(table continued on next page)

Table 11-8 *(continued)*
Stream and Unit Reconciliation Solution for Ammonia Example Case A. No Gross Errors Present in Measurements

STRM	VBL	STAT	UOM	TAG NAME	STANDARD DEVIATION	MT-STAT	MEASURED VALUE	CALC VALUE
S20								
	RATE	U	M3/HR					.237
	TEMP	U	C					–8.15
	PRES	F	ATM					150.000
	X1	F	MOL%					.0000
	X2	F	MOL%					.0000
	X3	F	MOL%					.0000
	X4	F	MOL%					.0000
	X5	F	MOL%					100.0000
S21								
	RATE	U	M3/HR					100.237
	TEMP	U	C					–21.79
	PRES	F	ATM					150.000
	X1	U	MOL%					.0000
	X2	U	MOL%					.0000
	X3	U	MOL%					.0000
	X4	U	MOL%					.0000
	X5	U	MOL%					100.0000
S22								
	RATE	M	KG/HR	F22	11.250	1.50	375.000	365.382
	TEMP	M	C	T22	1.20	.73	–30.00	–29.12
	PRES	F	ATM					1.200
	X1	F	MOL%					.0000
	X2	F	MOL%					.0000
	X3	F	MOL%					.0000
	X4	F	MOL%					.0000
	X5	F	MOL%					100.0000
S23								
	RATE	U	KG/HR					365.382
	TEMP	U	C					–29.12
	PRES	F	ATM					1.200
	X1	F	MOL%					.0000
	X2	F	MOL%					.0000
	X3	F	MOL%					.0000
	X4	F	MOL%					.0000
	X5	F	MOL%					100.0000

Table 11-8 *(continued)*
Stream and Unit Reconciliation Solution for Ammonia Example
Case A. No Gross Errors Present in Measurements

STRM	VBL	STAT	UOM	TAG NAME	STANDARD DEVIATION	MT-STAT	MEASURED VALUE	CALC VALUE
S3								
	RATE	M	M3/HR	F03	.500	1.50	100.000	100.000
	TEMP	U	C					–21.82
	PRES	F	ATM					150.000
	X1	F	MOL%					.0000
	X2	F	MOL%					.0000
	X3	F	MOL%					.0000
	X4	F	MOL%					.0000
	X5	F	MOL%					100.0000
S4								
	RATE	M	M3/HR	F04	3075.000	.30	1.03E+05	1.02E+05
	TEMP	M	C	T04	2.00	.31	135.00	134.86
	PRES	F	ATM					150.000
	X1	M	MOL%	H2-4	1.0000	1.31	58.0000	58.9875
	X2	M	MOL%	N2-4	1.0000	.36	21.0000	21.3170
	X3	M	MOL%	C1-4	1.0000	.09	13.0000	12.9218
	X4	M	MOL%	AR-4	1.0000	.28	4.0000	4.2499
	X5	M	MOL%	NH3-4	1.0000	.76	2.0000	2.5238
S5								
	RATE	M	M3/HR	F05	1650.000	.81	5.50E+04	5.60E+04
	TEMP	U	C					134.86
	PRES	F	ATM					150.000
	X1	U	MOL%					58.9875
	X2	U	MOL%					21.3170
	X3	U	MOL%					12.9218
	X4	U	MOL%					4.2499
	X5	U	MOL%					2.5238
S6								
	RATE	M	M3/HR	F06	450.000	1.11	1.50E+04	1.51E+04
	TEMP	U	C					134.86
	PRES	F	ATM					150.000
	X1	U	MOL%					58.9875
	X2	U	MOL%					21.3170
	X3	U	MOL%					12.9218
	X4	U	MOL%					4.2499
	X5	U	MOL%					2.5238

(table continued on next page)

Table 11-8 *(continued)*
Stream and Unit Reconciliation Solution for Ammonia Example
Case A. No Gross Errors Present in Measurements

STRM	VBL	STAT	UOM	TAG NAME	STANDARD DEVIATION	MT-STAT	MEASURED VALUE	CALC VALUE
S7								
	RATE	M	M3/HR	F07	1000.000	1.11	3.00E+04	3.06E+04
	TEMP	U	C					134.86
	PRES	F	ATM					150.000
	X1	U	MOL%					58.9875
	X2	U	MOL%					21.3170
	X3	U	MOL%					12.9218
	X4	U	MOL%					4.2499
	X5	U	MOL%					2.5238
S8								
	RATE	U	M3/HR					5.60E+04
	TEMP	M	C	T08	5.00	.47	430.00	428.98
	PRES	F	ATM					150.000
	X1	U	MOL%					58.9875
	X2	U	MOL%					21.3170
	X3	U	MOL%					12.9218
	X4	U	MOL%					4.2499
	X5	U	MOL%					2.5238
S9								
	RATE	U	M3/HR					3.06E+04
	TEMP	U	C					401.79
	PRES	F	ATM					150.000
	X1	U	MOL%					58.9875
	X2	U	MOL%					21.3170
	X3	U	MOL%					12.9218
	X4	U	MOL%					4.2499
	X5	U	MOL%					2.5238

HEAT EXCHANGER DUTY AND UA VALUES

HEAT EXNGR	DUTY (M*KCAL/HR)	UA (KCAL/HR-C)
E1	5.51592	36208.1
E2	.122487	2056.84
E3	5.78119	81187.1
E4	2.85656	20601.9

Table 11-8 *(continued)*
Stream and Unit Reconciliation Solution for Ammonia Example Case A. No Gross Errors Present in Measurements

SPLITTERS: SPLIT FRACTIONS

UNIT	STRM	STAT	CALCULATED VALUE
SP1			
	S16	U	.98191
	S17	U	.01809
SP2			
	S5	U	.55063
	S6	U	.14869
	S7	U	.30068

REACTOR UNITS: EXTENT OF REACTION AND DUTY

UNIT	VBL	STAT	UOM	CALCULATED VALUE
R1				
	EX1	U	KG-MOL/HR	1.22E+02
	DUTY	U	M*KCAL/HR	.11449
R2				
	EX1	U	KG-MOL/HR	91.67165
	DUTY	F	M*KCAL/HR	.00000

FLASH UNITS

UNIT	VBL	STAT	UOM	TAGNAME	STANDARD DEVIATION	MEASURED VALUE	CALC VALUE
F1							
	TEMP	U	C				–21.82
	PRES	F	ATM				150.000
	DUTY	F	M*KCAL/HR				.00000
F2							
	TEMP	U	C				–8.15
	PRES	M	ATM	F2-P	.500	150.000	150.000
	DUTY	F	M*KCAL/HR				.00000

NOTATION

STRM : STREAM ID
VBL : VARIABLE NAME
STAT : VARIABLE STATUS IN THE MODEL (M=MEASURED, U=UNMEASURED, F=FIXED)
UOM : UNIT OF MEASURE
MT-STAT : MEASUREMENT TEST STATISTIC
NR : NON-REDUNDANT MEASUREMENT

Table 11-9
Reconciliation Solution for the Measured Variables in the Ammonia Example
Case A. No Gross Errors Present in Measurements

TAGNAME	UOM	MEASURED VALUE	RECONCILED VALUE	STANDARD DEVIATION	DETECTABILITY FACTOR	MT-STAT	RE-MARK
AR-14	MOL%	5.0000	4.6916	1.0000	.833	.3580	
AR-19	MOL%	5.0000	5.3069	1.0000	.833	.3755	
AR-4	MOL%	4.0000	4.2499	1.0000	.909	.2815	
C1-14	MOL%	15.0000	14.2649	1.0000	.833	.8556	
C1-19	MOL%	15.0000	16.1357	1.0000	.833	1.3998	
C1-4	MOL%	13.0000	12.9218	1.0000	.909	.0889	
F03	M3/HR	100.0000	100.0002	.5000	.001	1.5045	
F04	M3/HR	102500.0000	101649.4149	3075.0000	.909	.2974	
F05	M3/HR	55000.0000	55971.6878	1650.0000	.714	.8107	
F06	M3/HR	15000.0000	15114.1287	450.0000	.227	1.1110	
F07	M3/HR	30000.0000	30563.5985	1000.0000	.500	1.1110	
F10	M3/HR	102500.0000	101649.4149	3075.0000	.909	.2974	
F11	M3/HR	96000.0000	96187.9483	2880.0000	.909	.0709	
F13	M3/HR	92500.0000	92078.4929	2774.9999	.909	.1637	
F14	M3/HR	92500.0000	92078.4929	2999.9999	.909	.1497	
F16	M3/HR	92400.0000	90412.9081	2999.9999	.909	.7056	
F17	M3/HR	1600.0000	1665.5848	48.0000	.769	1.8404	
F19	M3/HR	1500.0000	1472.4771	45.0000	.714	.8443	
F2-P	ATM	150.0000	150.0000	.5000			NR
F22	KG/HR	375.0000	365.3821	11.2500	.555	1.5045	

TAGNAME	UOM	MEASURED VALUE	RECONCILED VALUE	STANDARD DEVIATION	DETECTABILITY FACTOR	MT-STAT	RE-MARK
H2-14	MOL%	49.0000	49.5273	1.0000	.769	.6674	
H2-19	MOL%	57.0000	56.0226	1.0000	.714	1.3363	
H2-4	MOL%	58.0000	58.9875	1.0000	.769	1.3087	
N2-14	MOL%	18.0000	18.3357	1.0000	.833	.3934	
N2-19	MOL%	21.0000	20.7403	1.0000	.833	.3216	
N2-4	MOL%	21.0000	21.3170	1.0000	.909	.3623	
NH3-14	MOL%	13.0000	13.1805	1.0000	.714	.2612	
NH3-19	MOL%	2.0000	1.7945	1.0000	.526	.3869	
NH3-4	MOL%	2.0000	2.5238	1.0000	.714	.7584	
P02	ATM	150.0000	149.9982	.5000	.002	1.5045	
T01	C	–23.0000	–22.4639	1.5000	.238	1.5045	
T04	C	135.0000	134.8639	2.0000	.222	.3053	
T08	C	430.0000	428.9847	5.0000	.435	.4688	
T10	C	380.0000	378.1588	5.0000	.526	.6867	
T11	C	465.0000	466.8580	5.0000	.555	.6867	
T12	C	390.0000	388.1953	5.0000	.526	.6713	
T13	C	450.0000	453.6328	5.0000	.667	1.0711	
T14	C	290.0000	290.0654	5.0000	.667	.0198	
T18	C	–15.0000	–15.2442	1.0000	.161	1.5045	
T22	C	–30.0000	–29.1232	1.2000	1.000	.7307	

Table 11-10
Summary of Calculation Results for the Ammonia Example
Case A. No Gross Errors Present in Measurements

NUMBER OF ITERATIONS = 5
MEASURED VARIABLES = 40 (1 NON-REDUNDANT)
UNMEASURED VARIABLES = 59 (0 UNOBSERVABLE)
FIXED VARIABLES = 41 (29 FIXED BY USER)
NUMBER OF EQUATIONS = 82
DEGREE OF REDUNDANCY = 20

GLOBAL TEST (.950 CONFIDENCE LEVEL)

GT STATISTIC = 10.96 CRITICAL VALUE = 31.40

*** MEASUREMENTS PASSED THE GLOBAL TEST ***

MEASUREMENT TEST (.950 CONFIDENCE LEVEL)

CRITICAL VALUE = 3.16

*** ALL MEASUREMENTS PASSED THE MEASUREMENT TEST ***

PRINCIPAL COMPONENT MEASUREMENT (PCM) TEST

(.950 JOINT, .997 INDIVIDUAL CONFIDENCE LEVELS)

CRITICAL VALUE = 3.02

*** ALL PRINCIPAL COMPONENTS PASSED THE PCM TEST ***

(text continued from page 344)

Table 11-9 contains the reconciliation results for all measured variables. Table 11-10 indicates general run data (number of iterations until convergence, number of equations, and number of variables for each category—measured, unmeasured, fixed, the number of nonredundant and unobservable variables, the degrees of redundancy and a summary of results from statistical tests. All results were generated with DATACON™, a product of Simulation Sciences, Inc. (an Invensys Company). The thermodynamic properties were calculated with the Soave-Redlich-Kwong (SRK) method for all components.

No gross errors were simulated in Case A (only random errors are present in measured data). All three statistical tests that were used, the global test (GT), the measurement test (MT), and the principal component measurement test (PCMT), properly indicated that there are no gross errors in measurements.

In Case B, three gross errors were simulated in the following measurements:

F07 (magnitude = 10000 m^3/hr, ratio $\delta/\sigma = 10$);
C1-4 (magnitude = 5 mol%, ratio $\delta/\sigma = 5$);
T22 (magnitude = 3.6 Deg. C, ratio $\delta/\sigma = 3$).

They have different detectability factors, as indicated in Table 11-9. To increase the chance of detection and correct identification, a higher ratio of the gross error magnitude δ to the corresponding standard deviation σ was used for the measurements with lower detectability factor. The lower the detectability, the higher the ratio δ/σ.

Table 11-11 shows the reconciliation results for Case B for all measured variables. No error elimination was used in this run. Table 11-12 indicates various run data and the summary results from the GT and MT. The GT indicates the existence of gross errors, while the MT declares 7 measurements in gross error. In addition to the three true gross errors, four other gross errors were found by the MT. Serial elimination was used for better error identification.

Table 11-13 shows the summarized results from a first run with serial elimination calculations. In this run, the MT was used in addition to GT. We notice that in the first elimination step there were two measured variables sharing the same MT statistic (the largest MT statistic value). This particular algorithm used the detectability factor as a tie breaker in the elimination process. Since F07 has a higher detectability factor (0.5105) than F06 (0.2297), F07 was chosen to be eliminated first. This turned out to be the right choice. Subsequently, F06 was not found in gross error anymore. Next, C1-4 and T22 were also eliminated and no more gross errors were found by both GT and MT. The estimated values for the three eliminated measurements are very close to the values reported in Table 11-9 for the no gross error case.

Table 11-14 shows a similar run, this time using the PCMT for gross error identification. Initially, both GT and PCMT indicate existence of gross errors. The elimination path and final results, however, are somewhat different from the run with the MT. In the first elimination pass, F05 was found to be the major contributor to the largest inflated principal component. It is true that F05, F06, and F07 are model-related to each other (they are all outlet streams of the splitter SP2) and the smearing effect of the gross error in F07 is easier.

In this case, the calculation of the contribution shares to the first largest principal component that failed the PCMT, indicates that F05 should be first eliminated. Subsequently, an associated temperature, T13, also has to be eliminated and properly adjusted in order to satisfy the heat balance for exchanger E3. The last two eliminated measurements, T22 and C1-4, are true gross errors. The MT and PCMT tests usually detect correctly gross errors in measured variables with relatively higher detectability factor. The outcome of the two types of tests for gross errors in measured variables with relatively lower detectability factors could be the same for both tests (no gross error detection or wrong gross error identification) or one test can perform better than another. It is not clear which test performs consistently better. More analysis and comparison of the two type of tests for the ammonia example can be found in Jordache and Tilton [31].

Table 11-11
Reconciliation Solution for the Measured Variables in the Ammonia Example
Case B. Gross Errors Present in Three Measurements
No Gross Error Elimination Applied

TAGNAME	UOM	MEASURED VALUE	RECONCILED VALUE	STANDARD DEVIATION	MT-STAT	RE-MARK
AR-14	MOL%	5.0000	4.3708	1.0000	.7306	
AR-19	MOL%	5.0000	4.9102	1.0000	.1095	
AR-4	MOL%	4.0000	3.9918	1.0000	9.215E-03	
C1-14	MOL%	15.0000	15.4917	1.0000	.5727	
C1-19	MOL%	15.0000	17.4036	1.0000	2.9590	
C1-4	MOL%	18.0000	14.1485	1.0000	4.3908	ERR?
F03	M3/HR	100.0000	100.0004	.5000	2.9101	
F04	M3/HR	102500.0000	98184.7437	3075.0000	1.5076	
F05	M3/HR	55000.0000	59426.0003	1650.0000	3.7092	ERR?
F06	M3/HR	15000.0000	15632.9692	450.0000	6.1226	ERR?
F07	M3/HR	20000.0000	23125.7742	1000.0000	6.1226	ERR?
F10	M3/HR	102500.0000	98184.7437	3075.0000	1.5076	
F11	M3/HR	96000.0000	93945.4424	2880.0000	.7756	
F13	M3/HR	92500.0000	89671.7007	2774.9999	1.0989	
F14	M3/HR	92500.0000	89671.7007	2999.9999	1.0048	
F16	M3/HR	92400.0000	87994.0266	2999.9999	1.5649	
F17	M3/HR	1600.0000	1677.6741	48.0000	2.1875	
F19	M3/HR	1500.0000	1493.3735	45.0000	.2046	
F2-P	ATM	150.0000	150.0000	.5000		NR
F22	KG/HR	375.0000	356.3566	11.2500	2.9101	
H2-14	MOL%	49.0000	49.5736	1.0000	.7259	
H2-19	MOL%	57.0000	55.6916	1.0000	1.7817	

(table continued on next page)

Table 11-11 *(continued)*
Reconciliation Solution for the Measured Variables in the Ammonia Example
Case B. Gross Errors Present in Three Measurements
No Gross Error Elimination Applied

TAGNAME	UOM	MEASURED VALUE	RECONCILED VALUE	STANDARD DEVIATION	MT-STAT	RE-MARK
H2-4	MOL%	58.0000	58.2810	1.0000	.3740	
N2-14	MOL%	18.0000	18.1509	1.0000	.1769	
N2-19	MOL%	21.0000	20.3910	1.0000	.7516	
N2-4	MOL%	21.0000	20.9124	1.0000	.1004	
NH3-14	MOL%	13.0000	12.4129	1.0000	.8485	
NH3-19	MOL%	2.0000	1.6036	1.0000	.7465	
NH3-4	MOL%	2.0000	2.6662	1.0000	.9663	
P02	ATM	150.0000	149.9964	.5000	2.9101	
T01	C	–23.0000	–21.9490	1.5000	2.9101	
T04	C	135.0000	134.7517	2.0000	.5537	
T08	C	430.0000	421.9775	5.0000	3.4451	ERR?
T10	C	380.0000	382.2519	5.0000	.8444	
T11	C	465.0000	462.7234	5.0000	.8444	
T12	C	390.0000	391.8733	5.0000	.6986	
T13	C	450.0000	461.2256	5.0000	3.3358	ERR?
T14	C	290.0000	286.6337	5.0000	1.0263	
T18	C	–15.0000	–15.4760	1.0000	2.9101	
T22	C	–33.6000	–29.1232	1.2000	3.7306	ERR?

NOTATION
ERR? : CANDIDATE FOR GROSS ERROR
NR : NON-REDUNDANT

Table 11-12
Summary of Calculation Results for the Ammonia Example
Case B. Gross Errors Present in Three Measurements
Measurement Test Used for GED

NUMBER OF ITERATIONS = 5
MEASURED VARIABLES = 40 (1 NON-REDUNDANT)
UNMEASURED VARIABLES = 59 (0 UNOBSERVABLE)
FIXED VARIABLES = 41 (29 FIXED BY USER)
NUMBER OF EQUATIONS = 82
DEGREE OF REDUNDANCY = 20

GLOBAL TEST (.950 CONFIDENCE LEVEL)

GT STATISTIC = 80.86 CRITICAL VALUE = 31.40

*** DID NOT PASS THE GLOBAL TEST ***

MEASUREMENT TEST (.950 CONFIDENCE LEVEL)

CRITICAL VALUE = 3.16

*** 7 MEASUREMENTS FAILED THE MEASUREMENT TEST***

STRM/UNIT VBL	UOM	TAG-NAME	MEASUREMENT VALUE	CALCULATED VALUE	MT-STAT
S6					
RATE	M3/HR	F06	15000.0000	15632.9692	6.1226
S7					
RATE	M3/HR	F07	20000.0000	23125.7742	6.1226
S4					
X3	MOL%	C1-4	18.0000	14.1485	4.3908
S22					
TEMP	C	T22	–33.6000	–29.1232	3.7306
S5					
RATE	M3/HR	F05	55000.0000	59426.0003	3.7092
S8					
TEMP	C	T08	430.0000	421.9775	3.4451
S13					
TEMP	C	T13	450.0000	461.2256	3.3358

Table 11-13
Summary of Calculation Results for the Ammonia Example
Case B. Gross Errors Present in Three Measurements
Serial Elimination of Gross Errors Applied
Measurement Test Used for GED

INITIAL DATA RECONCILIATION

*** 7 MEASUREMENTS FAILED THE MEASUREMENT TEST (SEE TABLE 11-12) ***

TAGNAME	DETECTABILITY FACTOR
F06	0.2297
F07	0.5105

*** MEASUREMENT F07 WILL BE DELETED IN THE NEXT PASS ***

PASS 1 OF SERIAL ERROR ELIMINATION

MEASURED VARIABLES = 39 (2 NON-REDUNDANT)
UNMEASURED VARIABLES = 60 (0 UNOBSERVABLE)
FIXED VARIABLES = 41 (29 FIXED BY USER)
NUMBER OF EQUATIONS = 82
DEGREE OF REDUNDANCY = 19

SUMMARY OF ELIMINATED MEASUREMENTS

PASS	STRM/UNIT VARIABLE	UOM	TAGNAME	STANDARD DEVIATION	MEASUREMENT VALUE	CALCULATED VALUE
1	S7					
	RATE	M3/HR	F07	1000.0000	19999.9992	32039.1433

GLOBAL TEST (.950 CONFIDENCE LEVEL)

GT STATISTIC = 43.42 CRITICAL VALUE = 30.10

*** DID NOT PASS THE GLOBAL TEST ***

Table 11-13 *(continued)*
Summary of Calculation Results for the Ammonia Example
Case B. Gross Errors Present in Three Measurements
Serial Elimination of Gross Errors Applied
Measurement Test Used for GED

MEASUREMENT TEST (.950 CONFIDENCE LEVEL)

CRITICAL VALUE = 3.14

*** 2 MEASUREMENTS FAILED THE MEASUREMENT TEST ***

STRM/UNIT VBL	UOM	TAG-NAME	MEASUREMENT VALUE	CALCULATED VALUE	MT-STAT
S4					
X3	MOL%	C1-4	18.0000	14.0417	4.5065
S22					
TEMP	C	T22	–33.6000	–29.1232	3.7306

*** MEASUREMENT C1-4 WILL BE DELETED IN THE NEXT PASS ***

PASS 2 OF SERIAL ERROR ELIMINATION

MEASURED VARIABLES = 38 (2 NON-REDUNDANT)
UNMEASURED VARIABLES = 61 (0 UNOBSERVABLE)
FIXED VARIABLES = 41 (29 FIXED BY USER)
NUMBER OF EQUATIONS = 82
DEGREE OF REDUNDANCY = 18

SUMMARY OF ELIMINATED MEASUREMENTS

PASS	STRM/UNIT VARIABLE	UOM	TAGNAME	STANDARD DEVIATION	MEASUREMENT VALUE	CALCULATED VALUE
1	S7					
	RATE	M3/HR	F07	1000.0000	19999.9992	32202.9716
2	S4					
	X3	MOL%	C1-4	1.0000	18.0000	12.8745

GLOBAL TEST (.950 CONFIDENCE LEVEL)

GT STATISTIC = 23.10 CRITICAL VALUE = 28.90

*** MEASUREMENTS PASSED THE GLOBAL TEST ***

(table continued on next page)

Table 11-13 *(continued)*
Summary of Calculation Results for the Ammonia Example
Case B. Gross Errors Present in Three Measurements
Serial Elimination of Gross Errors Applied
Measurement Test Used for GED

MEASUREMENT TEST (.950 CONFIDENCE LEVEL)

CRITICAL VALUE = 3.13

*** 1 MEASUREMENT FAILED THE MEASUREMENT TEST ***

STRM/UNIT VBL	UOM	TAG-NAME	MEASUREMENT VALUE	CALCULATED VALUE	MT-STAT
S22					
TEMP	C	T22	–33.6000	–29.1232	3.7306

*** MEASUREMENT T22 WILL BE DELETED IN THE NEXT PASS ***

PASS 3 OF SERIAL ERROR ELIMINATION

MEASURED VARIABLES = 37 (2 NON-REDUNDANT)
UNMEASURED VARIABLES = 62 (0 UNOBSERVABLE)
FIXED VARIABLES = 41 (29 FIXED BY USER)
NUMBER OF EQUATIONS = 82
DEGREE OF REDUNDANCY = 17

SUMMARY OF ELIMINATED MEASUREMENTS

PASS	STRM/UNIT VARIABLE	UOM	TAGNAME	STANDARD DEVIATION	MEASUREMENT VALUE	CALCULATED VALUE
1	S7					
	RATE	M3/HR	F07	1000.0000	19999.9992	32203.0128
2	S4					
	X3	MOL%	C1-4	1.0000	18.0000	12.8743
3	S22					
	TEMP	C	T22	1.2000	–33.6000	–29.1232

GLOBAL TEST (.950 CONFIDENCE LEVEL)

GT STATISTIC = 9.18 CRITICAL VALUE = 27.60

*** MEASUREMENTS PASSED THE GLOBAL TEST ***

MEASUREMENT TEST (.950 CONFIDENCE LEVEL)

CRITICAL VALUE = 3.12

*** ALL MEASUREMENTS PASSED THE MEASUREMENT TEST ***

*** ALL VARIABLES ARE WITHIN BOUNDS ***

Table 11-14
Summary of Calculation Results for the Ammonia Example
Case B. Gross Errors Present in Three Measurements
Serial Elimination of Gross Errors Applied
Principal Component Measurement Test Used for GED

INITIAL DATA RECONCILIATION

PRINCIPAL COMPONENT MEASUREMENT TEST

(.950 JOINT, .997 INDIVIDUAL CONFIDENCE LEVELS)

CRITICAL VALUE = 3.02

*** 3 PRINCIPAL COMPONENTS FAILED THE PCM TEST ***

MAJOR CONTRIBUTING MEASUREMENTS TO THE FAILED PRINCIPAL COMPONENTS AND SHARES

OUTLIER PC #	PC SCORE	STRM/UNIT VARIABLE, SHARE %	STRM/UNIT VARIABLE, SHARE %	STRM/UNIT VARIABLE, SHARE %	STRM/UNIT VARIABLE, SHARE %
10	4.513	S5 RATE 28	S8 TEMP –20	S16 RATE 17	S13 RATE 14
		S10 RATE 12	S4 RATE 12	S14 RATE 11	S12 TEMP –8
		S7 RATE 7	S11 TEMP 7	S10 TEMP 7	S11 RATE 6
		S14 TEMP 5	S13 TEMP 4		
4	3.731	S22 TEMP 100			
13	3.049	S4 X3 74	S14 X4 10	S14 X5 8	S14 X1 –7
		S19 X2 5	S19 X1 4		

*** MEASUREMENT F05 WILL BE DELETED IN THE NEXT PASS ***

(table continued on next page)

Table 11-14 *(continued)*
Summary of Calculation Results for the Ammonia Example
Case B. Gross Errors Present in Three Measurements
Serial Elimination of Gross Errors Applied
Principal Component Measurement Test Used for GED

PASS 1 OF SERIAL ERROR ELIMINATION

MEASURED VARIABLES = 39 (1 NON-REDUNDANT)
UNMEASURED VARIABLES = 60 (0 UNOBSERVABLE)
FIXED VARIABLES = 41 (29 FIXED BY USER)
NUMBER OF EQUATIONS = 82
DEGREE OF REDUNDANCY = 19

SUMMARY OF ELIMINATED MEASUREMENTS

PASS	STRM/UNIT VARIABLE	UOM	TAGNAME	STANDARD DEVIATION	MEASUREMENT VALUE	CALCULATED VALUE
1	S5					
	RATE	M3/HR	F05	1650.0000	54999.9996	63681.9602

GLOBAL TEST (.950 CONFIDENCE LEVEL)

GT STATISTIC = 66.85 CRITICAL VALUE = 30.10

*** DID NOT PASS THE GLOBAL TEST ***

PRINCIPAL COMPONENT MEASUREMENT TEST

(.950 JOINT, .997 INDIVIDUAL CONFIDENCE LEVELS)

CRITICAL VALUE = 3.00

*** 3 PRINCIPAL COMPONENTS FAILED THE PCM TEST ***

Table 11-14 *(continued)*
Summary of Calculation Results for the Ammonia Example
Case B. Gross Errors Present in Three Measurements
Serial Elimination of Gross Errors Applied
Principal Component Measurement Test Used for GED

MAJOR CONTRIBUTING MEASUREMENTS TO THE LARGEST PRINCIPAL COMPONENT AND SHARES

OUTLIER PC #	PC SCORE	STRM/UNIT, VARIABLE, SHARE %	STRM/UNIT, VARIABLE, SHARE %	STRM/UNIT, VARIABLE, SHARE %	STRM/UNIT, VARIABLE, SHARE %
2	4.348	S13 TEMP 38	S8 TEMP 30	S14 TEMP 16	S11 TEMP 6
		S10 TEMP 6	S12 TEMP 5		
4	3.731	S22 TEMP 100			
11	3.243	S4 X3 59	S19 X3 15	S19 X1 10	S4 X1 8
		S4 X5 5	S14 X3 3	S14 X4 –3	

*** MEASUREMENT T13 WILL BE DELETED IN THE NEXT PASS ***

PASS 2 OF AUTOMATIC ERROR ELIMINATION

MEASURED VARIABLES	= 38 (1 NON-REDUNDANT)
UNMEASURED VARIABLES	= 61 (0 UNOBSERVABLE)
FIXED VARIABLES	= 41 (29 FIXED BY USER)
NUMBER OF EQUATIONS	= 82
DEGREE OF REDUNDANCY	= 18

(table continued on next page)

Table 11-14 *(continued)*
**Summary of Calculation Results for the Ammonia Example.
Case B. Gross Errors Present in Three Measurements.
Serial Elimination of Gross Errors Applied.
Principal Component Measurement Test Used for GED.**

SUMMARY OF ELIMINATED MEASUREMENTS

PASS	STRM/UNIT VARIABLE	UOM	TAGNAME	STANDARD DEVIATION	MEASUREMENT VALUE	CALCULATED VALUE
1	S5					
	RATE	M3/HR	F05	1650.0000	54999.9996	65494.1018
2	S13					
	TEMP	C	T13	5.0000	450.0000	481.3397

GLOBAL TEST (.950 CONFIDENCE LEVEL)

GT STATISTIC = 49.74 CRITICAL VALUE = 28.90

*** DID NOT PASS THE GLOBAL TEST ***

PRINCIPAL COMPONENT MEASUREMENT TEST

(.950 JOINT, .997 INDIVIDUAL CONFIDENCE LEVELS)

CRITICAL VALUE = 2.98

*** 2 PRINCIPAL COMPONENTS FAILED THE PCM TEST ***

MAJOR CONTRIBUTING MEASUREMENTS TO
THE LARGEST PRINCIPAL COMPONENT AND SHARES

OUTLIER PC #	PC SCORE	STRM/UNIT, VARIABLE, SHARE %	STRM/UNIT, VARIABLE, SHARE %	STRM/UNIT, VARIABLE, SHARE %	STRM/UNIT, VARIABLE, SHARE %
3	3.731	S22 TEMP 100			
10	3.043	S4 X3 58	S19 X3 14	S19 X1 11	S4 X1 8
		S4 X5 5	S14 X5 4	S14 X4 –3	S14 X3 3

*** MEASUREMENT T22 WILL BE DELETED IN THE NEXT PASS ***

Table 11-14 *(continued)*
Summary of Calculation Results for the Ammonia Example
Case B. Gross Errors Present in Three Measurements
Serial Elimination of Gross Errors Applied
Principal Component Measurement Test Used for GED

PASS 3 OF SERIAL ERROR ELIMINATION

MEASURED VARIABLES = 37 (1 NON-REDUNDANT)
UNMEASURED VARIABLES = 62 (0 UNOBSERVABLE)
FIXED VARIABLES = 41 (29 FIXED BY USER)
NUMBER OF EQUATIONS = 82
DEGREE OF REDUNDANCY = 17

SUMMARY OF ELIMINATED MEASUREMENTS

PASS	STRM/UNIT VARIABLE	UOM	TAGNAME	STANDARD DEVIATION	MEASUREMENT VALUE	CALCULATED VALUE
1	S5					
	RATE	M3/HR	F05	1650.0000	54999.9996	65494.1276
2	S13					
	TEMP	C	T13	5.0000	450.0000	481.3396
3	S22					
	TEMP	C	T22	1.2000	–33.6000	–29.1232

GLOBAL TEST (.950 CONFIDENCE LEVEL)

GT STATISTIC = 35.82 CRITICAL VALUE = 27.60

*** DID NOT PASS THE GLOBAL TEST ***

PRINCIPAL COMPONENT MEASUREMENT TEST

(.950 JOINT, .997 INDIVIDUAL CONFIDENCE LEVELS)

CRITICAL VALUE = 2.97

*** 1 PRINCIPAL COMPONENTS FAILED THE PCM TEST ***

(table continued on next page)

Table 11-14 *(continued)*
Summary of Calculation Results for the Ammonia Example
Case B. Gross Errors Present in Three Measurements
Serial Elimination ofGross Errors Applied
Principal Component Measurement Test Used for GED

MAJOR CONTRIBUTING MEASUREMENTS TO THE LARGEST PRINCIPAL COMPONENT AND SHARES

OUTLIER PC #	PC SCORE	STRM/UNIT, VARIABLE, SHARE %	STRM/UNIT, VARIABLE, SHARE %	STRM/UNIT, VARIABLE, SHARE %	STRM/UNIT, VARIABLE, SHARE %
9	3.048	S4 X3 58	S19 X3 14	S19 X1 11	S4 X1 8
		S4 X5 5	S14 X5 4	S14 X4 –3	S14 X3 3

*** MEASUREMENT T22 WILL BE DELETED IN THE NEXT PASS ***

PASS 4 OF AUTOMATIC ERROR ELIMINATION

MEASURED VARIABLES = 36 (1 NON-REDUNDANT)
UNMEASURED VARIABLES = 63 (0 UNOBSERVABLE)
FIXED VARIABLES = 41 (29 FIXED BY USER)
NUMBER OF EQUATIONS = 82
DEGREE OF REDUNDANCY = 16

SUMMARY OF ELIMINATED MEASUREMENTS

PASS	STRM/UNIT VARIABLE	UOM	TAGNAME	STANDARD DEVIATION	MEASUREMENT VALUE	CALCULATED VALUE
1	S5 RATE	M3/HR	F05	1650.0000	54999.9996	65795.7594
2	S13 TEMP	C	T13	5.0000	450.0000	481.4463
3	S22 TEMP	C	T22	1.2000	–33.6000	–29.1232
4	S4 X3	MOL%	C1-4	1.0000	18.0000	12.8642

GLOBAL TEST (.950 CONFIDENCE LEVEL)

GT STATISTIC = 15.42 CRITICAL VALUE = 26.30

*** MEASUREMENTS PASSED THE GLOBAL TEST ***

*** ALL PRINCIPAL COMPONENTS PASSED THE PCM TEST ***

*** ALL VARIABLES ARE WITHIN BOUNDS ***

SUMMARY

- Steady-state process unit data reconciliation and gross error detection technology is widely used in chemical, petrochemical and other related industrial processes.
- On-line data reconciliation is important for enhancing the accuracy in process optimization and advanced process control.
- Steady-state detection is necessary in order to increase the accuracy in the reconciled values and to provide meaningful gross error detection. If the process is not operated at steady-state for a longer period of time, dynamic data reconciliation should be applied.
- Proper component and thermodynamic characterization and accurate compositions are very important for a successful data reconciliation and gross error detection.
- Rigorous model enables merging the data reconciliation and parameter estimation into one problem which can be solved simultaneously.
- Plant-wide material and utilities reconciliation is an important tool for production (or yield) accounting.
- The most challenging problem in production accounting is the estimation of various leaks and losses. With enough redundancy, data reconciliation can provide reasonable estimates for the magnitudes of materials that are not accounted for.
- Imposing bounds on variables is often necessary to ensure a feasible solution for the data reconciliation problem. To solve a bounded problem, an NLP-based software is needed.
- The existent gross error detection methods do not accurately detect gross errors all the time. Some methods are better than others, but their overall performance depends upon the model accuracy and the level of data redundancy.

REFERENCES

1. Ravikumar, V. R., S. Narasimhan, S. R. Singh, and M. O. Garg. "RAGE—The State of the Art Package for Plant Data Reconciliation and Gross Error Detection," presented at the International Symposium on Automation and Control Systems, New Dehli, India, 1992.
2. Chi, Y. S., T. A. Clinkscales, K. A. Fenech, A. V. Gokhale, C. Jordache, and V. L. Rice. "On-line, Closed Loop Control and Optimization of an FCCU Using a Self Adapting Dynamic Model," presented at the AIChE Spring National Meeting, Houston, Tex., 1993.
3. Bagajewicz, M., and S. L. Mullick. "Reconciliation of Plant Data. Applications and Future Trends," presented at the AIChE Spring National Meeting, Houston, Tex., 1995.
4. Charpentier, V., L. J. Chang, G. M. Schwenzer, and M. C. Bardin. "An On-Line Data Reconciliation System for Crude and Vacuum Units," presented at the NPRA Computer Conference, Houston, Tex., 1991.
5. Leung, G., and K. H. Pang. "A Data Reconciliation Strategy: From On-Line Implementation to Off-Line Applications," presented at the AIChE Spring National Meeting, Orlando, Fla., 1990.
6. Scott, M. D., J. M. Tiessen, and S. L. Mullick. "Reactor Integrated Rigorous on-line model (ROMTM) for a Multi-Unit Hydrotreater-Catalytic Reformer Complex Optimization," presented at the NPRA Computer Conference, Anaheim, Calif., 1994.
7. Chiari, M., G. Bussari, M. G. Grottoli, and S. Pierucci. "On-line Data Reconciliation and Optimization: Refinery Applications." *Computers Chem. Engng.* 21 (Suppl.,1997): S1185–S1190.
8. Nair, P., and C. Jordache. "Rigorous Data Reconciliation is Key to Optimal Operations." *Control* (Oct. 1991): 118–123.
9. Tamura, K. I, T. Sumioshi, G. D. Fisher, and C. E. Fontenot. "Optimization of Ethylene Plant Operations Using Rigorous Models," presented at the AIChE Spring National Meeting, Houston, Tex., 1991.
10. Sanchez, M. A., A. Bandoni, and J. Romagnoli. "PLADAT—A Package for Process Variable Classification and Plant Data Reconciliation." *Computers Chem. Engng.* (Suppl. 1992): S499–S506.
11. Natori, Y., M. Ogawa, and V. S. Verneuil. "Application of Data Reconciliation and Simulation to a Large Chemical Plant." Proceedings of Large Chemical Plants 8th International Symposium, Antwerp, Belgium, 1992, pp. 103–113.

12. Christiansen, L. J., N. Bruniche-Olsen, J. M. Carstensen, and M. Schroeder. "Performance Evaluation of Catalytic Processes." *Computers Chem. Engng.* 21 (Suppl., 1997): S1179–S1184.

13. Holly, W., R. Cook, and C. M. Crowe. "Reconciliation of Mass Flow Rate Measurements in a Chemical Extraction Plant." *The Canadian Jl. of Chem. Engng.* 67 (1989): 595–601.

14. Dempf, D., and T. List. "On-line Data Reconciliation in Chemical Plant." *Computers Chem. Engng.* 22 (Suppl., 1998): S1023–S1025.

15. Placido, J., and L. V. Loureiro. "Industrial Application of Data Reconciliation." *Computers Chem. Engng.* 22 (Suppl., 1998): S1035–S1038.

16. Ravikumar, V., S. Narasimhan, M. O. Garg, and S. R. Singh. "RAGE—A Software Tool for Data Reconciliation and Gross Error Detection," in *Foundations of Computer-Aided Process Operations* (edited by D.W.T. Rippin, J. C. Hale, and J. F. Davis). Amsterdam: CACHE/Elsevier, 1994, 429–436.

17. Stephenson, G. R., and C. F. Shewchuck. "Reconciliation of Process Data with Process Simulation." *AIChE Journal* 32 (1986): 247–254.

18. Meyer, M., B. Koehret, and M. Enjalbert. "Data Reconciliation on Multicomponent Network Process." *Computers Chem. Engng.* 17 (no. 8, 1993): 807–817.

19. Narasimhan, S., R.S.H. Mah, and A. C. Tamhane. "A Composite Statistical Test for Detecting Changes in Steady State." *AIChE Journal* 32 (1986): 1409–1418.

20. Narasimhan, S., C. S. Kao, and R.S.H. Mah. "Detecting Changes in Steady State Using the Mathematical Theory of Evidence." *AIChE Journal* 33(1990): 1930–1932.

21. CEP Software Directory, a Supplement to *Chem. Engng. Progress,* published by the American Institute of Chemical Engineers. 1998.

22. Tong, H., and D. Bluck. "An Industrial Application of Principal Component Test to Fault Detection and Identification," presented at the IFAC Conference, 1998.

23. Reagan E., B. Tilton, and S. Salumenek. "Yield Accounting and Data Integration," presented at the NPRA Computer Conference, Atlanta, Ga., 1996.

24. Grosdidier, P. "Understand Operation Information Systems." *Hydrocarbon Processing* (Sept.1998): 67–78.

25. Veverka, V. V., and F. Madron. *Material and Energy Balancing in Process Industries: From Microscopic Balances to Large Plants.* Amsterdam: Elsevier, 1997.

26. Kelly, J. "Practical Issues in the Mass Reconciliation of Large Plant-Wide Flowsheets," presented at the AIChE Spring National Meeting, Houston, Tex., 1999.

27. Crowe, C. M. "Recursive Identification of Gross Errors in Linear Data Reconciliation." *AIChE Journal* 35 (1989): 869–872.

28. Tilton, B. "Tools for Plant Performance Analysis," presented at SIMTECH 97, San Diego, Calif., 1997.

29. Felix, B., and S. Wilson. "A Methodology for Loss Detection: Integrating Loss Detection with the Yield Accounting System in an Open Simulation Environment," presented at SIMTECH 97, San Diego, Calif., 1997.

30. DATACON Workbook. Brea, Calif.: Simulation Sciences, Inc., 1996.

31. Jordache, C., and B. Tilton. "Gross Error Detection by Serial Elimination: Principal Component Measurement Test versus Univariate Measurement Test," presented at the AIChE Spring National Meeting, Houston, Tex., March 1999.

Appendix A

Basic Concepts in Linear Algebra

This Appendix provides a ready reference of linear algebraic concepts used in this book. The book by Noble and Daniel [1] can be referred to for a more rigorous understanding starting from the notion of a field. The book by Strang [2] provides a lucid treatment of this subject and is especially recommended for those more interested in applying the concepts. We also assume that the readers are familiar with notions such as addition of vectors or matrices, matrix multiplication, multiplication of vectors and matrices by a scalar, and so on.

VECTORS AND THEIR PROPERTIES

We begin with the concept of a *real vector,* which is defined as an ordered set of *real numbers.* Examples of three vectors are given below:

$$\mathbf{a} = \begin{bmatrix} 1.0 \\ 2.0 \\ 2.5 \end{bmatrix} \qquad \mathbf{b} = \begin{bmatrix} 0 \\ 3 \\ 5 \end{bmatrix} \qquad \mathbf{c} = \begin{bmatrix} 0.25 & 2.6 & 3.2 \end{bmatrix}$$

where **a, b** are column vectors and **c** is a row vector. Unless otherwise mentioned, when we use the term vector we imply a column vector. The zero vector, denoted by **0,** is a special vector whose elements are all zero, and the *i*th unit vector denoted by $\mathbf{e}_i$ is a vector where the element in row *i* is 1 and all other elements are 0.

By suitably adding vectors which have the same number of elements, other vectors can be formed. A *linear combination of two or more vectors* is a vector which is formed by multiplying each vector by a real number (scalar) and adding them. For example, consider a linear combination of the above vectors **a** and **b** represented by the vector $\mathbf{d} = \alpha_1\mathbf{a} + \alpha_2\mathbf{b}$. If we choose the scalars $\alpha_1 = 1.5$, and $\alpha_2 = 2.0$, then the vector **d** is given by

$$\mathbf{d} = \begin{bmatrix} 1.50 \\ 9.00 \\ 13.75 \end{bmatrix}$$

Given a set of *n* vectors, $S = \{\mathbf{a}_1, \mathbf{a}_2, \ldots, \mathbf{a}_n\}$, we can generate all possible linear combinations of vectors in this set, $\alpha_1\mathbf{a}_1 + \alpha_2\mathbf{a}_2 + \ldots \alpha_n\mathbf{a}_n$, by choosing all possible values for the scalars α_i. We refer to the collection of vectors thus generated as a *vector space spanned* by the vectors in set S and denote it as **V**(S). It should be noted that the zero vector is a member of this space.

A set of vectors S is said to be linearly independent, if a linear combination of the vectors in this set equals **0** only for the case when all the scalars α_i are equal to 0 and not for any other choice of the scalar values. If a set of vectors is linearly dependent, then there is a vector in this set (whose scalar multiplying factor (α_i is nonzero) which can be expressed as a linear combination of the other vectors in this set. We can delete this vector and again check if the remaining vectors are linearly independent. If not, we can repeat this procedure until we are left with a set of vectors that are linearly independent.

The set of vectors which remain form a *minimal set of linearly independent vectors* which span the vector space **V**(S). This minimal set of vectors is said to form a *basis* set for **V**(S). For example, if we consider the set S consisting of vectors **a**, **b**, and **d** defined above, then this forms a linearly dependent set because any vector in this set can be expressed as a linear combination of the other two vectors in this set. We can choose to delete vector **d** from this set, in which case we are left with the two vectors **a** and **b** which can be verified to be linearly independent. Thus, the vectors **a** and **b** form a basis set for the vector space spanned the three vectors **a**, **b**, and **d.** Another basis set for the same vector space is **a** and **d**.

There can be many different choice of a basis set for a vector space, but the number of vectors in every basis set is the same, and is denoted as the *dimension* of the vector space. It must be borne in mind that the number of elements in any vector in a basis set (components of a vector) need not be equal to the dimension of the vector space spanned by the basis. This is illustrated clearly by the vector space spanned by the basis set **a** and **b**, where each of these vectors has 3 components but the dimension of the vector space spanned by them is only 2. Note that we often speak of a vector having n elements as an n-dimensional vector. This only implies that the vector having n elements is a member of the n-dimensional space of vectors. We use the notation $\Re^n$ for the n-dimensional real vector space.

MATRICES AND THEIR PROPERTIES

A real matrix of order $m \times n$ is an ordered set of elements consisting of m rows of n elements each. Each row of a matrix can be regarded as an n-dimensional row vector and each column of the matrix can be regarded as an m-dimensional column vector. Thus, an $m \times n$ matrix can either be considered as an ordered set of m row vectors each of dimension n, or as a set of n column vectors each of dimension m.

Two special matrices are the zero matrix denoted by $\mathbf{0}$, whose elements are 0, and the identity matrix of order $n \times n$, denoted by the symbol **I**, whose column i is the unit vector $\mathbf{e}_i$. It should be noted that we do not explicitly denote the dimensions of the matrix in the notation because it is usually clear from the context.

There are four important vector spaces associated with every matrix as defined below:

(1) **Row Space**: the space spanned by the rows.
(2) **Column Space**: the space spanned by the columns. This space is also known as the *range space* of a matrix.
(3) **Null Space**: the space spanned by all vectors **x** which satisfy $\mathbf{Ax} = \mathbf{0}$, where **x** is a vector belonging to $\Re^n$.
(4) **Left Null Space**: the null space of the transpose of a matrix.

There are some important properties that link these vector spaces. The *rank* of a matrix is equal to the dimension of its row space, which is also equal to the dimension of its column space. This immediately implies that the rank of a matrix $r \leq min(m, n)$. A matrix of order $n \times n$ is known as a square matrix of order n. If the rank of such a matrix is n, then the matrix is known as a nonsingular matrix and its inverse exists.

The following equality can also be proved:

$$r + \dim N(\mathbf{A}) = n \tag{A-1}$$

where $N(\mathbf{A})$ is the *null space* of matrix **A**. From the above equation, it follows that if the rank of a matrix is equal to n (which implies that the columns of a matrix are linearly independent), then the dimension of the null space of the matrix is 0. The only vector which satisfies $\mathbf{Ax} = \mathbf{0}$ in this case is the $\mathbf{0}$ vector. In general, we are interested in obtaining a vector $\mathbf{x}$ which is the solution of the linear set of equations $\mathbf{Ax} = \mathbf{b}$. In other words, we wish to express the vector **b** as a linear combination of the columns of **A**. This is possible only if **b** is a member of the column space of **A**. Furthermore, the solution is *unique* if the null space of **A** has dimension 0. This property is used in obtaining the solution of the unmeasured variables in data reconciliation discussed in Chapter 3.

In general, we can express the solution vector **x** as

$$\mathbf{x} = \mathbf{x}_r + \mathbf{x}_n \tag{A-2}$$

where $\mathbf{x}_r$ belongs to the column space of **A** and $\mathbf{x}_n$ belongs to the null space of **A**. This is known as the *range and null space decomposition* (RND), which is used in the RND-SQP nonlinear constrained optimization algorithm discussed in Chapter 5.

The *eigenvalues* of a square matrix **A** of order n are the n roots of its characteristic equation:

$$|\mathbf{A} - \lambda \mathbf{I}| = 0 \tag{A-3}$$

The set of these roots is denoted by $\lambda(\mathbf{A}) = \{\lambda_1, \lambda_2, \ldots, \lambda_n\}$. If we define the *trace* of matrix **A** by

$$\mathrm{tr}(\mathbf{A}) = \sum_{i=1}^{n} a_{ii} \tag{A-4}$$

then

$$\mathrm{tr}(\mathbf{A}) = \lambda_1 + \lambda_2 + \ldots + \lambda_n \tag{A-5}$$

The nonzero vectors $\mathbf{x}_i$ of size *n* that satisfy the equation

$$\mathbf{A}\mathbf{x}_i = \lambda_i \mathbf{x}_i \tag{A-6}$$

are referred to as *eigenvectors.* The eigenvalues and eigenvectors of certain matrices are used to build the principal component tests in Chapter 7.

REFERENCES

1. Noble, B., and J. Daniel. *Applied Linear Algebra.* Englewood Cliffs, N.J.: Prentice-Hall, 1977.
2. Strang, G. *Linear Algebra and its Applications,* 3rd ed. Orlando, Fla.: Harcourt Brace Jovanovich, 1988.
3. Golub, G. H., and C. F. Van Loan. *Matrix Computations.* Baltimore: Johns Hopkins University Press, 1996.

Appendix B

Graph Theory Fundamentals

Graph theory deals with problems related to topological properties of figures. It is also useful for analyzing problems concerning discrete objects and their interrelationships. In this appendix, we define some of the important concepts of graph theory used in the book and illustrate them using examples. Some facts are simply stated without proofs and we direct the interested reader to the book by Deo [1] for these proofs and additional concepts and theorems.

GRAPHS, PROCESS GRAPHS, AND SUBGRAPHS

A *graph* consists of a set of *nodes,* V, and a set of *edges,* E. Each edge is associated with a pair of nodes, which it joins. An example of a graph is shown in Figure B-1, which has six nodes drawn as circles and eight edges shown as lines. Each edge is said to be *incident* on the nodes with which it is associated. The *degree* of a node is the number of edges incident on it. A *process graph* is a graph which is simply obtained from a process flowsheet by adding an additional node called the environment node to which all process feeds and products are connected.

For example, the graph in Figure B-1 is the process graph of a simplified ammonia process whose flowsheet is shown in Figure 10-1. The nodes of a process graph correspond to process units and the edges of the process graph correspond to streams that interconnect the units.

Thus, if the process contains n units and e streams, then the corresponding process graph contains $n+1$ nodes and e edges. For ease of reference, we number or label the edges and nodes of the graph using the same numbers or labels as used in the process flowsheet, except for the environment node which is labelled as *E*. If the directions of the edges are ignored, as in Figure B-1, then an undirected graph is obtained; otherwise, the graph is directed. In this text, we are only concerned with undirected graphs.

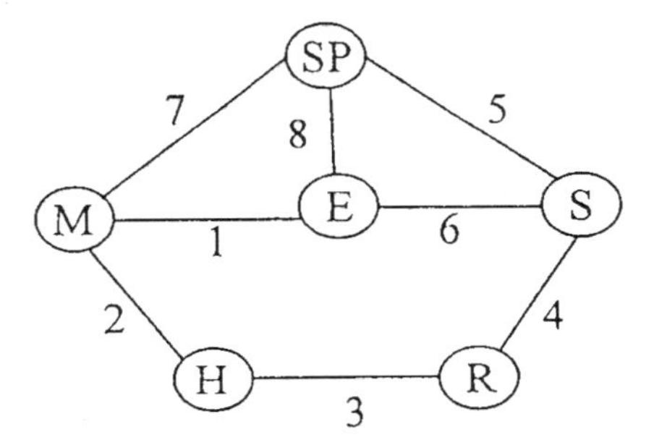

Figure B-1. A graph.

A *subgraph* of a graph consists of a subset of nodes and edges of the graph. Each edge of the subgraph joins the same two nodes as it does in the graph. In other words, if an edge is part of a subgraph, then the end nodes with which it is associated in the graph should also be part of the subgraph. The graph in Figure B-2 is a subgraph of the graph shown in Figure B-1.

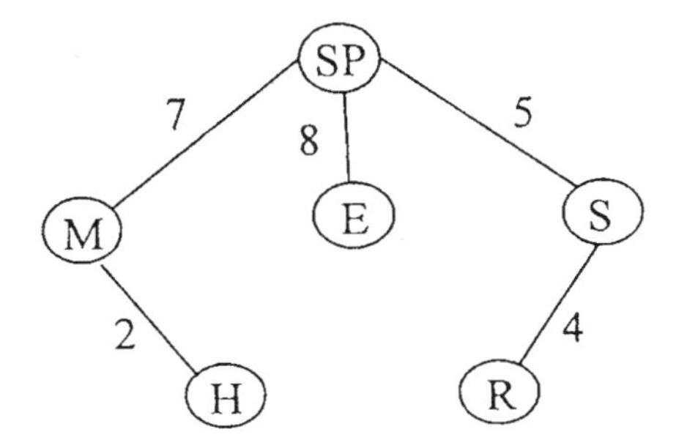

Figure B-2. A subgraph of the graph in Figure B-1.

PATHS, CYCLES, AND CONNECTIVITY

A *path* between two nodes (denoted as the initial and terminal nodes of the path) is a finite alternating sequence of edges and nodes such that each edge in the sequence is incident on the two nodes preceding and succeeding it, and no node appears more than once in this sequence. A path is called a *cycle* if the initial and terminal nodes are the same. For example, in Figure B-1, the alternating sequence of nodes and edges, E-1-M-2-H-3-R, is a path between initial node E and terminal node R, while the sequence E-6-S-5-SP-8-E is a cycle. A graph is *connected* if there exists a path between every pair of nodes of the graph. The graph in Figure B-1 is a connected graph as is generally the case for all process graphs.

SPANNING TREES, BRANCHES, AND CHORDS

A connected subgraph of the graph which does not contain any cycles and which includes all nodes of the graph is called a *spanning tree* of the graph. An edge of the graph that is part of the spanning tree is called a *branch,* while edges of the graph not part of the spanning tree are called *chords.* Figure B-2 is a spanning tree of the graph of Figure B-1. Corresponding to this spanning tree, edges 2, 4, 5, 7, and 8 are branches while the remaining edges 1, 3, and 6 are chords.

It should be noted that branches and chords are defined with respect to the specified spanning tree of a graph. If a different spanning tree of the graph is chosen then, accordingly, different edges of the graph are classified as branches or chords. It can be proved that a spanning tree contains n branches and e-n chords where n is the number of units in the process flowsheet (or one less than the number of nodes of the process graph), and e is the number of streams or edges of the graph.

GRAPH OPERATIONS

A graph can be modified by operations such as deletion of edges or nodes and by merging of nodes. The *deletion of an edge* from a graph results in a subgraph which contains all nodes and all edges except the deleted edge. For example, the spanning tree shown in Figure B-2 can be obtained from the graph in Figure B-1 by deleting edges 1, 3, and 6. The *deletion of a node* from a graph results in a subgraph which contains all the nodes of the graph except the deleted node, and contains all edges of the graph except the edges which are incident on the deleted node.

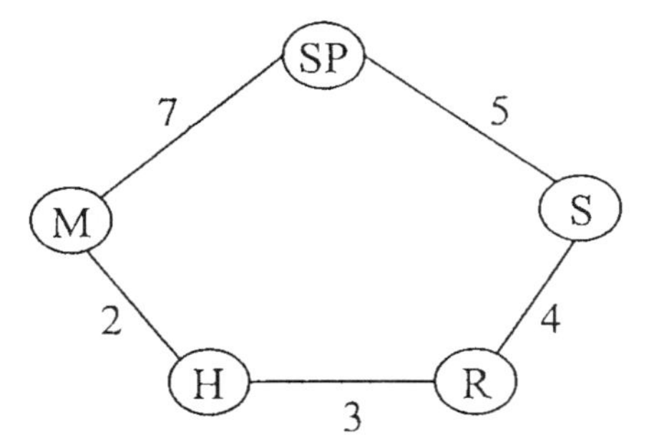

Figure B-3. Subgraph formed by deleting node E from graph in Figure B-1.

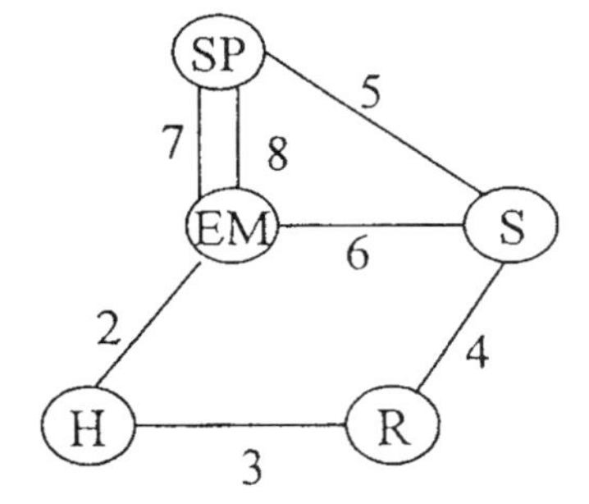

Figure B-4. Graph formed by merging nodes E and M of graph in Figure B-1.

The subgraph shown in Figure B-3 can be obtained from the graph of Figure B-1 by deleting the node *E.* The *merging of two nodes* of a graph results in a modified graph obtained by replacing the two merged nodes by a new node and deleting the edges incident on both these nodes. Edges which are incident on only one of the two merged nodes in the original graph are now incident on the new node of the modified graph. The graph in Figure B-4 is obtained from graph in Figure B-1 by merging nodes *E* and *M.* The new merged node in Figure B-4 is denoted as *EM.*

CUTSETS, FUNDAMENTAL CUTSETS, AND FUNDAMENTAL CYCLES

A *cutset* of a graph is a set of edges of a graph whose deletion disconnects the graph, but the deletion of a proper subset of the edges of a cutset does not disconnect the graph. The set of edges [2, 5, 6] is a cutset of the graph in Figure B-1 since the deletion of this set of edges disconnects the graph into two node sets one containing *M, E,* and *SP,* and the other containing *H, R,* and *S.* On the other hand, the set of edges [1, 2, 5, 6] is not a cutset although the removal of this set of edges disconnects the graph since its proper set of edges [2, 5, 6] is a cutset.

There is a correspondence between cutsets and flow balances that can be written for a process. A flow balance can be written around every unit of a process, which will involve the flows of streams that enter or exit this unit. It can be verified that the edges corresponding to these streams form a cutset of the process graph. Thus, corresponding to every cutset consisting of all edges incident on a node, a flow balance can be written. Flow balances can also be written corresponding to other cutsets which are essentially linear combinations of flow balances around individual process units.

Thus, corresponding to the cutset [2, 5, 6] of the graph in Figure B-1, the flow balance equation involving the flows of streams 2, 5, and 6 is a linear combination of the flow balances around process units *H, R,* and *S* of the process. It should be noted that the direction of the streams should be taken into account when writing the flow balances corresponding to cutsets of the process graph.

A cutset of the graph which contains only one branch of a spanning tree of the graph and zero or more chords is called a *fundamental cutset* corresponding to the spanning tree. For example, edge set [1, 3, 7] is a fundamental cutset of the graph in Figure B-1, corresponding to the spanning tree shown in Figure B-2. However, although the set of edges [2, 5, 6] is also a cutset of the graph in Figure B-1, it is not a fundamental cutset with respect to the spanning tree of Figure B-2 because it contains two branches, 2 and 5 of the spanning tree. With respect to every branch of a spanning tree of a graph, a fundamental cutset can be identified. The fundamental cutsets corresponding to the spanning tree, Figure B-2, of the graph in Figure B-1, are [2, 3], [4, 3], [5, 3, 6], [7, 1, 3], and [8, 1, 6], where the branch in each fundamental cutset is indicated by an underscore.

A concept, which is complementary to a fundamental cutset, is that of a fundamental cycle with respect to a spanning tree of a graph. A *fundamental cycle* with respect to a spanning tree of a graph is a cycle of the graph formed by exactly one chord and one or more branches. The cycle *E-1-M-7-SP-8-E,* which consists of edges [1, 7, 8], is a fundamental cycle of the graph of Figure B-1 with respect to the spanning tree B-2, which consists of chord *1* and branches *7* and *8.* For each chord of a spanning tree of a graph a fundamental cycle can be identified. The fundamental cycles with respect to the spanning tree, Figure B-2, of the graph in Figure B-1 are [1, 7, 8], [6, 5, 8], and [3, 4, 5, 7, 2], where the chords are indicated by an underscore.

Fundamental cutsets are complementary to fundamental cycles in the sense that if a chord c_j occurs in the fundamental cutset of a branch b_i, then branch b_i occurs in the fundamental cycle of chord c_j. This may be verified from the fundamental cycles and fundamental cutsets with respect to the spanning tree of Figure B-2 listed in the preceding paragraph. This property can be used to identify the fundamental cycles with respect to a spanning tree given the fundamental cutsets with respect to the same spanning tree.

Fundamental cycles (or fundamental cutsets) can be used to generate new spanning trees of a graph starting from a given spanning tree. The technique known as an *elementary tree transformation* (ETT) involves the interchange of a chord with a branch. In this technique, we add to the spanning tree a chord and delete a branch belonging to the fundamental cycle with respect to the original spanning tree formed by the chord which has been added. For example, the spanning tree shown in Figure B-5 is a new spanning tree of the graph in Figure B-1 obtained from the spanning tree in Figure B-2 by adding chord *1,* and deleting branch *7,* which belongs to the fundamental cycle formed by chord *1.* The new spanning tree differs from the initial spanning tree in respect of one chord and one branch, and is also referred to as the neighbor of the initial spanning tree.

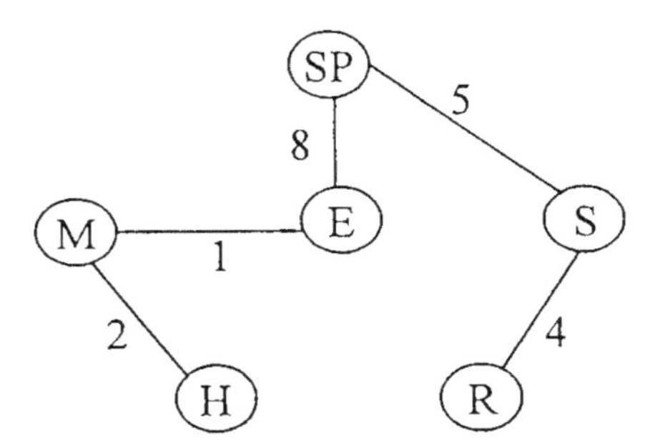

Figure B-5. Spanning tree formed by ETT of spanning tree in Figure B-2.

REFERENCE

1. Deo, N. *Graph Theory with Applications to Engineering and Computer Science.* Englewood Cliffs, N.J.: Prentice-Hall, 1974.

Appendix C

Fundamentals of Probability and Statistics

RANDOM VARIABLES AND PROBABILITY DENSITY FUNCTIONS

Probability is a mathematical theory dealing with the laws of *random events*. For example, the result of a physical or chemical experiment is a random event. The measured or inferred value obtained at the end of experiment is a *random variable* which lies within a specified interval with a certain probability.

It is easier to understand the behavior of random variables if we analyze a discrete event. The rolling of a pair of dice provides a good example of a random variable. It is impossible to predict the outcome of an individual roll; however, it is more likely that the summation number for the pair of dice is a 7 rather than a 12. This is because, of the 36 possible rolls, there is only one way to roll a 12—namely (6,6), while there are six ways to roll a 7—namely (6,1), (5,2), (4,3), (3,4), (2,5), (1,6).

Let us assume that we roll the dice thousands of times and record how many times each roll occurred. Then, the probability function for each roll R, i.e., P(R), or so-called *probability density function* (PDF, or p.d.f.) is:

$$P(R) = \frac{\text{Number of times roll R occurred}}{\text{Total number of rolls}} \qquad \text{(C-1)}$$

Figure C-1 shows the graph obtained by plotting P(R) for all possible rolls R. The probability of rolling a given value R in a single throw of the dice is the area under its rectangle. For example, you have a 4 in 36 chance of rolling a 9. The probability of rolling 3 or 12 in a single throw of the dice is the total area under their rectangles, 2/36+1/36=3/36.

The PDF graph in Figure C-1 is discontinuous, because rolling a pair of dice produces only discrete values, i.e., the resulting value must be an integer between 2 and 12 inclusively. Integrals of the PDF are quite useful because they determine the probability of occurrence of a group of events. For example, we can obtain P $(5 \le R \le 9)$ as the area of the plot which lies between R=5 and R=9, inclusively, i.e., 2(4/36)+ 2(5/36) +6/36 = 24/36, or 0.6667. Therefore, 66.67% of the rolls will have values $5 \le R \le 9$, or, in other words, the probability of getting a roll R such that $5 \le R \le 9$ is 66.67%. Another useful quantity is the probability that the roll is greater (or smaller) than a particular value. For example, P(R>9) = P(R=10)+P(R=11)+P(R=12) = 3/36+2/26+1/36 = 6/36.

Most *errors in plant measurements* are random variables. But unlike the rolls of a pair of dice in the previous example, they are continuous variables, not results of discrete events. This means that there is an infinity of possible "discrete" values for the events associated with continuous variables. For that reason, for continuous random variables, the integral

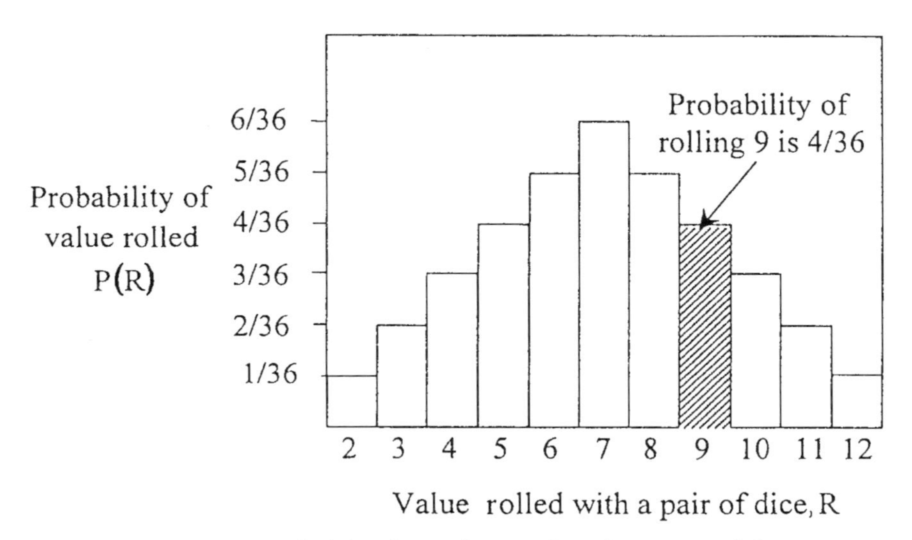

Figure C-1. Probability density function for rolling a pair of dice.

probabilities are of special practical interest. For example, we can make statements such as "there is a 95% chance that the true flow rate for stream S10 lies between 5,000 and 6,000 BPD." Or, "there is a 2.5% probability that the true flow rate in stream S10 exceeds 6,100 BPD." In order to calculate such probabilities, a continuous probability density function is required. For reasons specified in Chapter 2, the most widely used density function for continuous random variables in physical and chemical sciences is the *normal distribution function.*

The normal distribution is also known as the *Gaussian distribution,* and its PDF is described by the formula:

$$F(X) = \frac{1}{\sigma\sqrt{2\pi}} \exp\left(-\frac{(X-\mu)^2}{2\sigma^2} \right) \tag{C-2}$$

where μ is the mean value of the random variable X, and σ is its standard deviation. Since in practice we expect the errors to be zero on the average, $\mu = 0$ for the measurement error density function. Figure C-2 shows a Gaussian PDF with a mean of zero and a standard deviation of 1. This function is a continuous analog of the dice rolling density function. The normal distribution with zero mean and standard deviation of 1 is also known as *standard normal distribution.*

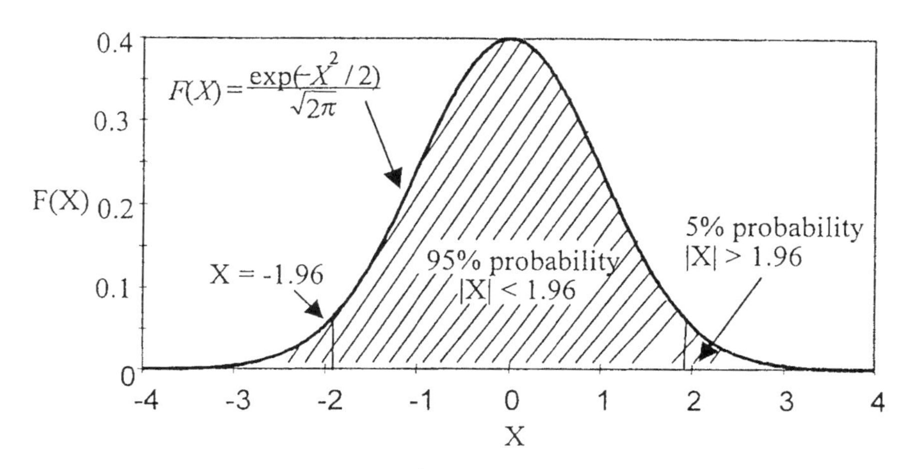

Figure C-2. Normal distribution density function.

The most important properties of the normal distribution PDF are as follows:

1. The maximum value of $F(X)$ occurs at the mean, μ.
2. The standard deviation σ determines the width (or the *skewness*) of the curve. For a very accurate instrument (small σ), the density function will look like a sharp peak centered at zero. On the contrary, for an inaccurate instrument (large σ), the PDF will look rather flat.
3. The $1/\sigma\sqrt{2\pi}$ factor normalizes the density so that $\int_{-\infty}^{\infty} F(X)dX = 1$.
4. It is symmetric about the mean.
5. The probability of a measurement error lying between X_1 and X_2 is:

$$P(X_1 \le X \le X_2) = \int_{X_1}^{X_2} F(X)dX \qquad \text{(C-3)}$$

   This probability is equivalent to the area under the curve between X_1 and X_2.
6. Similarly, the probability that a measurement error (its absolute value) is greater than a particular value X* is

$$P(|X| > X^*) = 1 - P(-X^* \le X \le X^*) = 1 - \int_{-X^*}^{X^*} F(X)dX \qquad \text{(C-4)}$$

   This probability is equivalent to the area under the curve outside the interval ($-X^*$ and X^*).

   As illustrated in Figure C-2, 95% of the random errors should lie within 1.96 standard deviations. Analytically, this means:

$$P(-1.96\sigma < X < 1.96\sigma) = \int_{-1.96\sigma}^{1.96\sigma} F(X)dX = 0.95 \qquad \text{(C-5)}$$

   This is often called the *95% confidence interval.* The 99% confidence interval occurs within 2.58 standard deviations of the mean. Note that these figures are true when there is only one measured variable. For multiple measured variables, the threshold is recalculated based on the rules given in Chapter 7 (see Sidak's rule).

   Another distribution of interest for the statistical applications in this book is the *chi-square* (χ^2) *distribution.* If R_1, R_2, . . . R_ν are independent random variables described by a standard normal distribution, then a chi-square random variable is defined by:

$$\chi^2(\nu) = \sum_{i=1}^{\nu} R_i^2 \tag{C-6}$$

The integer ν is usually known as the *number of degrees of freedom.* The probability density function for a χ^2 for different degrees of freedom ν is illustrated in Figure C-3.

The probability distribution function for the chi-square distribution is described analytically by the following formula:

$$F(\chi^2) = \frac{(\chi^2)^{(\nu-2)/2} \exp(-\chi^2/2)}{(2^{\nu/2})[\Gamma(\nu/2)]} \tag{C-7}$$

where Γ is the gamma function. The most important properties of the chi-square distribution are:

1. The mean value of $\chi^2(\nu)$ is ν.

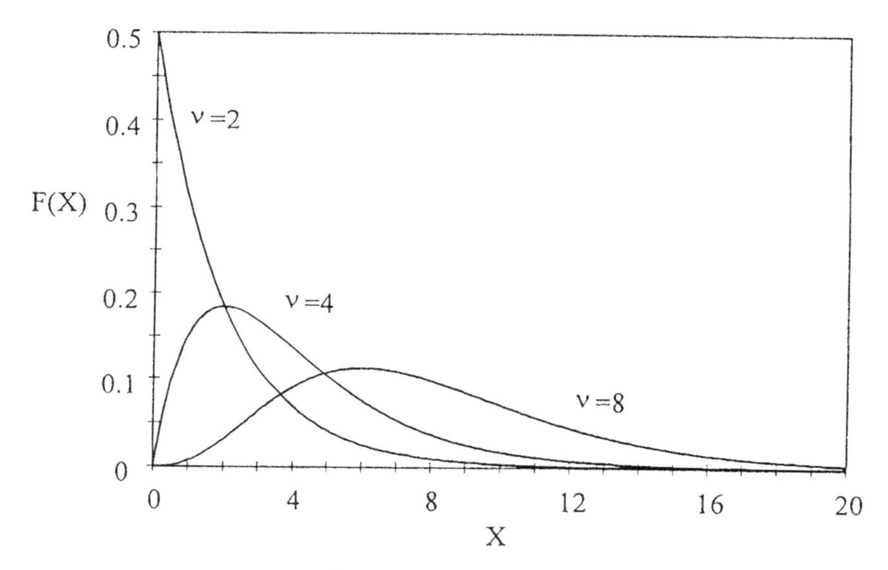

Figure C-3. Chi-square distribution density function.

2. As ν approaches infinity, the chi-square density approaches the normal distribution. The $\chi^2(8)$ curve in Figure C-3 is starting to illustrate this behavior.

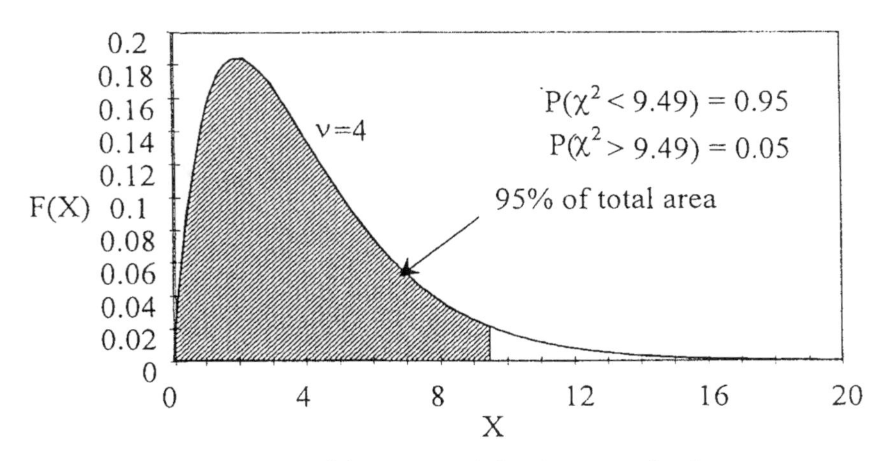

Figure C-4. Confidence intervals for chi-square distribution.

Confidence intervals can also be constructed for the χ^2 distribution. For example, Figure C-4 shows a 95% confidence region for a χ^2 distribution function with 4 degrees of freedom. In this particular case, 95% of the random χ^2 variables should lie between 0 and 9.49.

STATISTICAL PROPERTIES OF RANDOM VARIABLES

The statistical properties for random variables were indirectly mentioned as part of the analytical description of the probability density functions above. Now we are going to define them in a more general framework. There are two basic properties for the random variables (also known in statistics as *moments*): the ***mean*** and the ***variance*** of the random variable.

The **mean** value of a random variable X, μ_X, is defined as the *expected value* of X. For a continuous variable, it can be expressed analytically as:

$$\mu_x = E(X) = \int_{-\infty}^{\infty} F[X] dX \qquad \text{(C-8)}$$

The expected value defined above can also be defined as the first moment about zero. In general, the first moment about a constant value δ is given by $E[X-\delta]$. If δ is equal to the expected value of X, the central first moment is obtained, and the corresponding distribution is called the *central distribution.* Otherwise, a *noncentral distribution* is obtained.

The mean (expected value) of a random variable Z whose distribution is a joint distribution of other random variables, $X_1, X_2 \ldots X_m$ (i.e., a *multivariate distribution*) is defined as

$$\mu_z = E[Z] = \int_{-\infty}^{+\infty} \cdots\cdots \int_{-\infty}^{+\infty} f(X_1, X_2 \ldots X_m)\Phi(X_1, X_2 \ldots X_m)dX_1 \ldots dX_m \tag{C-9}$$

where $Z = f(X_1, X_2 \ldots X_m)$ and $\Phi(X_1, X_2 \ldots X_m)$ is the *joint probability density function* of the random variables $X_1, X_2 \ldots X_m$.

If $f(X_1, X_2 \ldots X_m)$ is a linear function, i.e.,

$$f(X_1, X_2 \ldots X_m) = \sum_{i=1}^{m} c_i X_i \tag{C-10}$$

then the mean value of Z is also linear:

$$\mu_z = \sum_{i=1}^{m} c_i \mu_i \tag{C-11}$$

The **variance** of a random variable X, var(X), is defined as the *second moment* about zero, i.e.,

$$\mathrm{Var}(X) = E[X - E(X)]^2 = E(X^2) - [E(X)]^2 \tag{C-12}$$

The relationship between the variance and the standard deviation of a random variable is given by:

$$\mathrm{Var}(X) = \sigma_X^2 \tag{C-13}$$

The variance of the multivariate random variable $Z = f(X_1, X_2 \ldots X_m)$ is defined as

$$\sigma_Z^2 = \int_{-\infty}^{+\infty} \cdots \int_{-\infty}^{+\infty} \left[f(X_1, X_2 \ldots X_m) - \mu_Z\right]^2 \Phi(X_1, X_2 \ldots X_m) dX_1 \ldots dX_m \quad (C-14)$$

If $f(X_1, X_2 \ldots X_m)$ is a linear function and the errors of the primary random variables $X_1, \ldots X_m$ are mutually independent random variables, (C-13) reduces to:

$$\sigma_Z^2 = \sum_{j=1}^{m} c_j^2 \sigma_j^2 \quad (C-15)$$

since the variance of a constant is zero.

A practical application of the above definitions is the derivation of the mean vector and the covariance matrix of a vector of random variables which is a linear function of other vector of random variables. For example, let us assume a linear equation in vector form such as:

$$\mathbf{y} = \mathbf{A}\mathbf{x} + \mathbf{c} \quad (C-16)$$

Let $E(\mathbf{x})=\mathbf{0}$ and $Cov(\mathbf{x}) = \mathbf{Q}$, where $Cov(\mathbf{x})$ means the covariance matrix of vector of random variables $\mathbf{x}$. Then, the expected value of $\mathbf{y}$ is:

$$E(\mathbf{y}) = E(\mathbf{A}\mathbf{x} + \mathbf{c}) = \mathbf{A}E(\mathbf{x}) + \mathbf{c} = \mathbf{c} \quad (C-17)$$

$$Cov(\mathbf{y}) = E\left\{[\mathbf{y} - E(\mathbf{y})][\mathbf{y} - E(\mathbf{y})]^T\right\} = \mathbf{A}E\left(\mathbf{x}\mathbf{x}^T\right)\mathbf{A}^T = \mathbf{A}\mathbf{Q}\mathbf{A}^T \quad (C-18)$$

These results for linear transformations of vector of random variables are used in many derivations throughout this book.

HYPOTHESIS TESTING

Hypothesis testing is a very important statistical tool for making decisions about random variables. The procedure uses information from a random sample of data to test the truth or falsity of a statement. The basic statement about a random variable is usually called the *null hypothesis,* denoted by H_0. The opposite hypothesis about same random variable is called the *alternative hypothesis,* here denoted by H_1.

The decision to accept or reject the null hypothesis is based on a statistical test. The test statistic (the value of the statistical test for given data) is first calculated with the data in the random sample. A decision criterion (a threshold of the statistical test) is used to make the decision about the hypothesis H_0. Two kinds of errors may be made at this point. If the null hypothesis is rejected when it is actually true, a *Type I error* is made. Alternatively, when the null hypothesis is accepted when it is actually false, then a *Type II error* is made. The probabilities of occurrence of Type I and Type II errors are as follows:

$$\alpha = \mathrm{P}(\text{Type I error}) = \mathrm{P}(\text{Reject } H_0 \mid H_0 = \text{true}) \qquad \text{(C-19)}$$

$$\gamma = \mathrm{P}(\text{Type II error}) = \mathrm{P}(\text{Reject } H_1 \mid H_1 = \text{true}) \qquad \text{(C-20)}$$

The *power of the test* is often used to evaluate a particular statistical test, and it is defined as:

$$\text{Power} = \mathrm{P}(\text{Accept } H_1 \mid H_1 = \text{true}) = 1 - \gamma \qquad \text{(C-21)}$$

In this book, the hypothesis testing is used to test the null hypothesis:

H_0 : there is no gross error in process data,

versus the alternative hypothesis:

H_1 : there is at least a gross error in process data,

or, more specifically:

H_{1j} : there is a gross error in measurement j.

The choice of the test threshold depends on the statistical test that is used for hypothesis testing. If the statistical test follows a standard normal distribution, such as some of the statistical tests in Chapter 7, a threshold $Z_{1-\alpha/2}$ is used, at a chosen α level of significance. The value α is used to control the probability of Type I error at value α.

For multiple tests, as in the case of multiple measurements in the plant, the probability of Type I error is higher than α. An upper bound β can be designed, as explained in Chapter 7. Let z_j be the test statistic for measurement *j*. If $|z_j| > Z_{1-\alpha/2}$, then the null hypothesis H_0 is rejected and hypothesis H_{1j} is accepted. This means that z_j is outside the $(-Z_{1-\beta/2}, +Z_{1-\beta/2}$ confidence interval for a standard normal distribution. This is similar to value X being outside the interval (–1.96, +1.96) for $\alpha = 0.05$ in Figure C-1.

On the other hand, if a global test described in Chapter 7 is used to test the null hypothesis H_0 against the global alternative hypothesis H_1, the threshold for the test is $\chi^2_{1-\alpha,\nu}$ at α chosen level of significance. As in Figure C-4, if the test statistic is greater than $\chi^2_{1-\alpha,\nu}$, the null hypothesis is rejected and a gross error is declared in the measurement set.

REFERENCES

1. Wadsworth, H. M. *Handbook of Statistical Methods for Engineers and Scientists.* New York: McGraw-Hill, 1990.
2. Hines, W. W., and D. C. Montgomery. *Probability and Statistics in Engineering and Management Science.* New York: John Wiley & Sons, 1980.
3. DATACON Workbook. Brea, Calif.: Simulation Sciences, Inc., 1996.

Index

Author Index

The Authors

Shankar Narasimhan is an Associate Professor of Chemical Engineering with the Indian Institute of Technology, Madras. He holds a B.Tech degree in Chemical Engineering from the IIT Madras, and M.S. and Ph.D. degrees in Chemical Engineering, both from Northwestern University. Dr. Narasimhan has published over 20 papers and he is well-known in the area of data reconciliation and gross error detection. His areas of expertise are the optimal design of sensor networks, fault diagnosis, and the application of artificial intelligence tools. Dr. Narasimhan has also led the development of data reconciliation and gross error detection package, RAGE, jointly with the R&D center of Engineers India Limited, New Delhi, and is the recipient of the Engineers India Ltd. Research Award instituted at IIT, Kanpur.

Cornelius Jordache (also known as Corneliu Iordache) is currently a corporate data reconciliation specialist with Simulation Sciences, Inc., an Invensys company. He holds a diploma in Industrial Chemistry from the Polytechnic Institute of Bucharest, and M.S. and Ph.D. degrees in Chemical Engineering, both from Northwestern University. Prior to his study in the U.S., he was an Assistant Professor with the Polytechnic Institute of Bucharest. Dr. Jordache has published and presented almost 20 papers and has co-authored a textbook on chemical reactors. He has extensive experience in the design and development of industrial data reconciliation software and has recently led the development of the gross error detection part of the ROMeo™ system, a new on-line product of Simulation Sciences, Inc., designed for using rigorous models for enhanced on-line optimization.